# Homogenization

## Methods and Applications

Translations of
MATHEMATICAL
MONOGRAPHS

Volume 234

# Homogenization

## Methods and Applications

G. A. Chechkin
A. L. Piatnitski
A. S. Shamaev

Translated by
Tamara Rozhkovskaya

American Mathematical Society
Providence, Rhode Island

А. Л. Пятницкий, Г. А. Чечкин, А. С. Шамаев

УСРЕДНЕНИЕ МЕТОДЫ И ПРИЛОЖЕНИЯ

Новосибирск, «Тамара Рожковская», 2007



Translated from the Russian by T. Rozhkovskaya

2000 *Mathematics Subject Classification.* Primary 35Bxx.

For additional information and updates on this book, visit
**www.ams.org/bookpages/mmono-234**

**Library of Congress Cataloging-in-Publication Data**

Chechkin, G. A. (Gregori A.), 1966–
[Usrednenie. English]
Homogenization : methods and applications / G. A. Chechkin, A. L. Piatnitski, A. S. Shamaev.
p. cm. — (Translations of mathematical monographs ; v. 234)
Includes bibliographical references and index.
ISBN-13: 978-0-8218-3873-0 (alk. paper)
ISBN-10: 0-8218-3873-3 (alk. paper)
1. Homogenization (Differential equations) I. Piatnitskii, A. L. (Andrei L.), 1955– II. Shamaev, A. S. (Aleksei S.), 1956– III. Title.

QA377.C43 2007
515′.35—dc22 2007060746



10 9 8 7 6 5 4 3 2 1 12 11 10 09 08 07

# Contents

# Preface

The purpose of this monograph is to present some recent directions of homogenization theory with particular emphasis on differential operators with rapidly oscillating coefficients, boundary value problems with rapidly alternating boundary conditions, equations in perforated domains, and other topics developed intensively during the last decades.

To explain the goal and ideas of homogenization theory for differential operators, we consider the boundary value problem for an equation with rapidly oscillating periodic coefficients:

$$-\operatorname{div}(A(x/\varepsilon)\nabla u_\varepsilon) = f \text{ in } \Omega, \quad u_\varepsilon = 0 \text{ on } \partial\Omega, \tag{0.1}$$

where $\Omega$ is a bounded domain in $\mathbb{R}^d$, $A(\xi)$ is a matrix with periodic measurable entries satisfying the ellipticity and boundedness conditions

$$\alpha I \leq A \leq \alpha^{-1} I, \quad \alpha > 0,$$

and $I$ is the identity matrix.

The boundary value problem (0.1) simulates the simplest stationary processes in a strongly inhomogeneous medium, i.e., a medium whose characteristics may sharply change depending on the spatial variables and the parameter $\varepsilon$ that characterizes, in a certain sense, the inhomogeneity scale. Such processes are quite difficult to study even with modern supercomputers because the use of finite-difference methods requires that the step be much less than $\varepsilon$, which leads to a huge amount of computations for small $\varepsilon$. It seems reasonable to try to replace (0.1) with another problem that is independent of small $\varepsilon$ (i.e., it is the so-called "homogenized problem") and, at the same time, has a solution which is "close" to the solution of (0.1) for small $\varepsilon$. The question is how the "homogenized problem" and "closeness" should be understood. Answers are given by homogenization theory. We give a (nonrigorous) formulation of one of such homogenization results obtained first for a periodic inhomogeneous medium. There is a constant matrix $\widehat{A}$ whose entries depend only on the matrix $A(\xi)$ such that for any $f$ the solutions $u_\varepsilon(x)$ converge to the solution of the problem

$$-\operatorname{div}(\widehat{A}\nabla u_0) = f \quad \text{in } \Omega, \quad u_0 = 0 \quad \text{on } \partial\Omega,$$

as $\varepsilon \to 0$; moreover, the entries of $\widehat{A}$ are determined by solving some auxiliary problem in the class of periodic functions.

The first proof of a homogenization theorem was obtained by De Giorgi and Spagnolo [**122, 123, 53, 54, 55**]. Shortly thereafter, Bakhvalov and then Lions established the same result by another method based on the asymptotic expansion technique, where a solution is represented in the form

$$u(x, \xi) + \varepsilon u_1(x, \xi) + \cdots, \quad \xi = x/\varepsilon,$$

with $u$ and $u_1$ periodic with respect to $\xi$ (cf. [**8, 9, 79**]). Using the technique of asymptotic analysis, it is possible not only to obtain a homogenized equation and establish the convergence of $u_\varepsilon$ to the solution of the homogenized equation, but also to estimate the convergence rate. The asymptotic expansion method still remains one of the basic tools of homogenization theory.

Another approach to homogenization theory, based on the notion of compensated compactness, was developed by Murat [**91**] and Tartar [**124**].

In 1989, Nguetseng [**100**] introduced the notion of "two-scale convergence", which provides a new approach to homogenization problems. This approach, developed by Allaire [**3**], turns out to be very effective for problems of more complicated structure than in the case of a standard homogenization model. In particular, in applications there are homogenization problems where the solutions $u_\varepsilon(x)$ do not have a limit in the classical sense, but it can be proved that the norms $\|u_\varepsilon(x) - v(x, \frac{x}{\varepsilon})\|$ converge to zero in some function space. Here, the function $v(x, \xi)$ is periodic with respect to $\xi$ and the variable $x$ belongs to the domain $\Omega$, whereas $\xi$ belongs to a periodicity cell. In this situation, the $u_\varepsilon$ do not have the classical limit, and the weak limit cannot be viewed as a satisfactory approximation of $u_\varepsilon$ for small $\varepsilon$. However, the asymptotic behavior of the solution can be characterized by the so-called *two-scale* limit, i.e., a function $v(x, \xi)$ that is periodic with respect to $\xi$.

In applications, there are situations that could be characterized as the "partial oscillating behavior of $u_\varepsilon(x)$." For example, assume that $\Omega$ is divided into several domains (depending, in general, on $\varepsilon$) and the solution strongly converges to a limit in some of these domains, and oscillates in the remaining domains. This situation corresponds to materials or media consisting of several phases with quite different physical and mechanical properties of each phase (for example, a composite material consisting of a hard frame and a soft fill material; a medium consisting of a viscous compressible or incompressible fluid and elastic hard inclusions; a conducting material consisting of a well-conducting phase and a material whose properties are close to those of a dielectric). Then by a homogenized problem it is natural to understand a boundary value problem for a function $v(x, \xi)$ of two variables. Such a problem can be reduced to simpler problems of mathematical physics with unknown functions depending only on $x$ or only on $\xi$, and the two-scale limit can be expressed as an algebraic combination of such functions. Such a procedure is usually referred to as the procedure of "asymptotic expansion." Recent research shows that an "asymptotic expansion" does not always hold even if the problem for the two-scale limit is well posed. However, the analysis of the homogenized problem can be informative even if the "asymptotic expansion" procedure is not applicable.

The literature on homogenization and related topics is quite extensive. We would like to draw the attention of the reader to the following books: [**4, 89, 85, 17, 50, 137, 138, 107, 102, 103, 10, 113, 6, 66, 67, 116, 115, 47, 44, 30, 15**]. Nevertheless, the authors run the risk of writing a new book with the hope that the presented material will be of interest for experts in homogenization theory as well as for nonexperts. This book was designed as an introduction to homogenization theory (so that modern methods are explained by rather simple examples and the rigorous proofs are accompanied by a number of exercises of different levels) and, at the same time, as a monograph combining little-known recent achievements with classical results of homogenization theory.

We list the main topics covered in this monograph.

*Operators with random coefficients.* Numerous works on homogenization of operators with random coefficients were published (cf., for example, [**137, 138**] and the references therein). In particular, for a random elliptic differential operator in the divergence form with stationary ergodic coefficients, the existence of a limit differential operator with constant deterministic coefficients was established and it was shown that the classical homogenization results remain valid for almost all realizations of a random medium (cf. [**137, 138**]). A remarkable observation concerning "nonclassical averaging" was made recently. Even if a homogenization procedure does not yield a deterministic limit operator, it is reasonable to study the limit behavior of the entire family of solutions in the sense of the weak convergence of measures regarded as solutions in a suitable function space. In such cases, the limit measure is usually determined by solving a stochastic partial differential equation.

*Homogenization in $p$-connected domains.* Recently, homogenization problems in $p$-connected domains (cf., for example, [**132**] and [**24**]) have been extensively studied. The "$p$-connectedness" is a generalization of the well-known notion of connectedness of a set, and it was introduced to describe processes in strongly inhomogeneous domains that are not connected from the geometric point of view, for example, electric current in a family of disconnected domains such that the intersection of the closures of these domains has positive capacity. Using the notion of $p$-connectedness, which is weaker than the usual geometric connectedness, it is possible to construct homogenization theory for differential operators in $p$-connected domains which is similar to homogenization theory in perforated domains in some ways but is quite different in other ways.

*Appearance of a "term étrange."* As was shown by Marchenko and Khruslov [**85**] in the 1960s, after homogenization of differential operators in perforated domains, a zeroth order term, called a "term étrange," can appear in the homogenized equation if the concentration (or the volume density) of holes tends to zero (this is similar to the appearance of a potential in the Schrödinger equation). This interesting phenomenon was investigated by many authors (cf., for example, [**45**]). We demonstrate the appearance of a "term étrange" by considering a simple example of the Dirichlet problem for the Laplace equation and then discuss other similar situations where a potential arises in the limit equation. We study in detail a problem with boundary conditions of the third type (the Fourier boundary conditions) on a perforation surface.

*Boundary homogenization.* One of the sections is devoted to the analysis of the behavior of solutions to boundary value problems in domains with oscillating boundary and to problems with rapidly oscillating type of boundary conditions (the so-called "boundary homogenization"). We pay particular attention to the method of matching asymptotic expansions, in spite of the fact that this method, suggested by Il′in [**69, 70**], is not directly related to the homogenization of differential operators and is well presented in the literature (cf., for example, [**71**]). However, recent research shows that the method of matching asymptotic expansions is an extremely effective tool in the study of boundary homogenization problems (cf., for example, [**60, 62, 63**]). Therefore, we find it reasonable to illustrate the main ideas of the matching method by a simple example and to show how the method works for boundary homogenization problems. In our opinion, this direction of homogenization theory is very promising for further developments.

*Homogenization of equations of hydrodynamics.* We consider a range of homogenization problems for equations of hydrodynamics, for example, the Darcy law relating pressure with speed of filtering fluid. To derive the Darcy law, we consider the boundary value problem for the Stokes equations with adherence condition on the boundary of cavities and then pass to a limit with respect to the characteristic size of inhomogeneity. This is done using two different methods. The first consists in passing to the limit in the corresponding integral identity and is similar to well-known classical homogenization methods. In the second case, the two-scale convergence technique is used. Unlike the weak limit, the two-scale limit is a function of two variables. To demonstrate features and advantages of the two-scale limit method, we start with a simple example of elliptic equations with rapidly oscillating coefficients and then proceed with the Stokes equations in a perforated domain.

*Domains with rapidly oscillating boundary.* To analyze the asymptotic behavior of solutions to boundary value problems in domains with rapidly oscillating boundary, for a model problem we take the initial-boundary value problem for a parabolic equation in a perforated domain with the boundary that periodically oscillates in time. An interesting phenomenon is observed in this case: first order terms describing translations appear in the limit equation although there are no translations in the original problem, which corresponds to the divergence form of the prelimit equation. To study this phenomenon, we introduce moving coordinates. Then the process can be described by an equation in the divergence form; this approach corresponds to the study of an effective diffusion process against the background of a large translation. The use of moving coordinates can also be helpful in the study of other problems, for example, the fluid filtration problem in domains with boundary alternating in time.

The book also contains various auxiliary information: the solvability theory of the main boundary value problems for the elasticity system and for the Stokes equations, some important inequalities, properties of function spaces, fundamental theorems in probability theory such as ergodic theorems, the central limit theorem for processes with mixing and the classical Bogolyubov averaging method.

The book is based on special lecture courses on modern aspects in mathematical physics, asymptotic methods, and homogenization of differential operators given by the authors at Lomonosov Moscow State University. The lectures were significantly revised and extended for this publication. The authors would like to express their deep gratitude to Professors V. V. Zhikov and P. P. Gadyl′shin for the extensive discussions and comments and, in particular, for the valuable additional information about the two-scale convergence, $p$-connectedness, and matching of asymptotic expansions.

*G. A. Chechkin*
*A. L. Piatnitski*
*A. S. Shamaev*

Moscow, 15.11.2005

CHAPTER 1

# Related Topics

## 1. Some facts of functional analysis

**1.1. Weak derivatives and Sobolev spaces.** Let $\Omega$ be a domain (an open connected set) in $\mathbb{R}^d$. The domain $\Omega$ is assumed to be bounded unless otherwise stated. Denote by $L_p(\Omega)$, $1 \le p < \infty$, the space of functions defined in $\Omega$ and $p$th-power summable in the sense of Lebesgue. Functions equal everywhere except for a set of zero $d$-dimensional measure are identified, i.e., such functions are regarded as a single element of $L_p(\Omega)$. The space $L_p(\Omega)$ equipped with the norm

$$\|u\|_{L_p(\Omega)} = \left( \int_\Omega |u|^p \, dx \right)^{1/p}$$

is a *Banach space* (i.e., a complete linear normed space). Thus, an element of $L_p(\Omega)$ is an equivalence class of functions different only on a set of zero measure.

Introduce the space $L_\infty(\Omega)$ of essentially bounded (i.e., bounded by a constant almost everywhere) real-valued functions in $\Omega$ that are Lebesgue measurable. The space $L_\infty(\Omega)$ is equipped with the norm

$$\|u\|_{L_\infty(\Omega)} = \operatorname*{ess\,sup}_{x\in\Omega} |u(x)| = \inf_{A\subset\Omega,\ |A|=0} \ \sup_{x\in\Omega\setminus A} |u(x)|,$$

where $|A|$ denotes the measure of a set $A \subset \Omega$, i.e., "ess sup" means the supremum up to a set of zero measure. In the case of a bounded domain $\Omega$, we have

$$\|u\|_{L_\infty(\Omega)} = \lim_{p\to\infty} \|u\|_{L_p(\Omega)},$$

which justifies the notation $L_\infty(\Omega)$.

DEFINITION 1.1. A function $v_i \in L_1(\Omega)$ is called the *weak derivative of $u \in L_1(\Omega)$ directed along the $x_i$-axis* if

$$\int_\Omega u \frac{\partial}{\partial x_i} \varphi \, dx = -\int_\Omega v_i \varphi \, dx \quad \forall \varphi \in C_0^\infty(\Omega), \ i = 1, \dots, d.$$

DEFINITION 1.2. The space of functions in $L_p(\Omega)$ such that their weak derivatives of order up to and including $l$ belong to $L_p(\Omega)$ is called a *Sobolev space* and is denoted by $W_p^l(\Omega)$. It is a Banach space with the norm

$$\|u\|_{W_p^l(\Omega)} = \left( \int_\Omega \sum_{0\le l_1+\dots+l_d\le l} \left| \frac{\partial^{l_1+\dots+l_d} u}{\partial x_1^{l_1} \cdots \partial x_d^{l_d}} \right|^p dx \right)^{1/p}.$$

In the case $p = 2$, the spaces $L_2(\Omega)$ and $W_2^l(\Omega)$ are *Hilbert spaces* (i.e., complete linear spaces equipped with inner products) with the inner products

$$(u, v)_{L_2(\Omega)} = \int_\Omega u(x) v(x) dx$$

and

$$(u,v)_{W_2^l(\Omega)} = \sum_{0\le l_1+\dots+l_d\le l} \int_\Omega \frac{\partial^{l_1+\dots+l_d}u}{\partial x_1^{l_1}\cdots\partial x_d^{l_d}} \frac{\partial^{l_1+\dots+l_d}v}{\partial x_1^{l_1}\cdots\partial x_d^{l_d}}\,dx$$

respectively. Denote by $H^1(\Omega)$ the completion of $C^\infty(\overline{\Omega})$ in the $W_2^1(\Omega)$-norm

$$\|u\| = \left(\int_\Omega (u^2 + |\nabla_x u|^2)\,dx\right)^{1/2}$$

and by $\overset{\circ}{H}{}^1(\Omega)$ the completion of $C_0^\infty(\Omega)$ in the $W_2^1(\Omega)$-norm.

If the boundary of $\Omega$ is sufficiently smooth, the spaces $H^1(\Omega)$ and $W_2^1(\Omega)$ coincide (the proof of this fact can be found, for example, in [**88**]), and we will not distinguish between them.

The Sobolev spaces can be defined for any real numbers $l$ and $p$. In particular, for $p=2$, $l=s$, and $\Omega\equiv\mathbb{R}^d$, this can be done with the help of the Fourier transform. We recall that the *Fourier transform*, defined by the formula

$$\widehat{u}(\lambda) = \frac{1}{(\sqrt{2\pi})^d}\int_{\mathbb{R}^d} u(x)e^{-i\lambda x}\,dx,$$

is an isometric linear operator in $L_2(\mathbb{R}^d)$. Using the Fourier transform, we can define the Hilbert space $H^s(\mathbb{R}^d)$ as follows. For functions in the *Schwartz class*, i.e., smooth functions that, together with all partial derivatives, rapidly decrease at infinity, we introduce the $H^s(\mathbb{R}^d)$-norm

$$\|u\|_{H^s(\mathbb{R}^d)} = \|(1+|\lambda|^2)^{s/2}\widehat{u}(\lambda)\|_{L_2(\mathbb{R}^d)}.$$

The space $H^s(\mathbb{R}^d)$ is the completion of the Schwartz class in this norm.

Let $\Omega$ be an arbitrary domain in $\mathbb{R}^d$. Denote by $\|u\|_{H^s(\Omega)}$ the infimum of the norms $\|\widetilde{u}\|_{H^s(\mathbb{R}^d)}$, where $\widetilde{u}(x)$ is an extension of $u(x)$ from $\Omega$ to the space $\mathbb{R}^d$.

We introduce an equivalent norm in the space $H^s(\Omega)$. Denote by $W_p^l(\Omega)$, $p\ne 1$, $k<l<k+1$, $k=0,1,2,\dots$, the closure of $C^\infty(\overline{\Omega})$ in the norm

$$\|v\|_{W_p^l(\Omega)} \equiv \left(\|v\|^p_{W_p^k(\Omega)} + \int_{\Omega\times\Omega} \frac{|D^{(k)}v(y) - D^{(k)}v(x)|^p}{|y-x|^{d+p(l-k)}}\,dx\,dy\right)^{1/p}.$$

The equivalence of this norm and the norm defined via the Fourier transform is proved, for example, in [**80**].

The *dual* of $\overset{\circ}{H}{}^1(\Omega)$ (i.e., the set of all linear continuous functionals on $\overset{\circ}{H}{}^1(\Omega)$) is denoted by $H^{-1}(\Omega)$. The value of $f\in H^{-1}(\Omega)$ at $u\in\overset{\circ}{H}{}^1(\Omega)$ is denoted by $[f,u]$ and the norm is defined by the formula

$$\|f\|_{H^{-1}(\Omega)} = \sup|\,[f,u]\,|,$$

where the supremum is taken over all $u\in\overset{\circ}{H}{}^1(\Omega)$ such that $\|u\|_{\overset{\circ}{H}{}^1(\Omega)} = 1$.

DEFINITION 1.3. We say that $\nabla u\in(H^{-1}(\Omega))^d$ is the *weak gradient* of a function $u\in L_2(\Omega)$ if

$$[(\nabla u)_i,\varphi] = -\int_\Omega u\frac{\partial}{\partial x_i}\varphi\,dx \quad \forall\varphi\in\overset{\circ}{H}{}^1(\Omega),\ i=1,\dots,d.$$

DEFINITION 1.4. The *weak divergence* $\operatorname{div} u \in H^{-1}(\Omega)$ of a vector $u \in (L_2(\Omega))^d$ is defined by the formula

$$[\operatorname{div} u, \varphi] = -\int_\Omega (u, \nabla_x \varphi)\, dx \quad \forall \varphi \in \overset{\circ}{H}{}^1(\Omega),$$

where $(\cdot, \cdot)$ is the Euclidean inner product.

We have

$$\|\operatorname{div} u\|_{H^{-1}(\Omega)} = \sup_{\|v\|_{H^1(\Omega)}=1} \int_\Omega (u, \nabla v)\, dx \leq \|u\|_{L_2(\Omega)}.$$

DEFINITION 1.5. A skew-symmetric matrix $\operatorname{curl} u$ with entries in $H^{-1}(\Omega)$ is called the *weak curl* of a vector $u \in (L_2(\Omega))^d$ if

$$[(\operatorname{curl} u)_{ij}, \varphi] = -\int_\Omega \left( u_i \frac{\partial \varphi}{\partial x_j} - u_j \frac{\partial \varphi}{\partial x_i} \right) dx \quad \forall \varphi \in \overset{\circ}{H}{}^1(\Omega), \quad i, j = 1, \dots, d.$$

**1.2. Strong, weak, and weak* convergence.** Let $B$ be a Banach space equipped with the norm $\|\cdot\|_B$. Denote by $B'$ the dual of $B$. The value of $f \in B'$ at $u \in B$ is denoted by $[f, u]$. The norm of $f$ in $B'$ is defined by the formula

$$\|f\|_{B'} = \sup_{\|u\|_B=1} |[f, u]|.$$

DEFINITION 1.6. A sequence $u_\varepsilon \in B$ *strongly converges* to $u_0 \in B$ as $\varepsilon \to 0$ if $\|u_\varepsilon - u_0\|_B \to 0$ as $\varepsilon \to 0$. In this case, we write

$$u_\varepsilon \to u_0 \quad \text{strongly in } B \quad \text{as } \varepsilon \to 0.$$

DEFINITION 1.7. A sequence $u_\varepsilon \in B$ *weakly converges* to $u_0 \in B$ if $[f, u_\varepsilon] \to [f, u_0]$ as $\varepsilon \to 0$ for all $f \in B'$. In this case, we write

$$u_\varepsilon \rightharpoonup u_0 \quad \text{weakly in } B \quad \text{as } \varepsilon \to 0.$$

It is easy to check that the strong convergence implies the weak convergence.

EXERCISE 1.1. Give an example of a weakly, but not strongly converging sequence in a Banach space.

DEFINITION 1.8. A sequence $f_\varepsilon \in B'$ *weak*-converges* to $f_0 \in B'$ if $[f_\varepsilon, u] \to [f_0, u]$ as $\varepsilon \to 0$ for all $u \in B$. In this case, we write

$$f_\varepsilon \overset{*}{\rightharpoonup} f_0 \quad \text{weak}^* \text{ in } B' \quad \text{as } \varepsilon \to 0.$$

If $B$ is a *reflexive* space (i.e., $B'' \equiv B$), then the weak convergence coincides with the weak* convergence.

We recall that the spaces $L_p(\Omega)$ and $L_q(\Omega)$ $(1/p + 1/q = 1)$, $1 < p < \infty$, are dual, and consequently the spaces $L_p(\Omega)$ are reflexive. The spaces $W_p^l(\Omega)$, $1 < p < \infty$, are also reflexive for all $l \in \mathbb{R}$. Furthermore, $L_\infty(\Omega)$ is the dual of $L_1(\Omega)$, but neither of these spaces is reflexive.

EXERCISE 1.2. Prove that $L_1(\Omega)$ is not the dual of $L_\infty(\Omega)$.
*Hint.* The space $L_\infty(\Omega)$ is not separable.

A Hilbert space is identified with its dual because of the Riesz representation theorem (cf., for example, [**130**]).

**1.3. Coerciveness and $G$-convergence.** Let $B$ be a separable reflexive Banach space. Denote by $\mathcal{L}(B, B')$ the set of linear bounded operators acting from $B$ to $B'$. The norm of an operator $A \in \mathcal{L}(B, B')$ is defined by the formula

$$\|A\|_{\mathcal{L}(B,B')} = \sup_{x \in B, \|x\|_B = 1} \|Ax\|_{B'}.$$

DEFINITION 1.9. An operator $A$ is said to be *coercive* if there are constants $\kappa_1, \kappa_2$, $0 < \kappa_2 \leq \kappa_1 < \infty$, such that

1) $\|A\|_{\mathcal{L}(B,B')} \leq \kappa_1$,
2) $[Au, u] \geq \kappa_2 \|u\|_B^2$ for any $u \in B$.

THEOREM 1.1. *Let $A$ be a coercive operator. Then for any $f \in B'$ the equation $Au = f$ has a unique solution.*

PROOF. It suffices to show that the operator $A$ is bijective. Since

$$\kappa_2 \|u\|_B^2 \leq [Au, u] \leq \|Au\|_{B'} \|u\|_B,$$

we have

$$\kappa_2 \|u\|_B \leq \|Au\|_{B'}. \tag{1.1}$$

We show that $A$ is *injective* (i.e., $\operatorname{Ker} A = \{0\}$). Let $Au_1 = Au_2$. Then

$$0 = \|A(u_1 - u_2)\|_{B'} \geq \kappa_2 \|u_1 - u_2\|_B.$$

Therefore, $u_1 = u_2$. Consequently, the kernel of $A$ consists of the zero element and the operator $A$ is injective.

We show that the image of $A$, denoted by $\operatorname{Im} A$, is closed. Suppose that $Au_n$ is a Cauchy sequence in $B'$. By (1.1), $u_n$ is a Cauchy sequence in $B$. Since the space $B$ is complete, there exists $u_* \in B$ such that $u_n \to u_*$ in $B$ as $n \to +\infty$. By Definition 1.9, $\|Au\|_{B'} \leq \kappa_1 \|u\|_B$. Therefore,

$$\|A(u_n - u_*)\|_{B'} \leq \kappa_1 \|u_n - u_*\|_B \to 0 \quad \text{as } n \to +\infty.$$

Consequently, there exists the limit $Au_* \in B'$ of the Cauchy sequence $Au_n$ as $n \to +\infty$. Hence $\operatorname{Im} A$ is closed.

We show that $A$ is *surjective* (i.e., $\operatorname{Im} A = B'$). Assume the contrary. Let $\operatorname{Im} A \neq B'$. By the Hahn–Banach theorem, for the closed subspace $\operatorname{Im} A$ of the Banach space $B'$ there exists a nonzero linear functional vanishing on this subspace, i.e., there exists a nonzero element $z \in B'' \equiv B$ such that $[Au, z] = 0$ for any $u \in B$. Substituting $z$ for $u$ in the last equality, we find

$$0 = [Az, z] \geq \kappa_2 \|z\|_B^2,$$

which implies $z = 0$. We arrive at a contradiction. Hence $A$ is surjective and, consequently, $A$ is bijective. □

DEFINITION 1.10 (cf. [**122, 123, 53, 54, 55, 84**]). A sequence of operators $A_\varepsilon$ *$G$-converges* to an operator $A_0$ as $\varepsilon \to 0$ if $[g, A_\varepsilon^{-1} f] \to [g, A_0^{-1} f]$ as $\varepsilon \to 0$ for all $f, g \in B'$.

In this case, we write

$$A_\varepsilon \xrightarrow{G} A_0 \quad \text{as } \varepsilon \to 0.$$

It is easy to see that the $G$-convergence implies the weak convergence of the inverse operators and can be interpreted as the weak convergence of solutions to the corresponding equations: the solutions of the equations $A_\varepsilon u_\varepsilon = f$ weakly converge in $B$ to the solution of the equation $A_0 u_* = f$.

THEOREM 1.2. *Let $A_\varepsilon$ be a sequence of coercive operators with common constants $\kappa_1$ and $\kappa_2$. Then the sequence $A_\varepsilon$ is G-compact (in the sense of the G-convergence), i.e., from any sequence $A_{\varepsilon'}$ it is possible to extract a subsequence $A_{\varepsilon''}$ that G-converges to a coercive operator $A_0$ as $\varepsilon'' \to 0$.*

PROOF. We show that the sequence $A_\varepsilon^{-1}$ is uniformly bounded with respect to $\varepsilon$ in the $\mathcal{L}(B', B)$-norm. By the definition of coerciveness, $[A_\varepsilon u, u] \geq \kappa_2 \|u\|_B^2$. Therefore, $\|u\|_B \leq \kappa_2^{-1} \|A_\varepsilon u\|_{B'}$ and, consequently,

$$\|A_\varepsilon^{-1}\|_{\mathcal{L}(B',B)} \leq \frac{1}{\kappa_2}.$$

As is known, it is possible to extract a weakly converging subsequence from any bounded sequence in a separable reflexive Banach space.

The spaces $B$ and $B'$ are separable. Let $f_j \in B'$, $j = 1, 2, \dots$, be everywhere dense. Define an operator $L$ on $\{f_j\}$ as follows. Since the norms $\|A_\varepsilon^{-1} f_j\|_B$ are bounded for any $j$, we can extract a weakly converging subsequence $A_{\varepsilon_n^1}^{-1} f_1$ from the bounded sequence $A_\varepsilon^{-1} f_1$:

$$A_{\varepsilon_1^1}^{-1} f_1, \dots, A_{\varepsilon_n^1}^{-1} f_1, \dots \rightharpoonup U_1 \quad \text{as } \varepsilon_n^1 \to 0.$$

We set $Lf_1 = U_1$. From $A_{\varepsilon_n^1}^{-1} f_2$ we extract a weakly converging subsequence

$$A_{\varepsilon_1^2}^{-1} f_2, \dots, A_{\varepsilon_n^2}^{-1} f_2, \dots \rightharpoonup U_2$$

and set $Lf_2 = U_2$. Continue this procedure by setting $Lf_j = U_j$, $j = 1, 2, \dots$. Choosing the Cantor diagonal $A_{\varepsilon_n^n}^{-1}$, for any $j$ and $g \in B'$ we find

$$\lim_{n\to\infty} [g, A_{\varepsilon_n^n}^{-1} f_j] = [g, Lf_j].$$

We show that $L$ is a bounded operator on $\{f_j\}$. Indeed, passing to the limit in the inequality $[f_j, A_{\varepsilon_n^n}^{-1} f_j] \geq \kappa_2 \|A_{\varepsilon_n^n}^{-1} f_j\|_B^2$ as $n \to \infty$ and using the semicontinuity of the norm under the weak limit passage, we find

$$\kappa_2 \|Lf_j\|_B^2 \leq [f_j, Lf_j] \leq \|f_j\|_{B'} \|Lf_j\|_B,$$

which implies

$$\|Lf_j\|_B \leq \frac{1}{\kappa_2} \|f_j\|_{B'}.$$

Thus, the required estimate is proved for a dense set. Finally, consider arbitrary $f \in B$. Since $\{f_j\}$ is dense, there exists a sequence $f_s$ such that $f_s \to f$ in $B'$ as $s \to +\infty$. Since the sequence $Lf_s$ is bounded, for some subsequence $s'$ we have $Lf_{s'} \rightharpoonup u^*$ weakly in $B$. We have

$$[g, A_{\varepsilon_n^n}^{-1} f] - [g, u^*] = [g, A_{\varepsilon_n^n}^{-1} f - A_{\varepsilon_n^n}^{-1} f_s + A_{\varepsilon_n^n}^{-1} f_s - Lf_s + Lf_s - u^*] .$$

We choose sufficiently large $s'$ such that $\|A_{\varepsilon_n^n}^{-1}(f - f_{s'})\|_B$ and $|[g, (Lf_{s'} - u^*)]|$ are small for all $n$. Then we choose sufficiently large $n$ such that $|[g, (A_{\varepsilon_n^n}^{-1} f_s - Lf_s)]|$ is small. Thus, the left-hand side can be made as small as we like provided that $n$ is chosen in a suitable way. Therefore,

$$\lim_{n\to\infty} [g, A_{\varepsilon_n^n}^{-1} f] = [g, u^*],$$

which implies that $u^*$ is independent of the choice of the subsequence $f_{s'}$. We set $Lf \equiv u^*$. Thus, $L$ is defined on the entire space $B'$.

EXERCISE 1.3. Prove that the operator $L$ is linear.

We prove that $L$ is coercive. As has already been shown, $\|L\|_{\mathcal{L}(B',B)} \le 1/\kappa_2$. Furthermore,

$$[f, A_{\varepsilon_n^n}^{-1} f] \ge \kappa_2 \|A_{\varepsilon_n^n}^{-1} f\|_B^2 \ge \kappa_2 \frac{1}{\kappa_1} \|A_{\varepsilon_n^n} A_{\varepsilon_n^n}^{-1} f\|_{B'}^2 \ge \frac{\kappa_2}{\kappa_1} \|f\|_{B'}^2.$$

Passing to the limit as $n \to \infty$, we find that

$$[f, Lf] \ge \frac{\kappa_2}{\kappa_1} \|f\|_{B'}^2.$$

Hence there exists the inverse operator $L^{-1}$ such that $\|L^{-1}\|_{\mathcal{L}(B,B')} \le \kappa_2$. We show that the inverse operator is also coercive, i.e., the second inequality in Definition 1.9 holds. We have $[f, A_{\varepsilon_n^n}^{-1} f] \ge \kappa_2 \|A_{\varepsilon_n^n}^{-1} f\|_B^2$. Passing to the limit as $n \to \infty$ and using the lower semicontinuity of the norm under the weak limit passage, we find $[f, Lf] \ge \kappa_2 \|Lf\|_B^2$ or $[L^{-1}u, u] \ge \kappa_2 \|u\|_B^2$. Thus, $L^{-1} = A_0$, where $A_0$ is a coercive operator. □

### 1.4. The Poincaré inequality.

DEFINITION 1.11. A domain $\Omega$ is said to be *star-shaped with respect to a ball* $Q$ if for any points $x \in Q$ and $y \in \Omega$ the segment joining $x$ and $y$ lies inside $\Omega$.

Denote by $\rho$ the diameter of a ball $Q$, by $\sigma_d$ the area of the unit sphere in $\mathbb{R}^d$, and by $M$ the diameter of a domain $\Omega$.

THEOREM 1.3. *Let $u \in H^1(\Omega)$, and let $\Omega$ be star-shaped with respect to $Q$. Then*

$$\int_\Omega u^2 dx \le C_1(\Omega, Q, d) \Big( \int_\Omega u\, dx \Big)^2 + C_2(\Omega, Q, d) \int_\Omega |\nabla u|^2 dx, \tag{1.2}$$

*where the positive constants $C_1(\Omega, Q, d)$ and $C_2(\Omega, Q, d)$ are independent of $u(x)$ and satisfy the estimates*

$$C_1(\Omega, Q, d) \le \frac{1}{|\Omega|}, \quad C_2(\Omega, Q, d) \le \frac{M^{d+1} \rho \sigma_d}{d|Q|}.$$

PROOF. Let $u(x) \in C^1(\overline{\Omega})$. By the Newton–Leibniz formula, for $x \in Q$ and $y \in \Omega$ we have

$$u(x) - u(y) = \int_0^\eta \frac{\partial u}{\partial r}\, dr, \tag{1.3}$$

where $r$ is the natural parameter of the segment joining the points $x$, $y$ and $\eta = |x-y|$. Let $(r, \theta_1, \dots, \theta_{d-1})$ denote the spherical coordinates with the origin $x$. Since the domain $\Omega$ is star-shaped with respect to $Q$, the boundary of $\Omega$ at every point $x \in Q$ can be described in the spherical coordinates as follows: $r = \eta(\theta_1, \dots, \theta_{d-1})$.

Consider the squared equality (1.3). Using the Cauchy–Schwarz–Bunyakovskii inequality, we find

$$u^2(x) + u^2(y) - 2u(x)u(y) = \Big( \int_0^\eta \frac{\partial u}{\partial r}\, dr \Big)^2 \le M \int_0^\eta \Big( \frac{\partial u}{\partial r} \Big)^2 dr.$$

We fix $x \in Q$ and integrate the last inequality with respect to $y \in \Omega$:

$$\begin{aligned}
&|\Omega|u^2(x) + \int_\Omega u^2(y)\,dy - 2u(x)\int_\Omega u(y)\,dy \\
&\quad \le \int_\Omega M \int_0^\eta \left|\frac{\partial u}{\partial r}\right|^2 dr\,dy = M\int_0^\eta \int_{S_1}\int_0^\eta \left|\frac{\partial u}{\partial r}\right|^2 dr\; s^{d-1} d\omega\,ds \\
&\quad \le \frac{M^{d+1}}{d}\int_{S_1}\int_0^\eta \left|\frac{\partial u}{\partial r}\right|^2 r^{1-d} r^{d-1}\,dr\,d\omega = \frac{M^{d+1}}{d}\int_\Omega \left|\frac{\partial u}{\partial r}\right|^2 |x-y|^{1-d}\,dy,
\end{aligned}$$

where $d\omega = \sin^{d-2}\theta_1 \sin^{d-3}\theta_2 \cdots \sin\theta_{d-2} d\theta_1 \cdots d\theta_{d-1}$. We have

$$|\Omega|u^2(x) + \int_\Omega u^2(y)\,dy \le 2u(x)\int_\Omega u(y)\,dy + \frac{M^{d+1}}{d}\int_\Omega |\nabla u|^2 |x-y|^{1-d}\,dy.$$

Using the inequality $2ab \le \varepsilon a^2 + \frac{b^2}{\varepsilon}$, we integrate with respect to $x \in Q$:

$$\begin{aligned}
&|\Omega|\int_Q u^2(x)\,dx + |Q|\int_\Omega u^2(y)\,dy \\
&\le 2\int_Q u(x)\,dx \int_\Omega u(y)\,dy + \frac{M^{d+1}}{d}\int_Q\int_\Omega |\nabla u|^2|x-y|^{1-d}\,dy\,dx \\
&\le \varepsilon\Big(\int_Q u(x)\,dx\Big)^2 + \frac{1}{\varepsilon}\Big(\int_\Omega u(y)\,dy\Big)^2 + \frac{M^{d+1}}{d}\int_Q\int_\Omega |\nabla u|^2|x-y|^{1-d}\,dy\,dx.
\end{aligned}$$

Taking $|\Omega|/|Q|$ for $\varepsilon$ and using the Cauchy–Schwarz–Bunyakovskii inequality, we estimate the first integral on the right-hand side of the last inequality as follows:

$$\varepsilon\Big(\int_Q u(x)\,dx\Big)^2 \le \varepsilon|Q|\int_Q u^2(x)dx = |\Omega|\int_Q u^2(x)dx.$$

Consequently,

$$\begin{aligned}
&|\Omega|\int_Q u^2(x)\,dx + |Q|\int_\Omega u^2(y)\,dy \\
&\le |\Omega|\int_Q u^2(x)\,dx + \frac{|Q|}{|\Omega|}\Big(\int_\Omega u(y)\,dy\Big)^2 + \frac{M^{d+1}}{d}\int_Q\int_\Omega |\nabla u|^2|x-y|^{1-d}\,dy\,dx.
\end{aligned}$$

We cancel the first terms on the left- and right-hand sides:

$$\int_\Omega u^2(y)\,dy \le \frac{1}{|\Omega|}\Big(\int_\Omega u(y)\,dy\Big)^2 + \frac{M^{d+1}}{d|Q|}\int_Q\int_\Omega |\nabla u|^2|x-y|^{1-d}\,dy\,dx. \tag{1.4}$$

Since the functions $|\nabla u|^2$ and $|x-y|^{1-d}$ are summable, we can apply the Fubini theorem to the last integral on the right-hand side of (1.4):

$$\begin{aligned}
\frac{M^{d+1}}{d|Q|}\int_Q\int_\Omega |\nabla u|^2|x-y|^{1-d}\,dy\,dx &= \frac{M^{d+1}}{d|Q|}\int_\Omega\Big(|\nabla u|^2\int_Q |x-y|^{1-d}\,dx\Big)dy \\
&= \frac{M^{d+1}}{d|Q|}\int_\Omega\Big(|\nabla u|^2\int_Q r^{1-d} r^{d-1}\,dr\,d\omega\Big)\,dy = \frac{M^{d+1}\rho\sigma_d}{d|Q|}\int_\Omega |\nabla u|^2\,dy.
\end{aligned}$$

Finally, we have

$$\int_\Omega u^2(y)\,dy \le \frac{1}{|\Omega|}\Big(\int_\Omega u(y)\,dy\Big)^2 + \frac{M^{d+1}\rho\sigma_d}{d|Q|}\int_\Omega |\nabla u|^2\,dy, \tag{1.5}$$

where $\sigma_d$ is the area of the unit $(d-1)$-dimensional sphere. Note that

$$C_1 = \frac{1}{|\Omega|}, \quad C_2 = \frac{M^{d+1}\rho\sigma_d}{d|Q|} = C(d)\frac{M^{d+1}}{\rho^{d-1}}.$$

Thus, the inequality (1.5) is proved for smooth functions. Since smooth functions are dense in $H^1(\Omega)$, the inequality (1.5) remains valid for functions in $H^1(\Omega)$. $\square$

**1.5. Integral estimates for rapidly oscillating functions with zero average.** Introduce the notation: $\Omega$ is a domain in $\mathbb{R}^d$ and $\square = (0,1) \times \cdots \times (0,1) = (0,1)^d$ (the direct product of intervals is taken $d$ times). Let $h(\xi)$ be periodic in $\xi_1, \ldots, \xi_d$ with period 1 relative to each variable.

We say that a function $h(x)$ is *Hölder continuous with exponent* $\alpha > 0$ in $\Omega$ if there exists a constant $L > 0$ such that for any $x, y \in \Omega$,

$$|h(x) - h(y)| \le L|x-y|^\alpha.$$

In the case $\alpha = 1$, the function $h(x)$ is said to be *Lipschitz continuous* and $L$ is called the *Lipschitz constant.*

The space $H^1_{\text{per}}(\square)$ is the closure of the set of smooth 1-periodic functions in the norm

$$\|u\|_{H^1_{\text{per}}(\square)} = \left( \int_\square (u^2 + |\nabla u|^2)\, d\xi \right)^{1/2}.$$

LEMMA 1.1. *Let $h(\xi)$ be a 1-periodic function of the form $h(\xi) = h_0(\xi) + \sum_{i=1}^d \frac{\partial}{\partial \xi_i} h_i(\xi)$, where $h_i \in L_\infty(\square)$, $i = 0, 1, \ldots, d$, and*

$$\langle h_0 \rangle \equiv \int_\square h_0(\xi)\, d\xi = 0. \tag{1.6}$$

*Then for $u \in \overset{\circ}{H}{}^1(\Omega)$, $v \in H^1(\Omega)$,*

$$\left| \int_\Omega h\left(\frac{x}{\varepsilon}\right) u(x)\, v(x)\, dx \right| \le K\varepsilon\, \|u\|_{H^1(\Omega)}\, \|v\|_{H^1(\Omega)}, \tag{1.7}$$

*where $K$ depends only on $h$ and $\Omega$.*

REMARK 1.1. This assertion was proved in [**12**] and, independently, in [**31**] (cf. also [**11**]) for smooth functions $h(x/\varepsilon)$ in the case $d = 2$. Similar inequalities were obtained in [**109**]–[**111**]. For Lipschitz continuous functions the proof is contained in [**41**].

PROOF. We need an important estimate obtained by De Giorgi [**52**] (cf. also [**65**, Theorem 8.24]). Note that it is possible to prove Lemma 1.1 without this estimate. However, it is reasonable to employ it in this proof because it will be useful below.

Suppose that $\Omega$ is a bounded domain, $\Omega_1 \Subset \Omega$, and

$$L[u] := \sum_{i,j=1}^d \frac{\partial}{\partial x_i}\left( a_{ij}(x) \frac{\partial u}{\partial x_j} \right) + \sum_{i=1}^d b_i(x) \frac{\partial u}{\partial x_i} + c(x) u = f \quad \text{in } \Omega,$$

where $a_{ij}(x)$, $b_i(x)$, and $c(x)$ are measurable functions such that $\lambda_1|\xi|^2 \le a_{ij}(x)\xi_i\xi_j \le \lambda_2|\xi|^2$, $|b_i(x)|, |c(x)| \le h$, $f(x) \in W^{-1}_{p,0}(\Omega)$, $p > d$, $W_p^{-1}(\Omega)$ is the space of linear bounded functionals on $W_p^1(\Omega)$, and $u \in H^1(\Omega)$. Then $u \in C^\alpha(\Omega_1)$, where $\alpha > 0$ depends only on $\lambda_1, \lambda_2$, and the following estimate holds:

$$\|u\|_{C^\alpha(\Omega_1)} \leq C_1 \|u\|_{L_2(\Omega)} + C_2 \|f\|_{W_p^{-1}(\Omega)}, \tag{1.8}$$

where the constants $C_1$ and $C_2$ depend only on $\Omega, \Omega_1$, $\lambda_1$, $\lambda_2$, and $h$.

EXERCISE 1.4. Under the condition $\langle h_0 \rangle = 0$, prove that the equation

$$\Delta_\xi \Psi_0(\xi) = h_0(\xi) \tag{1.9}$$

is solvable in the space $H^1_{\mathrm{per}}(\square)$; moreover, the solution is unique up to an additive constant. Furthermore, $\nabla \Psi_0 \in (L_\infty(\square))^d$ and

$$h(\xi) = \operatorname{div}(\nabla \Psi_0(\xi) + \widetilde{h}(\xi)), \quad (\nabla \Psi_0(\xi) + \widetilde{h}(\xi)) \in (L_\infty(\square))^d,$$

where $\widetilde{h}(\xi) = (h_1(\xi), \dots, h_n(\xi))$.

EXERCISE 1.5. Let $h_0(\xi)$ be a 1-periodic function in $L_\infty(\square)$ with zero average. Prove (without the estimate (1.8)) that there exists a 1-periodic, almost everywhere bounded function $\overline{U}$ such that $\nabla \overline{U} = h(\xi)$. Using this result, prove Lemma 1.1 without the estimate (1.8).

We return to the proof of Lemma 1.1. Substituting $h(\xi) = \operatorname{div}(\nabla \Psi_0(\xi) + \widetilde{h}(\xi))$, integrating by parts, and using the Cauchy–Schwarz–Bunyakovskii inequality, we find

$$\begin{aligned} &\left| \int_\Omega h\Big(\frac{x}{\varepsilon}\Big) u(x)\, v(x)\, dx \right| = \left| \int_\Omega \Big( \operatorname{div}_\xi (\nabla_\xi \Psi_0(\xi) + \widetilde{h}(\xi)) \Big)\Big|_{\xi = x/\varepsilon} u(x)\, v(x)\, dx \right| \\ &= \left| \int_\Omega \varepsilon \operatorname{div}_x \Big[ \nabla_\xi \Psi_0(\xi)\Big|_{\xi = x/\varepsilon} + \widetilde{h}\Big(\frac{x}{\varepsilon}\Big) \Big] u(x)\, v(x)\, dx \right| \\ &= \varepsilon \left| \int_\Omega \Big( \Big( \nabla_\xi \Psi(\xi)\Big|_{\xi = x/\varepsilon} + \widetilde{h}\Big(\frac{x}{\varepsilon}\Big) \Big), \nabla_x (uv) \Big)\, dx \right| \leq K\varepsilon \|u\|_{H^1(\Omega)} \|v\|_{H^1(\Omega)}, \end{aligned} \tag{1.10}$$

where we used the boundedness of the gradient of $\Psi_0$. □

REMARK 1.2. If $u$ and $v$ are not sufficiently smooth (for example, only of class $H^{1/2}$), we need estimates of the form (1.7). In this case, the interpolation inequalities allow us to derive weaker estimates similar to (1.7). We consider only one example and refer the interested reader to [**80**] for details.

Suppose that $p \in [0, 1]$ and $h \in L_\infty(\square)$. Using the bilinear form

$$\int_\Omega h\Big(\frac{x}{\varepsilon}\Big)\, u(x)\, v(x)\, dx, \quad u, v \in H^p(\Omega),$$

we construct an operator $T_\varepsilon : H^p(\Omega) \to H^{-p}(\Omega)$ by the formula

$$\langle T_\varepsilon u, v \rangle_p = \int_\Omega h\Big(\frac{x}{\varepsilon}\Big)\, u(x)\, v(x)\, dx.$$

From (1.10) it follows that $\|T_\varepsilon\|_{\mathcal{L}(H^1(\Omega), H^{-1}(\Omega))} \leq K\varepsilon$. It is obvious that

$$\|T_\varepsilon\|_{\mathcal{L}(L_2(\Omega), L_2(\Omega))} \leq K_1,$$

where the constant $K_1$ is independent of $\varepsilon$.

Using the interpolation technique [**80, 114**], we can derive the estimate

$$\|T_\varepsilon\|_{\mathcal{L}(H^p(\Omega), H^{-p}(\Omega))} \leq K_1^{1-p} K^p \varepsilon^p. \tag{1.11}$$

Setting $p = 1/2$ in (1.11), we obtain an inequality of type (1.7) for functions in $H^{1/2}$.

Lemma 1.1 generalizes, in particular, the following well-known result:

$$h\Big(\frac{x}{\varepsilon}\Big) \rightharpoonup \langle h\rangle \quad \text{weakly as } \varepsilon \to 0.$$

Lemma 1.1 yields an estimate for the convergence rate.

**1.6. The Rellich theorem.** Let $\Omega$ be a bounded domain in $\mathbb{R}^d$.

THEOREM 1.4 (the Rellich theorem). *Let $u_n \in \overset{\circ}{H}{}^1(\Omega)$ be uniformly bounded in $H^1(\Omega)$ with respect to $n$. Then there exists a function $u_0 \in \overset{\circ}{H}{}^1(\Omega)$ and a subsequence $u_m$ such that*

$$u_m \to u_0 \text{ strongly in } L_2(\Omega), \qquad \frac{\partial u_m}{\partial x_j} \rightharpoonup \frac{\partial u_0}{\partial x_j} \text{ weakly in } L_2(\Omega) \text{ as } m \to +\infty.$$

The Rellich theorem is valid for functions in $H^1(\Omega)$ if the boundary of $\Omega$ is sufficiently regular, for example, $\partial\Omega$ is Lipschitz. We prove the Rellich theorem in the simplest case, for functions in $\overset{\circ}{H}{}^1(\Omega)$. In this case, no regularity conditions on the boundary of $\Omega$ are required. The Rellich theorem for functions in $H^1(\Omega)$ can be easily deduced from Theorem 1.4 provided that we already have the following result concerning the extension of functions in $H^1(\Omega)$ to functions in $H^1(\widetilde{\Omega})$, where $\overline{\Omega} \subset \widetilde{\Omega}$ (cf., for example, [**88**]).

THEOREM 1.5 (on extension). *Suppose that $\Omega$ is a Lipschitz domain (with Lipschitz boundary) and $\overline{\Omega} \subset \widetilde{\Omega}$. Then there exists an extension operator $P : H^1(\Omega) \to \overset{\circ}{H}{}^1(\widetilde{\Omega})$ such that*

$$\|Pu\|_{L_2(\widetilde{\Omega})} \le C\|u\|_{L_2(\Omega)}, \qquad \|\nabla(Pu)\|_{L_2(\widetilde{\Omega})} \le C'\|\nabla u\|_{L_2(\Omega)},$$

*where the constants $C$ and $C'$ are independent of $u$.*

EXERCISE 1.6. Using Theorem 1.5, prove the assertion of the Rellich theorem for functions in $H^1(\Omega)$.

We formulate the Sobolev–Kondrashov compactness theorem asserting the compactness of the embedding operators $W_p^l(\Omega) \hookrightarrow L_q(S)$ and $W_p^l(\Omega) \hookrightarrow C(\overline{\Omega})$ (the proof can be found in [**120, 2**]).

THEOREM 1.6. *Suppose that $\Omega$ is an open smooth domain in $\mathbb{R}^d$, $S$ is a smooth $k$-dimensional surface, and $S \subset \overline{\Omega}$.*

1. *If $pl < d$, $q < \frac{pk}{d-pl}$, and $k > d - pl$, then the embedding operator $W_p^l(\Omega) \hookrightarrow L_q(S)$ is compact.*
2. *If $pl > d$ and $k = d$, then the embedding operator $W_p^l(\Omega) \hookrightarrow C(\overline{\Omega})$ is compact.*

PROOF OF THEOREM 1.4. Without loss of generality, we can assume that $\Omega \Subset Q$, where $Q$ is the unit cube in $\mathbb{R}^d$. We extend $u_n$ by zero to $Q\backslash\overline{\Omega}$ and then extend it as a periodic function to $\mathbb{R}^d$. We expand $u_n$ into the Fourier series

$$u_n = \sum_k a_k^n e^{i(k,x)},$$

where $k = (k_1, \dots, k_d)$, $(k, x) = \sum_{l=1}^{d} k_l x_l$, and differentiate with respect to $x_j$. The series

$$\frac{\partial u_n}{\partial x_j} = i \sum_k k_j a_k^n e^{i(k,x)}$$

converges in $L_2(\Omega)$. The sequence $a_k^n$ is uniformly bounded with respect to $n$ since the norms of $u_n(x)$ are uniformly bounded. We extract a converging subsequence $a_1^{n_1}$ from $a_1^n$ and denote the limit by $a_1$. Then we extract a converging subsequence $a_2^{n_2}$ from the bounded sequence $a_2^{n_1}$ and denote the limit by $a_2$ and so on. Finally, we extract the Cantor diagonal from the sequences $n_1, n_2, \dots, n_s, \dots$ and denote it by $m$. Thus, for any fixed $k$,

$$a_k^m \to a_k \quad \text{as } m \to +\infty.$$

Let us show that $u_m(x)$ is a Cauchy sequence. We have

$$\begin{aligned} \|u_{m_1} - u_{m_2}\|_{L_2(\Omega)}^2 &= \sum_k (a_k^{m_1} - a_k^{m_2})^2 \\ &= \sum_{|k|<M} (a_k^{m_1} - a_k^{m_2})^2 + \sum_{|k|\geq M} (a_k^{m_1} - a_k^{m_2})^2. \end{aligned} \tag{1.12}$$

The first sum on the right-hand side of (1.12) contains a finite number of terms. The functions $a_k^m$ converge for any $k$; therefore, for any $\varepsilon$, we can choose $M_0$ such that for all $m_1, m_2 > M_0$ the first sum is less than $\varepsilon/2$. We show that the second sum is also less than $\varepsilon/2$ if $M$ is chosen in a suitable way. Indeed, by the Parseval theorem, we have

$$\|u_m\|_{\overset{\circ}{H}{}^1(\Omega)}^2 = \sum_{j=1}^{d} \int_\Omega \left(\frac{\partial u_m}{\partial x_j}\right)^2 dx = \sum_{j=1}^{d} \sum_k k_j^2 (a_k^m)^2 < L.$$

Therefore,

$$\sum_{|k|\geq M} (a_k^m)^2 \leq \frac{1}{M^2} \sum_{|k|\geq M} (a_k^m)^2 |k|^2 \leq \frac{L}{M^2}.$$

Hence

$$\sum_{|k|\geq M} (a_k^{m_1} - a_k^{m_2})^2 \leq 2 \sum_{|k|\geq M} ((a_k^{m_1})^2 + (a_k^{m_1})^2) \leq \frac{4L}{M^2}.$$

For any $\varepsilon$ there exists $M$ such that $\frac{4L}{M^2} < \frac{\varepsilon}{2}$. Thus, $u_m(x)$ is a Cauchy sequence. Since $L_2(\Omega)$ is complete, there exists $u_0 \in L_2(\Omega)$ such that the functions $u_m(x)$ strongly converge to $u_0$ in $L_2(\Omega)$ as $m \to +\infty$. Since the sequence $\frac{\partial u_m}{\partial x_j}$ is uniformly bounded in $L_2(\Omega)$, there exists a function $v_j \in L_2(\Omega)$ and a subsequence $\frac{\partial u_{m'}}{\partial x_j}$ that weakly converges to $v_j(x)$ in $L_2(\Omega)$. Integrating by parts, it is easy to show that $v_j(x) = \frac{\partial u_0}{\partial x_j}$. □

EXERCISE 1.7. Prove the equality $v_j(x) = \frac{\partial u_0}{\partial x_j}$.

**1.7. The Lax–Milgram theorem.** Let $H$ be a Hilbert space over $\mathbb{R}$ equipped with the inner product $(\cdot,\cdot)_H$ and the corresponding norm $\|\cdot\|_H$. Let $[\cdot,\cdot]$ denote a bilinear form. The following assertion generalizes the Riesz representation theorem.

LEMMA 1.2 (the Lax–Milgram theorem). *If*
1) $|[u, v]| \leq \kappa_1 \|u\|_H \|v\|_H$, *and*
2) $[u, u] \geq \kappa_2 \|u\|_H^2$,
*then for any linear functional* $f(u)$ *on* $H$ *there exists a unique* $v \in H$ *such that* $f(u) = [v, u]$.

PROOF. We cannot apply the Riesz representation theorem directly since $[\cdot,\cdot]$ cannot be taken for a new inner product in $H$. Indeed, in the general case, the bilinear form $[\cdot,\cdot]$ is not symmetric.

For fixed $v$ the form $[v, u]$ can be regarded as a linear continuous functional on $H$. By the Riesz representation theorem, there exists $Av \in H$ such that $[v, u] = (Av, u)_H$. Thereby we can define an operator $A : H \to H$. It suffices to show that $A$ is bijective. We have

$$\|Av\|_H^2 = (Av, Av)_H = [v, Av] \leq \kappa_1 \|v\|_H \|Av\|_H, \tag{1.13}$$

i.e., $\|A\|_{\mathcal{L}(H,H)} \leq \kappa_1$. On the other hand,

$$\|Au\|_H \|u\|_H \geq (Au, u)_H = [u, u] \geq \kappa_2 \|u\|_H^2, \tag{1.14}$$

i.e., $\|A\|_{\mathcal{L}(H,H)} \geq \kappa_2$.

First we show that $A$ is injective, i.e., $\operatorname{Ker} A = \{0\}$. Let $Au_1 = Au_2$. From (1.14) it follows that $0 = \|A(u_1 - u_2)\|_H \geq \kappa_2 \|u_1 - u_2\|$. Consequently, $u_1 = u_2$. Thus, the kernel of the operator $A$ consists of the zero element.

Let us now show that $\operatorname{Im} A$ is closed. Let $\{Av_n\}$ be a Cauchy sequence in $H$. Taking into account (1.14), we conclude that $v_n$ is a Cauchy sequence in $H$. By the completeness of $H$, there exists $v_0 \in H$ such that $v_n \to v_0$ as $n \to \infty$. From (1.13) it follows that

$$\|Av_n - Av_0\|_H \leq \kappa_1 \|v_n - v_0\|_H \to 0 \quad \text{as } n \to +\infty,$$

i.e., $Av_n \to Av_0$ as $n \to +\infty$. This means that the limit of $Av_n$ is the image of $v_0$ and, consequently, the image of $A$ is closed.

It remains to show that $A$ is surjective, i.e., $\operatorname{Im} A = H$. Assume the contrary. Suppose that there exists $w \in H$ such that $w \perp \operatorname{Im} A$. Then $0 = (w, Aw)_H = [w, w] \geq \kappa_2 \|w\|_H^2$. Hence $w = 0$ and $A$ is surjective. Thus, $A$ is bijective.

By the Riesz representation theorem, for any linear continuous functional $f(u)$ there exists a unique $z \in H$ such that $f(u) = (z, u)_H$. For $z \in H$ there exists a unique $x \in H$ such that $z = Ax$. Therefore, $f(u) = (z, u)_H = (Ax, u)_H = [x, u]$. □

**1.8. Existence of a vector field with a given divergence.** Consider a bounded domain $\Omega \subset \mathbb{R}^d$ and recall the definition of a star-shaped domain (cf. Definition 1.11) for the sake of convenience.

DEFINITION 1.12. A domain $\Omega \subset \mathbb{R}^d$ is said to be *star-shaped with respect to a ball* $Q \Subset \Omega$ if for any points $x \in Q$ and $y \in \Omega$ the segment joining $x$ and $y$ lies inside $\Omega$.

DEFINITION 1.13. A domain $\Omega \subset \mathbb{R}^d$ is said to be *star-shaped with respect to a system* $\{Q_i\}$ *of balls* if $\Omega = \bigcup_{i=1}^m \Omega_i$, where each domain $\Omega_i$ is star-shaped with respect to some ball $Q_i \Subset \Omega_i$.

EXERCISE 1.8. Is it true that a Lipschitz domain is star-shaped with respect to a ball?

EXERCISE 1.9. Is it true that a Lipschitz domain is star-shaped with respect to some system of balls?

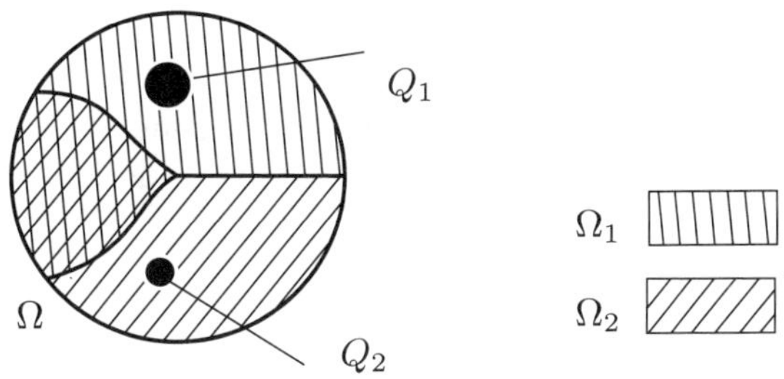


FIGURE 1.1

It is easy to see that the converse to Exercise 1.9 is not true. Indeed, the disk with a cut along the radius (cf. Figure 1.1) is not a Lipschitz domain. However, it can be represented as the union of two domains $\Omega_1$ and $\Omega_2$ that are star-shaped with respect to the balls $Q_1$ and $Q_2$ respectively.

DEFINITION 1.14. A domain is called a *generalized Lipschitz domain* if it is represented as the union of finitely many Lipschitz domains.

EXERCISE 1.10. Is Definition 1.14 equivalent to the definition of a domain that is star-shaped with respect to a finite system of balls?

EXERCISE 1.11. Give an example of a bounded domain which is not star-shaped with respect to any system of balls.

We will use the following important assertion.

THEOREM 1.7. *Let $\Omega$ be star-shaped with respect to a system of balls, and let a function $f(x) \in L_2(\Omega)$ have the zero average over $\Omega$, i.e., $\int_\Omega f(x)\,dx = 0$. Then there exists a (nonunique!) vector-valued function $\overline{U}(x) \in \overset{\circ}{H}{}^1(\Omega)$ such that* $\operatorname{div} \overline{U}(x) = f(x)$ *in $\Omega$ and*

$$\|\overline{U}\|_{\overset{\circ}{H}{}^1(\Omega)} \le C\|f\|_{L_2(\Omega)}, \tag{1.15}$$

*where the constant $C > 0$ is determined by the ratio of the diameters of the domains $\Omega_i$ and the balls $Q_i$ in the definition of a star-shaped domain and also by the volumes of $\Omega_i$ and their intersections.*

For Lipschitz domains this theorem was proved in [**76**]. The results of [**76**] were generalized in [**19**] to the case of star-shaped domains and $f \in L_p(\Omega)$. In [**19**], there is an explicit expression of the vector field $\overline{U}(x)$ in a star-shaped domain $\Omega$ with respect to some ball $Q$:

$$U(x) = \int_\Omega f(y)\left\{\frac{x-y}{|x-y|^d}\int_{|x-y|}^\infty q\left(y + \xi\frac{x-y}{|x-y|}\right)\xi^{d-1}\,d\xi\right\}dy, \tag{1.16}$$

where $q(x)$ is an arbitrary function in $C_0^\infty(Q)$ such that $\int_Q q(x)\,dx = 1$.

EXERCISE 1.12. Verify that $\overline{U}$ vanishes on the boundary of $\Omega$.

EXERCISE 1.13. Verify that $\overline{U}$ satisfies the equation $\operatorname{div} \overline{U}(x) = f(x)$.

To prove the estimate (1.15), we can use the explicit representation of $\overline{U}(x)$ in accordance with the Calderon–Zygmund theorem [**33**].

Based on the above representation, we establish the existence of a vector field $\overline{U}(x)$ with the required properties in the union of finitely many domains that are star-shaped with respect to balls. We need the following assertion.

EXERCISE 1.14. Let $\Omega = \bigcup_{i=1}^{m} \Omega_i$, where $\Omega_i$ are some domains, and let $f(x) \in L_p(\Omega)$ be such that $\int_\Omega f(x)\,dx = 0$. Then there exist functions $f_i(x) \in L_p(\Omega)$ such that

$$\operatorname{supp} f_i \Subset \Omega_i, \quad i = 1, \dots, m,$$

$$f(x) = \sum_{i=1}^{m} f_i(x), \quad \int_\Omega f_i(x)\,dx = 0, \quad i = 1, \dots, m.$$

*Hint.* For example, in the case $m = 2$, we can set

$$f_1(x) = \begin{cases} f(x) - \frac{\chi(x)}{\mu} \int_{\Omega_1} f(x)\,dx, & x \in \Omega_1, \\ 0, & x \in \Omega_2 \backslash \Omega_1. \end{cases}$$

$$f_2(x) = \begin{cases} (1 - \chi(x))f(x) - \frac{\chi(x)}{\mu} \int_{\Omega_2 \backslash \Omega_1} f(x)\,dx, & x \in \Omega_2, \\ 0, & x \in \Omega_1 \backslash \Omega_2, \end{cases}$$

where $\chi(x)$ is the characteristic function of the set $\Omega_1 \cap \Omega_2$ and $\mu$ is the measure of $\Omega_1 \cap \Omega_2$.

Now, the vector field $\overline{U}(x)$ can be represented as $\overline{U} = \sum_{i=1}^{m} \overline{U}_i$, where $\overline{U}_i(x)$, $i = 1, \dots, m$, are vector fields defined by (1.16) for the domains $\Omega_i$, $i = 1, \dots, m$, respectively.

As was shown in [**19**], the constant $C > 0$ in the estimate

$$\|\overline{U}\|_{\overset{\circ}{H}{}^1(\Omega)} \le C\|f\|_{L_2(\Omega)},$$

where $\Omega$ is star-shaped with respect to some ball $Q \Subset \Omega$, depends only on the ratio of the diameter of $\Omega$ to the diameter of $Q$. Thus, under a homothetic transformation of $\Omega$ with respect to $Q$, the constant $C$ in (1.15) remains unchanged.

Assume that $\Omega$ is the union of finitely many star-shaped domains $\Omega_i$ with respect to balls $Q_i$. Using the representation $f(x) = \sum_{i=1}^{m} f_i(x)$, $\operatorname{supp} f_i \subset \Omega_i$, it is easy to see that the constant $C > 0$ in the estimate (1.15) can be expressed in terms of the maximum constant $C_i > 0$ among all the constants in the estimates

$$\|\overline{U}_i\|_{\overset{\circ}{H}{}^1(\Omega_i)} \le C_i\|f_i\|_{L_2(\Omega_i)}$$

for $\Omega_i$, the volumes of $\Omega_i$, and the volumes of the pairwise intersections $\Omega_i \cap \Omega_j$.

EXERCISE 1.15. Suppose that $\Omega = \Omega_1 \cup \Omega_2$, where $\Omega_1$ and $\Omega_2$ are star-shaped with respect to balls $Q_1 \subset \Omega_1$ and $Q_2 \subset \Omega_2$ respectively. Let $C_1$ and $C_2$ be constants in (1.15) for $\Omega_1$ and $\Omega_2$. Derive an estimate for the constant $C$ in the inequality (1.15) for the domain $\Omega$ in terms of $C_1$, $C_2$, $|\Omega_1|$, $|\Omega_2|$, and $|\Omega_1 \cap \Omega_2|$.

EXERCISE 1.16. Let $\Omega$ be a domain star-shaped with respect to a system of balls. Prove that there exists a sequence of domains $\Omega_i$, $i = 1, 2, \dots$, such that $\Omega_{i+1} \supset \Omega_i$, $\Omega_i \Subset \Omega$, $\Omega = \bigcup_{i=1}^{\infty} \Omega_i$, and Theorem 1.7 is valid for each of the domains $\Omega_i$. Show that the constant in (1.15) is independent of $i = 1, 2, \dots$.

Exercise 1.16 will be used in the proof of the existence of a weak solution to the Stokes equations.

## 2. Some facts of ergodic theory and stochastic analysis

**2.1. Definitions and preliminaries.** Let $(\Omega, \mathcal{F}, \mu)$ be a *probability space*, i.e., a set $\Omega$ equipped with the $\sigma$-algebra $\mathcal{F}$ of subsets of $\Omega$ and with a $\sigma$-additive nonnegative measure $\mu$ on $\mathcal{F}$ such that $\mu(\Omega) = 1$.

DEFINITION 2.1. A *dynamical system with d-dimensional time* is a family of measurable mappings $T(x) : \Omega \to \Omega$, $x \in \mathbb{R}^d$, with the following properties:

1) *group property*,

$$T(x + y) = T(x)T(y) \quad \forall x, y \in \mathbb{R}^d, \quad T(0) = \mathrm{Id},$$

where Id is the identity mapping;

2) *isoperimetric property* (the transform $T(x)$ preserves the measure $\mu$ on $\Omega$),

$$T(x)\mathcal{U} \in \mathcal{F}, \quad \mu(T(x)\mathcal{U}) = \mu(\mathcal{U}) \quad \forall x \in \mathbb{R}^d, \ \mathcal{U} \in \mathcal{F};$$

3) for any measurable function $f(\omega)$ on $\Omega$ the function $f(T(x)\omega)$ is measurable on the Cartesian product $\Omega \times \mathbb{R}^d$, where the space $\mathbb{R}^d$ is equipped with the Borel $\sigma$-algebra of subsets.

Hereinafter, the Borel $\sigma$-algebra is denoted by $\mathcal{B}$.

Denote by $L_q(\Omega)$, $q \geq 1$, the space of measurable functions that are $q$-power summable with respect to the measure $\mu$, and by $L_\infty(\Omega)$ the space of essentially bounded measurable functions. In $L_2(\Omega)$, we consider the induced $d$-parametric group of operators $U(x)$, $x \in \mathbb{R}^d$,

$$(U(x)f)(\omega) = f(T(x)\omega), \quad f \in L_2(\Omega).$$

By condition 2) of Definition 2.1, the operator $U(x) : L_2(\Omega) \to L_2(\Omega)$ is unitary for every $x \in R^d$. Condition 3) of Definition 2.1 implies that the group $U(x)$ is *strongly continuous*, i.e., for any $f \in L_2(\Omega)$,

$$\lim_{x \to 0} \|U(x)f - f\|_{L_2(\Omega)} = 0.$$

DEFINITION 2.2. Let $f(\omega)$ be a measurable function on $\Omega$. By a *realization of* $f$ we mean the function $f(T(x)\omega)$ of variable $x \in \mathbb{R}^d$ with fixed $\omega \in \Omega$.

The following important assertion holds.

LEMMA 2.1. *Let $f \in L_q(\Omega)$. Then almost all realizations $f(T(x)\omega)$ belong to $L_q^{\mathrm{loc}}(\mathbb{R}^d)$.*

*If a sequence $f_k \in L_q(\Omega)$ converges in $L_q(\Omega)$ to a function $f$, then almost all realizations $f_k(T(x)\omega)$ converge on some subsequence to the realization $f(T(x)\omega)$ in $L_q^{\mathrm{loc}}(\mathbb{R}^d)$.*

PROOF. The first assertion immediately follows from the definitions and Fubini theorem. The second assertion follows from the equality

$$\int_{-a}^{a} \cdots \int_{-a}^{a} \left( \int_\Omega |f_k(T(x)\omega) - f(T(x)\omega)|^q d\mu \right) dx$$
$$= \int_\Omega \left( \int_{-a}^{a} \cdots \int_{-a}^{a} |f_k(T(x)\omega) - f(T(x)\omega)|^q dx \right) d\mu,$$

where $a = \text{const} > 0$, which is also a consequence of the Fubini theorem. Since the expression in parentheses under the integral sign on the right-hand side of the last equality converges to zero in $L_1(\Omega)$, we can extract a subsequence converging almost everywhere. □

DEFINITION 2.3. A measurable function $f(\omega)$ in $\Omega$ is said to be *invariant* if $f(T(x)\omega) = f(\omega)$ for any $x \in \mathbb{R}^d$ and almost all $\omega \in \Omega$.

DEFINITION 2.4. A dynamical system $T(x)$ is said to be *ergodic* if any invariant function is constant almost everywhere.

Let $f(x) \in L_1^{\mathrm{loc}}(\mathbb{R}^d)$.

DEFINITION 2.5. We say that a function $f(x)$ has the spatial average, denoted by $M(f)$, if the limit

$$M(f) = \lim_{\varepsilon \to 0} \frac{1}{|B|} \int_B f\Big(\frac{x}{\varepsilon}\Big)\, dx \tag{2.1}$$

exists for all bounded Borel sets $B \in \mathcal{B}$ and is independent of the choice of $B$.

Equivalently, we can write

$$M(f) = \lim_{t \to +\infty} \frac{1}{|B_t|} \int_{B_t} f(x)\, dx, \quad B_t = \{x \in \mathbb{R}^d \mid x/t \in B\}. \tag{2.2}$$

EXERCISE 2.1. Suppose that $f(x)$ has spatial average in $\mathbb{R}^d$ and the sequence $f(x/\varepsilon)$, $0 < \varepsilon \leq 1$, is bounded in $L_q(\mathcal{K})$ for some $q \geq 1$, where $\mathcal{K}$ is a compact set in $\mathbb{R}^d$. Prove the weak convergence

$$f\Big(\frac{x}{\varepsilon}\Big) \rightharpoonup M(f) \quad \text{in } L_q^{\mathrm{loc}}(\mathbb{R}^d) \quad \text{as } \varepsilon \to 0.$$

We need the following statement of the Birkhoff theorem (cf., for example, [**57, 137, 138**] and [**107**]).

THEOREM 2.1 (the Birkhoff theorem). *Let $f \in L_q(\Omega)$, $q \geq 1$. Then for almost all $\omega \in \Omega$ the realization $f(T(x)\omega)$ has spatial average $M(f(T(x)\omega))$. Furthermore, the average $M(f(T(x)\omega))$ is the conditional expectation of $f(\omega)$ with respect to the $\sigma$-algebra of invariant subsets. Consequently, $M(f(T(x)\omega))$ is an invariant function and*

$$E(f) \equiv \int_\Omega f(\omega)\, d\mu = \int_\Omega M(f(T(x)\omega))\, d\mu. \tag{2.3}$$

*In particular, if the dynamical system $T(x)$ is ergodic, then $E(f) = M(f(T(x)\omega))$ for almost all $\omega \in \Omega$.*

DEFINITION 2.6. A random field $f(x,\omega)$, $x \in \mathbb{R}^d$, $\omega \in \Omega$, is said to be *statistically homogeneous* if $f(x,\omega) = f(T(x)\omega)$, $f(\omega) \in L_2(\Omega)$, where $T(x)$ is a dynamical system in $\Omega$.

DEFINITION 2.7. 1. A vector-valued function $u(x) \in [L_{2,\mathrm{loc}}(\mathbb{R}^d)]^d$ is said to be *potential* if there exists a scalar function $V(x) \in H^1_{\mathrm{loc}}(\mathbb{R}^d)$ such that $\nabla V(x) = u(x)$, where the derivatives are understood in the sense of distributions.

2. A vector-valued function $\overline{u}(x) \in [L_{2,\mathrm{loc}}(\mathbb{R}^d)]^d$ is said to be *solenoidal* if $\mathrm{div}_x \overline{u}(x) = 0$ in the sense of distributions.

3. A random vector $f \in (L_2(\Omega))^d$ is said to be *potential* (*solenoidal*) if almost all realizations $f(T(x)\omega)$ of $f$ are potential (solenoidal) in $\mathbb{R}^d$.

### 2.2. The Weyl decomposition.

LEMMA 2.2 (the Weyl decomposition). *The following orthogonal decomposition holds:*

$$[L_2(\Omega)]^d = \mathcal{V}_2^{\text{pot}}(\Omega) \oplus \mathcal{V}_2^{\text{sol}}(\Omega) \oplus \mathbb{R}^d = \mathcal{V}_2^{\text{pot}}(\Omega) \oplus L_2^{\text{sol}}(\Omega), \tag{2.4}$$

*where* $\mathcal{V}_2^{\text{pot}}(\Omega) = \{f \in L_2^{\text{pot}}(\Omega), E(f) = 0\}$ *and* $\mathcal{V}_2^{\text{sol}}(\Omega) = \{f \in L_2^{\text{sol}}(\Omega), E(f) = 0\}$.

Our proof is close to that in [**137**]. We need some auxiliary assertions. We define the generator of a $d$-parametric group of unitary operators $U(x)$.

DEFINITION 2.8. By the *generator* $\partial_i$ of the group $U(x)$ of unitary operators we mean the operator

$$\partial_i f = \lim_{\Delta x_i \to 0} \frac{U(\Delta x_i) f(\omega) - f(\omega)}{\Delta x_i} = \lim_{\Delta x_i \to 0} \frac{f(T(\Delta x_i)\omega) - f(\omega)}{\Delta x_i} \quad \text{in } L_2(\Omega), \tag{2.5}$$

where $\Delta x_i = (0, \dots, \Delta x_i, \dots, 0)$. The operator $\partial_i$ is defined for functions $f \in L_2(\Omega)$ such that the limit (2.5) exists as $\Delta x_i \to 0$.

Since $U(x)$ is unitary, the operators $\partial_i$ are skew-symmetric, i.e.,

$$\int_\Omega \partial_i f g \, d\mu = - \int_\Omega f \partial_i g \, d\mu,$$

where $f$ and $g$ belong to the domain of $\partial_i$. In particular, this means that $\int_\Omega \partial_i f \, d\mu = 0$, $i = 1, \dots, d$.

We prove that the domain $D(\partial_i)$ of the operator $\partial_i$ is dense in $L_2(\Omega)$.

Let $f \in L_2(\Omega)$. Consider the regularized function

$$f^\delta(\omega) = \int_{\mathbb{R}^d} K^\delta(y) f(T(y)\omega) \, dy, \tag{2.6}$$

where $K^\delta(t) = \frac{1}{\delta^d} K(y/\delta)$ and $K(y) \in C_0^\infty(\mathbb{R}^d)$ is nonnegative, even, i.e., $K(y) = K(-y)$, and $\int_{\mathbb{R}^d} K(y) \, dy = 1$. Assume that the support of $K(y)$ is contained in the ball $\{t : |t| \le \rho\}$. It is easy to see that almost all realizations of $f^\delta$ are infinitely differentiable. Indeed,

$$f^\delta(T(x)\omega) = \int_{\mathbb{R}^d} K^\delta(y) f(T(x+y)\omega) \, dy = \int_{\mathbb{R}^d} K^\delta(y-x) f(T(y)\omega) \, dy \tag{2.7}$$

and the kernel $K^\delta$ is an infinitely differentiable function. We also note that (2.6) can be written in the form

$$f^\delta = \int_{\mathbb{R}^d} K^\delta(y) U(y) f \, dy.$$

Since $K(y)$ is even, the following equality holds (check it!):

$$\int_\Omega f^\delta g \, d\mu = \int_\Omega f \, g^\delta \, d\mu.$$

PROPOSITION 2.1. *For any* $f \in L_2(\Omega)$

$$\lim_{\delta \to 0} \|f - f^\delta\|_{L_2(\Omega)} = 0. \tag{2.8}$$

PROOF. Since $\int_{\mathbb{R}^d} K^\delta(t)\,dt \equiv 1$, we have

$$f^\delta - f = \int_{\mathbb{R}^d} K^\delta(t)(U(t)f - f)\,dt = \int_{\mathbb{R}^d} K(t)(U(\delta t)f - f)\,dt.$$

Since the group $U(x)$ is strongly continuous, we have

$$\|f^\delta - f\|_{L_2(\Omega)} \le \int_{|t|\le\rho} K(t)\|U(\delta t)f - f\|_{L_2(\Omega)}\,dt \le \sup_{|t|\le\rho} \|U(\delta t)f - f\|_{L_2(\Omega)} \to 0.$$

The proposition is proved. □

Let $f(\omega) \in L_2(\Omega)$. We prove that $f^\delta(\omega) \in D(\partial_i)$. The following equality holds (check it!):

$$\frac{f^\delta(T(\Delta x_i)\omega) - f^\delta(\omega)}{\Delta x_i} = \int_{\mathbb{R}^d} \frac{K^\delta(\Delta x - t) - K^\delta(-t)}{\Delta x_i}[U(t)f(\omega)]dt$$

(recall that $\Delta x_i = (0,\dots,\Delta x_i, 0,\dots,0)$). It is obvious that the right-hand side converges to a function in $L_2(\Omega)$ as $\Delta x_i \to 0$, i.e., $f^\delta(\omega) \in D(\partial_i)$.

The group $T(x)$ is strongly continuous and unitary because $i\partial_j$, $j = 1,\dots,d$, are closed self-adjoint operators (cf. [**57**]).

From (2.7) it follows that for any potential vector $f$ almost all realizations $f^\delta(T(x)\omega)$ of $f$ are also potential.

PROOF OF LEMMA 2.2. We represent an arbitrary vector $f \in (L_2(\Omega))^d$ in the form $f = p + s$, where $p \in \mathcal{V}_2^{\text{pot}}(\Omega)$ and $s \in (\mathcal{V}_2^{\text{pot}}(\Omega))^\perp$. It suffices to check that $s \in (L_2^{\text{sol}}(\Omega))$.

For an arbitrary function $u \in L_2(\Omega)$ we construct a vector in $\mathcal{V}_2^{\text{pot}}$ as follows. For a regularized function $u^{\delta_1}$ we consider $(\partial_1 u^{\delta_1},\dots,\partial_d u^{\delta_1})$. Recall that $\int_\Omega \partial_i u^{\delta_1}\,d\mu = 0$.

EXERCISE 2.2. Let $s \in (\mathcal{V}_2^{\text{pot}}(\Omega))^\perp$. Prove that $s^\delta \in (\mathcal{V}_2^{\text{pot}}(\Omega))^\perp$ for any $\delta > 0$.

Using Exercise 2.2, we see $s^\delta \in (\mathcal{V}_2^{\text{pot}}(\Omega))^\perp$. Therefore,

$$0 = \sum_{k=1}^d \int_\Omega s_k^\delta \partial_k(u^{\delta_1})\,d\mu = -\sum_{k=1}^d \int_\Omega u^{\delta_1}\partial_k s_k^\delta\,d\mu.$$

Passing to the limit as $\delta_1 \to 0$, we arrive at the equality

$$\sum_{k=1}^d \int_\Omega u\partial_k s_k^\delta\,d\mu = 0,$$

which holds for an arbitrary function $u \in L_2(\Omega)$. Hence $\sum_{k=1}^d \partial_k s_k^\delta = 0$ almost everywhere in $\Omega$. Then (2.5) implies $\operatorname{div}_x s^\delta(T(x)\omega) = 0$ for almost all $\omega \in \Omega$. It follows that $s^\delta \in L_2^{\text{sol}}(\Omega)$ and $s \in (L_2^{\text{sol}}(\Omega))^d$ (check it!). □

**2.3. Weak convergence of measures. The Prokhorov theorem.** Let $X$ be a complete separable metric space. Denote by $\mathcal{B}$ the Borel $\sigma$-algebra of subsets of $X$, i.e., the $\sigma$-algebra generated by all open sets in $X$. The set of all bounded continuous functions in $X$ is denoted by $C_B(X)$. Consider a sequence $\mu_n$ of nonnegative $\sigma$-additive measures on $X$.

DEFINITION 2.9. We say that a sequence of measures $\mu_n$ *weakly converges* to a measure $\mu$ if for every $f \in C_B(X)$ we have

$$\lim_{n\to\infty} \int_X f(x)\mu_n(dx) = \int_X f(x)\mu(dx).$$

Let $M$ be a family of nonnegative Borel measures on $X$.

DEFINITION 2.10. The family $M$ is said to be *relatively compact* if any sequence of measures $\mu_n$ in $M$ contains a weakly converging subsequence.

THEOREM 2.2 (the Prokhorov theorem). *Let $X$ be a complete separable metric space. A family $M$ of nonnegative measures is weakly relatively compact if and only if the following two conditions are satisfied:*

(1) *there exists $C > 0$ such that $\mu(X) \leq C$ for all $\mu \in M$;*

(2) *for any $\varepsilon > 0$ there exists a compact set $K_\varepsilon$ in $X$ such that $\mu(X \backslash K_\varepsilon) < \varepsilon$ for all $\mu \in M$.*

We recall the definition of the distribution of a random function and of the convergence in law. Let $(\Omega, \mathcal{F}, \mathbf{P})$ be a probability space. By a *random function* we mean a measurable mapping $\xi : \Omega \longrightarrow X$, where $X$ is a metric space equipped with the Borel $\sigma$-algebra. If for $X$ we take $\mathbb{R}^1$ or $\mathbb{R}^d$, then $\xi$ is usually called a *random variable* or a *random vector*, respectively.

DEFINITION 2.11. The *distribution of a random function* $\xi$ is a probability measure in $(X, \mathcal{B})$ defined for every $A \in \mathcal{B}$ by the equality $\mathcal{L}[\xi]\,(A) = \mathbf{P}\{\xi \in A\}$.

Let $\xi_n$ be random functions that take values in a metric space $X$ and are defined on the same or different probability spaces.

DEFINITION 2.12. We say that a sequence $\xi_n$ *converges in law* to $\xi$ as $n \to \infty$ if the distributions $\mathcal{L}[\xi_n]$ of $\xi_n$ weakly converge in $X$ to the distribution of $\xi$. In this case, we write

$$\xi_n \xrightarrow{\mathcal{L}} \xi.$$

We note that for probability measures in $(X, \mathcal{B})$, in particular, for the distributions of random functions, condition (1) in the Prokhorov theorem is automatically fulfilled. Therefore, for an arbitrary family of random functions taking values in a separable complete metric space, the necessary and sufficient condition for the relative compactness in law is the existence of a compact set $K_\varepsilon$, $\varepsilon > 0$, in $X$ such that $\mathbf{P}\,(\xi \notin K_\varepsilon) < \varepsilon$ for every $\xi$ in the family.

The Skorokhod theorem plays an important role in the analysis of weakly converging sequences of measures. Let $E$ be a separable Banach space.

THEOREM 2.3 (the Skorokhod theorem). *For an arbitrary sequence of probability measures $\mu_n$ on $\mathcal{B}(E)$ weakly converging to a measure $\mu$ there is a probability space $(\Omega, \mathcal{F}, \mathbf{P})$ and random functions $\eta, \eta_1, \ldots, \eta_k, \ldots$ that are defined in $(\Omega, \mathcal{F}, \mathbf{P})$, take values in $E$, and satisfy the relations*

$$\mathcal{L}[\eta_k] = \mu_k, \quad \mathcal{L}[\eta] = \mu, \quad \lim_{k\to\infty} \eta_k = \eta \quad \mathbf{P}\text{-}\textit{almost surely.}$$

**Central limit theorem.** Consider a stationary ergodic process $\xi_t$ with values in $\mathbb{R}^1$ or $\mathbb{R}^d$. Let $\xi_s$ be an arbitrary stationary random process on a probability space $(\Omega, \mathcal{F}, \mathbf{P})$. Denote by $\mathcal{F}_{\leq t}$ the $\sigma$-algebra generated by $\xi_s$ for all $s \leq t$, and let $\mathcal{F}_{\geq t}$ be the $\sigma$-algebra generated by $\xi_s$ for all $s \geq t$. In other words, $\mathcal{F}_{\leq t}$ ($\mathcal{F}_{\geq t}$) is the minimal $\sigma$-algebra, with respect to which the process $\xi_s$ is measurable for all $s \leq t$ ($s \geq t$). The existence of such a $\sigma$-algebra is a well-known fact in measure theory.

DEFINITION 2.13. The *strong mixing coefficient* of a process $\xi_s$ is defined by the formula

$$\kappa(\gamma) = \sup_{E_1 \in \mathcal{F}_{\leq t},\, E_2 \in \mathcal{F}_{\geq (t+\gamma)}} |\mathbf{P}(E_1)\mathbf{P}(E_2) - \mathbf{P}(E_1 \cap E_2)|,$$

where the supremum is taken over all measurable $E_1$ and $E_2$ with respect to $\mathcal{F}_{\leq t}$ and $\mathcal{F}_{\geq (t+\gamma)}$ respectively. Since the process is stationary, the expression on the right-hand side is independent of $t$. In particular, for $t$ we can take 0.

DEFINITION 2.14. The *uniform mixing coefficient* $\varphi(\gamma)$ is defined by the formula

$$\varphi(\gamma) = \sup_{\substack{E_1 \in \mathcal{F}_{\leq t},\, E_2 \in \mathcal{F}_{\geq (t+\gamma)} \\ \mathbf{P}(E_2) \neq 0}} \left| \mathbf{P}(E_1) - \frac{\mathbf{P}(E_1 \cap E_2)}{\mathbf{P}(E_2)} \right|,$$

and the *maximal correlation coefficient* $\rho(\gamma)$ is defined by the formula

$$\rho(\gamma) = \sup_{\eta_1,\, \eta_2} \left| \frac{\mathbf{E}((\eta_1 - \mathbf{E}\eta_1)(\eta_2 - \mathbf{E}\eta_2))}{\sqrt{\mathbf{E}\eta_1^2\, \mathbf{E}\eta_2^2}} \right|,$$

where $\eta_1 \in L_2(\Omega, \mathcal{F}_{\leq t}, \mathbf{P})$ and $\eta_2 \in L_2(\Omega, \mathcal{F}_{\geq (t+\gamma)}, \mathbf{P})$.

THEOREM 2.4. *Let $\xi_t \in (L_1(\Omega))^d$. Assume that at least one of the following conditions holds*:

$$\int_0^\infty \kappa(\gamma)\, d\gamma < \infty, \qquad \int_0^\infty \sqrt{\varphi(\gamma)} d\gamma < \infty, \qquad \int_0^\infty \rho(\gamma)\, d\gamma < \infty.$$

*Then for every $T > 0$ we have*

$$\frac{1}{\sqrt{t}} \int_0^{tT} \xi_\gamma d\gamma \xrightarrow{\mathcal{L}} \mathcal{N}(0, T\sigma\sigma^*),$$

*where*

$$\sigma\sigma^* = \mathbf{E} \int_0^\infty (\xi_\gamma \times \xi_0 + \xi_0 \times \xi_\gamma) d\gamma.$$

Hereinafter, $\mathcal{N}(\alpha, \beta)$ is the normal distribution with mean $\alpha$ and covariance $\beta$, whereas $\xi \times \eta$ denotes the Kronecker product of matrices: $(\xi \times \eta)_{ij} = \xi_i \eta_j$.

In many cases, this result can be significantly strengthened. Consider a mapping sending a trajectory $\{\xi_s, 0 \leq s < \infty\}$ to a continuous function in $[0, T]$ defined by the formula

$$Y_t(r) = \frac{1}{\sqrt{t}} \int_0^{tr} \xi_s ds, \quad 0 \leq r \leq T.$$

THEOREM 2.5. *Let $\xi_s \in L_\infty(\Omega)$. Let the assumptions of Theorem 2.4 be satisfied. Then, on each segment $[0, T]$,*

$$Y_t(\cdot) \xrightarrow{\mathcal{L}} \sigma W_\bullet \quad \text{as } t \to \infty$$

*in $(C[0, T])^d$, where $W_t$ is a standard $d$-dimensional Wiener process.*

**2.4. The simplest stochastic partial differential equations.** We briefly discuss the simplest stochastic partial differential equations with finite-dimensional multiplicative noise. Such equations appear after the homogenization of random parabolic operators with lower order terms (cf. Section 18 below).

Let $W_t$ be a standard Wiener process on a probability space $(\Omega, \mathcal{F}, \mathbf{P})$. We assume that it is measurable with respect to the right continuous filtration of the $\sigma$-algebra $\mathcal{F}_t$. The definition and properties of Wiener processes and stochastic integrals over Wiener processes can be found, for example, in [**101**].

On $[0,T] \times \Omega$, we construct the $\sigma$-algebra generated by the sets $(s,t] \times \mathcal{A}$, $0 \le s < t \le T$, $\mathcal{A} \in \mathcal{F}_s$, $\{0\} \times \mathcal{A}$, $\mathcal{A} \in \mathcal{F}_0$, and denote it by $\mathcal{G}^\infty$.

DEFINITION 2.15. A process $\xi_t$ with values in a Banach space $E$ is said to be *predictable* if $\xi_t$ regarded as a mapping from $([0,T]\times\Omega, \mathcal{G}^\infty)$ to $(E, \mathcal{B})$ is measurable, where $\mathcal{B}$ is the Borel $\sigma$-algebra.

In $\mathbb{R}^d$, we consider the uniformly elliptic differential operator

$$A = \frac{\partial}{\partial x_j}\left(a_{ij}(x)\frac{\partial}{\partial x_i}\right) + b_i(x)\frac{\partial}{\partial x_i} + c(x) \tag{2.9}$$

with measurable coefficients such that

$$a_{ij}(x) z_i z_j \ge \alpha |z|^2 \quad \forall z \in \mathbb{R}^d,$$
$$|a_{ij}(x)| \le \alpha^{-1}, \quad |b_i(x)| \le \alpha^{-1}, \quad |c(x)| \le \alpha^{-1},$$

where $\alpha > 0$. Define the operator $A$ in $L_2(\mathbb{R}^d)$ as the closure of (2.9) on $C_0^\infty(\mathbb{R}^d)$ in the space $L_2(\mathbb{R}^d)$.

It is known that the domain of the adjoint operator $A^*$ is dense in $L_2(\mathbb{R}^d)$.

Suppose that $u_0 \in L_2(\mathbb{R}^d)$ and $f(t,x) \in L_2\big((0,T)\times\mathbb{R}^d\big)$.

DEFINITION 2.16. A predictable process $u(t,\cdot)$ (henceforth, we omit the second argument in the notation) with values in $L_2(\mathbb{R}^d)$ is called a *solution* to the following boundary value problem for the stochastic partial differential equation:

$$du(t) = Audt + \lambda u dW_t + f(t),$$
$$u\big|_{t=0} = u_0$$

if

$$\mathbf{P}\Big\{\int_0^T \|u\|^2_{L_2(\mathbb{R}^d)}(t)dt < \infty\Big\} = 1$$

and for every $\varphi \in D(A^*)$,

$$(u(t),\varphi) = (u_0,\varphi) + \int_0^t (u(s), A^*\varphi)\, ds + \int_0^t (f(s),\varphi)\, ds + \lambda\int_0^t (u(s),\varphi)\, dW_s,$$

where

$$(u(t),\varphi) = \int_{\mathbb{R}^d} u(x,t)\varphi(x)\, dx.$$

As is shown in [**48**, Chapter 6], this problem has a unique solution.

In a similar way, we define a solution to the problem

$$du = Au\, dt + \Lambda_{ij}\frac{\partial u}{\partial x_i}\, dW_t^j + f(t),$$
$$u\big|_{t=0} = u_0,$$

where $W_t = \{W_t^i\}$ is a $d$-dimensional Wiener process and $\Lambda = \{\Lambda_{ij}\}$ is a $(d \times d)$-matrix.

For the well-posedness of the problem the so-called *superellipticity* condition is required:

$$\{a_{ij}\} - \frac{1}{2}\Lambda\Lambda^* \geq \kappa\,\mathrm{Id}, \quad \kappa > 0,$$

where the matrix $\Lambda$ was introduced before, $\{a_{ij}\}$ is the matrix of coefficients of the operator $A$, and Id is the identity matrix.

## 3. Elasticity system

**3.1. The main notions and statements of the problems.** Suppose that $\Omega$ is a bounded Lipschitz domain in $\mathbb{R}^d$ and $\Gamma$ is a subset of $\partial\Omega$ of positive $(d-1)$-dimensional measure. Consider the system of differential equations

$$L\overline{u} = \overline{f}, \tag{3.1}$$

where $\overline{u} = (u_1(x), \dots, u_d(x))$ is a vector-valued function in $\Omega$,

$$L\overline{u} \equiv \frac{\partial}{\partial x_k}\left(A^{kl}(x)\frac{\partial \overline{u}}{\partial x_l}\right), \quad x = (x_1, \dots, x_d), \qquad \frac{\partial \overline{u}}{\partial x_l} = \left(\frac{\partial u_1}{\partial x_l}, \dots, \frac{\partial u_d}{\partial x_l}\right),$$

$A^{kl}(x)$ are square $d \times d$-matrices with bounded measurable entries $a_{ij}^{kl}(x)$ such that

$$a_{ij}^{kl}(x) = a_{ji}^{lk}(x) = a_{il}^{kj}(x), \tag{3.2}$$

$$\kappa_1\eta_{ik}\eta_{ik} \leq a_{ij}^{kl}(x)\eta_{ik}\eta_{jl} \leq \kappa_2\eta_{ik}\eta_{ik}, \tag{3.3}$$

where $\kappa_1$ and $\kappa_2$ are positive constants, $(\eta_{ik})$ is an arbitrary symmetric $d \times d$-matrix with constant entries, and $x \in \Omega$. The components of the vector-valued function $\overline{f}(x) = (f_1(x), \dots, f_d(x))$ belong to the space $L_2(\Omega)$.

Hereinafter, we adopt the convention regarding summation with respect to repeating indices.

We write the elasticity system (3.1) in the coordinate form:

$$(L\overline{u})_i \equiv \frac{\partial}{\partial x_k}\left(a_{ij}^{kl}(x)\frac{\partial u_j}{\partial x_l}\right) = f_i(x), \quad i = 1, \dots, d. \tag{3.4}$$

The Lamé system, playing an important role in applications, is a special case of the system (3.4) with

$$a_{ij}^{kl} = \lambda\delta_{ik}\delta_{jl} + \mu\delta_{ij}\delta_{kl} + \mu\delta_{il}\delta_{kj}$$

or

$$(L\overline{u})_i = \sum_{k=1}^{d}\mu\frac{\partial^2 u_i}{\partial x_k^2} + (\lambda + \mu)\frac{\partial^2 u_k}{\partial x_k \partial x_i}, \quad i = 1, \dots, d,$$

i.e.,

$$(L\overline{u})_i = \mu\Delta u_i + (\lambda + \mu)\frac{\partial}{\partial x_i}\,\mathrm{div}\,\overline{u}, \quad i = 1, \dots, d. \tag{3.5}$$

The Lamé system is determined by two independent constants $\lambda$ and $\mu$. Note that 21 constants are required to determine a general linear elasticity system in the three-dimensional space provided that the symmetry condition (3.2) is satisfied (under the assumption that the coefficients are constant).

In the two-dimensional case, the matrices $A^{kl}$ in the Lamé system have the form

$$A^{11} = \begin{pmatrix} \lambda + 2\mu & 0 \\ 0 & \mu \end{pmatrix}, \quad A^{12} = \begin{pmatrix} 0 & \lambda \\ \mu & 0 \end{pmatrix}, \quad A^{21} = \begin{pmatrix} 0 & \mu \\ \lambda & 0 \end{pmatrix}, \quad A^{22} = \begin{pmatrix} \mu & 0 \\ 0 & \lambda + 2\mu \end{pmatrix}$$

or, in the coordinate notation,

$$(\lambda + 2\mu)\frac{\partial^2 u_1}{\partial x_1^2} + \mu \frac{\partial^2 u_1}{\partial x_2^2} + (\lambda + \mu)\frac{\partial^2 u_2}{\partial x_1 \partial x_2} = f_1(x),$$

$$(\lambda + 2\mu)\frac{\partial^2 u_2}{\partial x_2^2} + \mu \frac{\partial^2 u_2}{\partial x_1^2} + (\lambda + \mu)\frac{\partial^2 u_1}{\partial x_1 \partial x_2} = f_2(x).$$

We use the notation $e_{ij}(\overline{u}) \equiv \frac{1}{2}\left(\frac{\partial u_i}{\partial x_j} + \frac{\partial u_j}{\partial x_i}\right)$. The tensor $e_{ij}(\overline{u})$ is called the *strain tensor* and $\sigma_{ij}(\overline{u}) \equiv a_{ij}^{kl} e_{kl}(\overline{u})$ is referred to as the *stress tensor*. The elasticity system is written in the form

$$\frac{\partial}{\partial x_j}\sigma_{ij}(\overline{u}) = f_i(x), \quad i = 1, \dots, d. \tag{3.6}$$

In the three-dimensional case, the following notation is used: $\sigma_x = \sigma_{11}$, $\sigma_y = \sigma_{22}$, $\sigma_z = \sigma_{33}$, $\sigma_{xz} = \sigma_{13}$, $\sigma_{xy} = \sigma_{12}$, $\sigma_{yz} = \sigma_{23}$, $\varepsilon_x = e_{11}$, $\varepsilon_y = e_{22}$, $\varepsilon_z = e_{33}$, $\gamma_{xz} = e_{13}$, $\gamma_{xy} = e_{12}$, $\gamma_{yz} = e_{23}$. In this notation, the stress tensor takes the form

$$\begin{pmatrix} \sigma_x \\ \sigma_y \\ \sigma_z \\ \sigma_{xz} \\ \sigma_{xy} \\ \sigma_{yz} \end{pmatrix} = \begin{pmatrix} a_{11}^{11} & a_{12}^{12} & a_{13}^{13} & a_{13}^{11} & a_{12}^{11} & a_{13}^{12} \\ & a_{22}^{22} & a_{23}^{23} & a_{23}^{21} & a_{22}^{21} & a_{23}^{22} \\ & & a_{33}^{33} & a_{33}^{31} & a_{32}^{31} & a_{33}^{32} \\ & & & a_{33}^{11} & a_{32}^{11} & a_{33}^{12} \\ & & & & a_{22}^{11} & a_{23}^{12} \\ & & & & & a_{33}^{22} \end{pmatrix} \begin{pmatrix} \varepsilon_x \\ \varepsilon_y \\ \varepsilon_z \\ \gamma_{xz} \\ \gamma_{xy} \\ \gamma_{yz} \end{pmatrix},$$

where the matrix is symmetric. Such a form is convenient in applications because we can simultaneously observe all the coefficients of the systems in the form of a single matrix.

We write $L \in E(\kappa_1, \kappa_2)$ if our elasticity system satisfies (3.2) and (3.3).

EXERCISE 3.1. Prove that the Lamé system belongs to $E(2\mu, 2\mu + d\lambda)$.

EXERCISE 3.2. Using (3.2) and (3.3), prove the inequalities

$$\sum_{i,k=1}^{d} \kappa_1 \left(\frac{\partial u_i}{\partial x_k} + \frac{\partial u_k}{\partial x_i}\right)^2 \le a_{ij}^{kl} \frac{\partial u_i}{\partial x_k}\frac{\partial u_j}{\partial x_l} \le \sum_{i,k=1}^{d} \kappa_2 \left(\frac{\partial u_i}{\partial x_k} + \frac{\partial u_k}{\partial x_i}\right)^2$$

or

$$\kappa_1 e_{ik} e_{ik} \le \frac{1}{4} a_{ij}^{kl} \frac{\partial u_i}{\partial x_k}\frac{\partial u_j}{\partial x_l} \le \kappa_2 e_{ik} e_{ik}.$$

In our further considerations, an important role is played by the so-called *rigid displacements*, i.e., vector-valued functions of the form

$$\overline{\eta} = \overline{a} + A\overline{x},$$

where $\overline{a} = (a_1, \dots, a_d)$ is a constant vector and $A = (a_{ij})$ is a skew-symmetric matrix, $a_{ij} = -a_{ji}$. It is easy to check (do it!) that any rigid displacement is a solution to the homogeneous elasticity system $L\overline{u} = 0$. By the definition of a rigid displacement, we have $\frac{\partial \eta_i}{\partial x_j} = -\frac{\partial \eta_j}{\partial x_i}$. The converse assertion (cf. Exercise 3.3 below) is also valid.

EXERCISE 3.3. Suppose that $\overline{v}(x) \in H^1(\Omega)$ and $\frac{\partial v_i}{\partial x_j} = -\frac{\partial v_j}{\partial x_i}$, where the derivatives are understood as weak derivatives (i.e., $e_{ij}(\overline{v}) = 0$ in the sense of distributions). Then $\overline{v}(x)$ is a rigid displacement.

A solution to the elasticity system $L\overline{u} = \overline{f}$ can be understood in the classical sense if we additionally require that $\overline{u}(x) \in C^2(\Omega)$, $a_{ij}^{kl}(x) \in C^1(\Omega)$, and $\overline{f}(x) \in C(\Omega)$. These smoothness conditions can be weakened if we are interested in a weak solution to a boundary value problem for the elasticity system.

Let $\overline{f}$ and $\overline{\varphi}(x)$ be given vector-valued functions with components in $L_2(\Omega)$ and in $H^1(\Omega)$ respectively.

DEFINITION 3.1. By a *weak solution to the first boundary value problem* for the elasticity system $L\overline{u} = \overline{f}$ with the boundary data $\overline{\varphi}$ we mean a vector-valued function $\overline{u}(x) \in H^1(\Omega)$ such that $\overline{u} - \overline{\varphi} \in \overset{\circ}{H}{}^1(\Omega)$ and

$$\int_\Omega A^{kl} \frac{\partial \overline{u}}{\partial x_l} \frac{\partial \overline{v}}{\partial x_k}\, dx = -\int_\Omega (\overline{f}, \overline{v})\, dx \quad \forall\, \overline{v}(x) \in \overset{\circ}{H}{}^1(\Omega).$$

The weak statement corresponds to the classical problem

$$\begin{aligned} L\overline{u} &= \overline{f} \quad \text{in } \Omega, \\ \overline{u} &= \overline{\varphi} \quad \text{on } \partial\Omega, \end{aligned}$$

where $\overline{f}$ and $\overline{\varphi}$ are given vector-valued functions.

EXERCISE 3.4. Prove that a vector-valued function $\overline{u}(x) \in C^2(\Omega) \cap C(\overline{\Omega})$ satisfying the classical problem is a weak solution in the sense of Definition 3.1.

REMARK 3.1. If the boundary $\partial\Omega$ and functions $\varphi$, $f$ are sufficiently smooth, the converse assertion is also true: a weak solution is classical.

DEFINITION 3.2. By a *weak solution to the second boundary value problem* for the elasticity system $L\overline{u} = 0$ we mean a vector-valued function $\overline{u}(x) \in H^1(\Omega)$ such that

$$\int_\Omega A^{kl} \frac{\partial \overline{u}}{\partial x_l} \frac{\partial \overline{v}}{\partial x_k}\, dx = \int_{\partial\Omega} (\overline{\varphi}, \overline{v})\, ds - \int_\Omega (\overline{f}, \overline{v})\, dx \quad \forall\, \overline{v}(x) \in H^1(\Omega),$$

where $\overline{f}$ and $\overline{\varphi}(x)$ are given vector-valued functions with components in $L_2(\Omega)$ and $L_2(\partial\Omega)$ respectively.

The weak statement corresponds to the classical problem

$$\begin{aligned} L\overline{u} &= \overline{f} \quad \text{in } \Omega, \\ \sigma(\overline{u}) &= \overline{\varphi} \quad \text{on } \partial\Omega, \end{aligned}$$

where $\overline{f}$ and $\overline{\varphi}$ are given vector-valued functions, $(\sigma(\overline{u}))_i \equiv \sigma_{ij}(\overline{u}) n_j$, $\overline{n} = (n_1, \dots, n_d)$ is the outward normal to $\partial\Omega$, and $(\cdot)_i$ denotes the $i$th component of a vector-valued function.

Let $\Gamma$ be a measurable subset of $\partial\Omega$. The space $H^1(\Omega, \Gamma)$ of vector-valued functions is defined as the closure in the $H^1(\Omega)$-norm of the set of infinitely differentiable vector-valued functions $\overline{\varphi}(x)$ vanishing in some neighborhood of $\Gamma$.

Suppose that $\overline{f} \in (L_2(\Omega))^d$, $\overline{\psi} \in (H^1(\Omega))^d$, and $\overline{\varphi} \in (L_2(S))^d$, where $S = \partial\Omega \backslash \Gamma$. Let the sets $S$ and $\Gamma$ have positive $(d-1)$-dimensional measure.

DEFINITION 3.3. By a *weak solution to the mixed boundary value problem* (the *third boundary value problem*) for the elasticity system $L\overline{u} = 0$ we mean a vector-valued function $\overline{u}(x) \in (H^1(\Omega))^d$ such that $\overline{u}(x) - \overline{\psi}(x) \in (H^1(\Omega, \Gamma))^d$ and

$$\int_\Omega A^{kl} \frac{\partial \overline{u}}{\partial x_l} \frac{\partial \overline{v}}{\partial x_k}\, dx = \int_S (\overline{\varphi}, \overline{v})\, ds - \int_\Omega (\overline{f}, \overline{v})\, dx \quad \forall\, \overline{v}(x) \in (H^1(\Omega, \Gamma))^d.$$

The weak statement corresponds to the classical problem

$$\begin{aligned} L\overline{u} &= \overline{f} \quad \text{in } \Omega, \\ \overline{u} &= \overline{\psi} \quad \text{on } \Gamma, \\ \sigma(\overline{u}) &= \overline{\varphi} \quad \text{on } S, \end{aligned}$$

where $\overline{f}$, $\overline{\varphi}$, and $\overline{\psi}$ are given vector-valued functions.

Our next goal is to establish the existence and uniqueness of solutions to the above problems. For this purpose, we need the first and second Korn inequalities.

### 3.2. The first Korn inequality.

THEOREM 3.1 (the first Korn inequality). *Let $\Omega$ be a bounded domain. Then for any vector-valued function $\overline{u} \in \left(\overset{\circ}{H}{}^1(\Omega)\right)^d$*

$$\int_\Omega |\nabla \overline{u}|^2\, dx \le 2 \int_\Omega |e(\overline{u})|^2\, dx, \tag{3.7}$$

*where $|\nabla \overline{u}|^2 \equiv \sum_{i,j=1}^d \left(\frac{\partial \overline{u}_i}{\partial x_j}\right)^2$ and $|e(\overline{u})|^2 = e_{ik}(\overline{u}) e_{ik}(\overline{u})$.*

To simplify computations, we write $u, v, \dots$ instead of $\overline{u}, \overline{v}, \dots$.

PROOF OF THEOREM 3.1. Let $u(x) \in (C_0^\infty(\Omega))^d$. Then

$$\begin{aligned} \int_\Omega |e(u)|^2\, dx &= \frac{1}{4} \int_\Omega \sum_{i,j=1}^d \left( \frac{\partial u_i}{\partial x_j} + \frac{\partial u_j}{\partial x_i} \right)^2 dx \\ &= \frac{1}{2} \int_\Omega \sum_{i,j=1}^d \left[ \left( \frac{\partial u_i}{\partial x_j} \right)^2 + \frac{\partial u_i}{\partial x_j} \frac{\partial u_j}{\partial x_i} \right] dx \\ &= \frac{1}{2} \int_\Omega |\nabla u|^2\, dx - \frac{1}{2} \int_\Omega \sum_{i,j=1}^d \frac{\partial^2 u_i}{\partial x_j \partial x_i} u_j\, dx \\ &= \frac{1}{2} \int_\Omega |\nabla u|^2\, dx + \frac{1}{2} \int_\Omega \sum_{i,j=1}^d \frac{\partial u_i}{\partial x_i} \frac{\partial u_j}{\partial x_j}\, dx \\ &= \frac{1}{2} \int_\Omega |\nabla u|^2\, dx + \frac{1}{2} \int_\Omega \left( \frac{\partial u_1}{\partial x_1} + \dots + \frac{\partial u_d}{\partial x_d} \right)^2 dx \\ &\ge \frac{1}{2} \int_\Omega |\nabla u|^2\, dx, \end{aligned}$$

where we have used the divergence theorem. Note that integrals over the boundary of $\Omega$ vanish because $u(x) \in (C_0^\infty(\Omega))^d$.

To obtain the inequality (3.7) for $u \in \overset{\circ}{H}{}^1(\Omega)$, we can approximate $u(x)$ by functions in $(C_0^\infty(\Omega))^d$ and pass to the limit. □

### 3.3. The second Korn inequality.

THEOREM 3.2 (the second Korn inequality). *Let $\Omega$ be a bounded Lipschitz domain. Then for any vector-valued function $u \in (H^1(\Omega))^d$ we have*

$$\int_\Omega |\nabla u|^2\,dx \le C \int_\Omega (|u|^2 + |e(u)|^2)\,dx, \tag{3.8}$$

*where $C > 0$ is independent of $u(x)$.*

The proof of the second Korn inequality is more complicated than the proof of Theorem 3.1. We need two auxiliary inequalities.

LEMMA 3.1. *Let $\rho(x)$ be the distance from a point $x$ to $\partial\Omega$. Then any function $v(x)$ such that $\rho^2\Delta v \in L_2(\Omega)$ satisfies the inequality*

$$\|\rho\nabla v\|_{L_2(\Omega)} \le K\|v\|_{L_2(\Omega)} + \|\rho^2\Delta v\|_{L_2(\Omega)}, \tag{3.9}$$

*where $K > 0$ is independent of $v(x)$.*

LEMMA 3.2. *Suppose that $\Omega$ is a Lipschitz domain and $\rho(x)$ is the distance from a point $x$ to $\partial\Omega$. Then any function $v(x) \in C^\infty(\Omega) \cap L_2(\Omega)$ such that $\rho(x)\frac{\partial^2 v}{\partial x_i \partial x_j} \in L_2(\Omega)$ satisfies the inequality*

$$\|\nabla v\|_{L_2(\Omega)} \le K_1\Big\{\|v\|_{L_2(\Omega)} + \sum_{i,j=1}^d \Big\|\rho\frac{\partial^2 v}{\partial x_i \partial x_j}\Big\|_{L_2(\Omega)}\Big\}, \tag{3.10}$$

*where $K_1 > 0$ is independent of $v(x)$.*

Lemmas 3.1 and 3.2 will be proved later. Now, assuming that the statements of these lemmas hold, we prove the second Korn inequality.

PROOF OF THEOREM 3.2. It is easy to verify that

$$\frac{\partial^2 u_i}{\partial x_j^2} = 2\frac{\partial}{\partial x_j}e_{ij}(\overline{u}) - \frac{\partial}{\partial x_i}e_{jj}(\overline{u}), \tag{3.11}$$

$$\frac{\partial^2 u_i}{\partial x_k \partial x_l} = \frac{\partial}{\partial x_k}e_{il}(\overline{u}) + \frac{\partial}{\partial x_l}e_{ik}(\overline{u}) - \frac{\partial}{\partial x_i}e_{kl}(\overline{u}). \tag{3.12}$$

Define a vector-valued function $\overline{v}$ with components $v_i$ such that

$$\Delta v_i = \begin{cases} 0, & x \in \Omega^0\backslash\Omega, \\ \sum_{j=1}^d \Big(2\frac{\partial}{\partial x_j}e_{ij}(\overline{u}) - \frac{\partial}{\partial x_i}e_{jj}(\overline{u})\Big), & x \in \Omega, \end{cases}$$

where $\Omega^0$ compactly contains $\Omega$, i.e., $\Omega \Subset \Omega^0$, and

$$v_i = 0 \quad \text{on } \partial\Omega^0.$$

We set $\overline{w} = \overline{u} - \overline{v}$ in $\Omega$. Then $\Delta\overline{w} = 0$ in $\Omega$. We estimate $\overline{v}$ and $\overline{w}$ separately. Each component of $\overline{v}$ is a solution to the boundary value problem for the Poisson equation

$$\Delta V = f_i + \frac{\partial f_0}{\partial x_i} \quad \text{in } \Omega^0, \quad V = 0 \text{ on } \partial\Omega^0, \quad f_0, f_i \in L_2(\Omega^0).$$

This solution satisfies the estimate (cf., for example, [**88**])

$$\|V\|_{H^1(\Omega^0)} \le K_2\Big(\|f_0\|_{L_2(\Omega^0)} + \sum_{i=1}^d \|f_i\|_{L_2(\Omega^0)}\Big),$$

where the constant $K_2 > 0$ is independent of $f_0$ and $f_i$ (check it!). Therefore,

$$\int_\Omega |\overline{v}|^2\,dx + \int_\Omega |\nabla\overline{v}|^2\,dx \le K_3\|e(u)\|^2_{L_2(\Omega)}. \tag{3.13}$$

To estimate the vector-valued function $\overline{w}(x)$, we use the estimate in Lemma 3.2:

$$\|\nabla\overline{w}\|_{L_2(\Omega)} \le K_4\Big\{\|w\|_{L_2(\Omega)} + \sum_{i,j=1}^{d}\Big\|\rho\frac{\partial^2\overline{w}}{\partial x_i\partial x_j}\Big\|_{L_2(\Omega)}\Big\}.$$

Using the estimate in Lemma 3.1, the equality $\Delta\overline{w} = 0$, and the identity (3.12), we find

$$\begin{aligned}
\|\nabla\overline{w}\|_{L_2(\Omega)} &\le K_4\Big\{\|w\|_{L_2(\Omega)} \\
&\quad + \sum_{i,j,k=1}^{d}\Big\|\rho\Big(\frac{\partial}{\partial x_i}e_{kj}(\overline{w}) + \frac{\partial}{\partial x_j}e_{ki}(\overline{w}) - \frac{\partial}{\partial x_k}e_{ij}(\overline{w})\Big)\Big\|_{L_2(\Omega)}\Big\} \\
&\le K_5\{\,\|\overline{w}\|_{L_2(\Omega)} + \|\rho\nabla e(\overline{w})\|_{L_2(\Omega)}\} \\
&\le K_6\{\,\|\overline{w}\|_{L_2(\Omega)} + \|e(\overline{w})\|_{L_2(\Omega)} + \|\rho^2\Delta e(\overline{w})\|_{L_2(\Omega)}\} \\
&= K_6\{\,\|\overline{w}\|_{L_2(\Omega)} + \|e(\overline{w})\|_{L_2(\Omega)}\} \\
&\le K_6\{\|\overline{u} - \overline{v}\|_{L_2(\Omega)} + \|e(\overline{u} - \overline{v})\|_{L_2(\Omega)}\} \\
&\le K_7\{\|\overline{u}\|_{L_2(\Omega)} + \|e(\overline{u})\|_{L_2(\Omega)}\}.
\end{aligned} \tag{3.14}$$

Combining (3.13) with (3.14), we obtain the required inequality (3.8). □

PROOF OF LEMMA 3.1. Let $\Omega_{(\delta)}$ consist of points of $\Omega$ located at a distance greater than $\delta = \text{const} > 0$ from $\partial\Omega$. Integrating by parts, we find

$$\begin{aligned}
&\int_{\Omega_{(\delta)}} (\rho(x) - \delta)^2|\nabla v|^2\,dx \\
&\quad = -\int_{\Omega_{(\delta)}} 2(\rho(x) - \delta)\frac{\partial\rho}{\partial x_i}\frac{\partial v}{\partial x_i}v\,dx - \int_{\Omega_{(\delta)}} (\rho(x) - \delta)^2 v\Delta v\,dx \\
&\quad \le K_8\big\{\,\|v\|_{L_2(\Omega_{(\delta)})}\|\,(\rho - \delta)|\nabla v|\,\|_{L_2(\Omega_{(\delta)})} + \|v\|_{L_2(\Omega_{(\delta)})}\|\,(\rho - \delta)^2\Delta v\,\|_{L_2(\Omega_{(\delta)})}\big\}.
\end{aligned} \tag{3.15}$$

We "sidestep" from the boundary of $\Omega$ in the integration over $\Omega_{(\delta)}$ in order to provide the existence of the integral on the left-hand side of (3.15). Furthermore, the derivatives $\frac{\partial\rho}{\partial x_i}$ are bounded, since $\rho(x)$ is Lipschitz continuous and the weak derivatives of a Lipschitz function exist and are bounded almost everywhere (cf. Exercises 3.5 and 3.6).

EXERCISE 3.5. Prove that the distance $\rho(x)$ from a point $x \in \Omega$ to $\partial\Omega$ is a Lipschitz continuous function, i.e.,

$$|\rho(x_1) - \rho(x_2)| \le K_9|x_1 - x_2|, \quad x_1, x_2 \in \Omega,$$

where the constant $K_9$ is independent of $x_1, x_2 \in \Omega$. What is $K_9$?

EXERCISE 3.6. Prove that a Lipschitz continuous function $f(x)$ possesses the weak derivatives $\frac{\partial f}{\partial x_i}$ and these derivatives are bounded almost everywhere.

To continue (3.15), we use $ab \le \frac{a^2}{2\varepsilon} + \frac{\varepsilon b^2}{2}$ and $ab \le \frac{a^2}{2} + \frac{b^2}{2}$ for the terms on the right-hand side, where $a$, $b$, and $\varepsilon$ are arbitrary positive numbers. Taking $\varepsilon > 0$ small enough and simplifying the inequality, we find

$$\| (\rho - \delta)|\nabla v| \|_{L_2(\Omega_{(\delta)})} \le K_{10}\{ \|v\|_{L_2(\Omega_{(\delta)})} + \|(\rho-\delta)^2 \Delta v\|_{L_2(\Omega_{(\delta)})}\}.$$

Passing to the limit as $\delta \to 0$, we obtain (3.9). □

COROLLARY 3.1. *Let $\Omega'$ be a domain compactly contained in $\Omega$, i.e., $\Omega' \Subset \Omega$. Then*

$$\|\nabla v\|_{L_2(\Omega')} \le K(\Omega, \Omega')\{\|v\|_{L_2(\Omega)} + \|\Delta v\|_{L_2(\Omega)}\}. \tag{3.16}$$

PROOF OF LEMMA 3.2. For simplicity, we consider only the two-dimensional case. The case of higher dimensions is treated in a similar way.

We represent (cf. Figure 3.1) the domain $\Omega$ as the union of a smooth domain $\Omega'$ that is compactly contained in $\Omega$ and domains $\Omega_1, \dots, \Omega_N$ such that, after a suitable change of variables, the part of $\partial\Omega$ lying in $\overline{\Omega}_i$ is given by the equation $x_1 = \psi_i(x_2)$ with a Lipschitz continuous function $\psi_i(x_2)$, and $\Omega_i$ can be described as follows: $\Omega_i = \{(x_1, x_2) : 0 < x_1 < \psi_i(x_2), 0 < x_2 < x_2^0\}$.

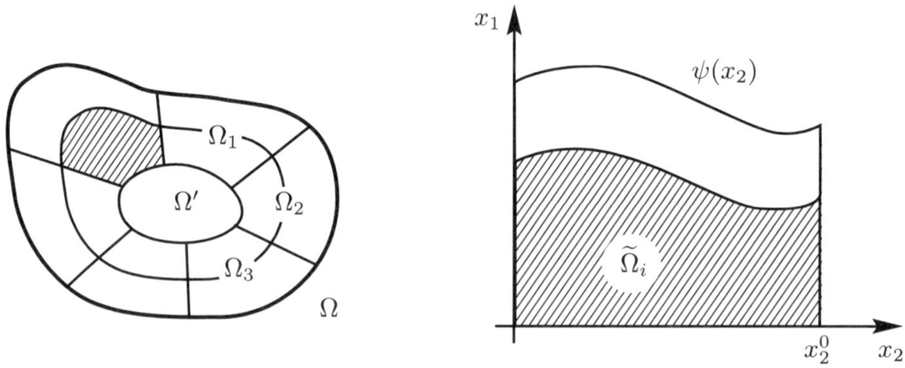


FIGURE 3.1

We also consider the domain

$$\widetilde{\Omega}_i = \left\{(x_1, x_2) : 0 < x_1 < \frac{\psi_i(x_2)}{2},\ 0 < x_2 < x_2^0\right\}$$

and denote $\widetilde{\Omega} = \bigcup_{i=1}^N \widetilde{\Omega}_i$. Using the elementary inequality (the proof will be given later)

$$\int_0^b f^2(\tau)\, d\tau \le K_{11}\Big( \int_{\frac{b}{2}}^b f^2(\tau)\, d\tau + \int_0^b \tau^2[f'(\tau)]^2\, d\tau\Big), \tag{3.17}$$

we have

(3.18)

$$\begin{aligned}\int_0^{\psi(x_2)} \left(\frac{\partial v}{\partial x_k}\right)^2 d\tau &\le \int_0^{\frac{\psi(x_2)}{2}} \left(\frac{\partial v}{\partial x_k}\right)^2 d\tau + \int_0^{\psi(x_2)} (\psi(x_2) - \tau)^2 \left(\frac{\partial^2 v}{\partial x_k \partial x_1}\right)^2 d\tau \\ &\le \int_0^{\frac{\psi(x_2)}{2}} \left(\frac{\partial v}{\partial x_k}\right)^2 d\tau + a^2 \int_0^{\psi(x_2)} \rho^2(\tau) \left(\frac{\partial^2 v}{\partial x_k \partial x_1}\right)^2 d\tau,\end{aligned}$$

where $k = 1, 2$, and $\rho(\tau)$ is the distance between the point corresponding to the parameter $\tau$ on the integration segment $(0, \psi(x_2))$ and the boundary $\partial\Omega_i \cap \partial\Omega$. Here, we used the inequality $|\psi(x_2) - \tau| \le a\rho(\tau)$ (where $a$ is a constant) for the Lipschitz continuous function $\psi(x_2)$. The proof is left to the reader.

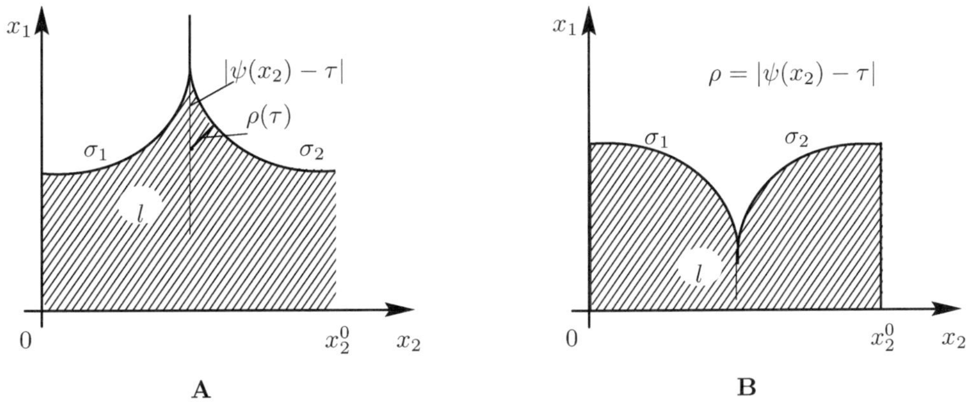


FIGURE 3.2

REMARK 3.2. The assumption that the domain is Lipschitz is sufficient to prove the inequality (3.18). In Figure 3.2, there are fragments of non-Lipschitz domains where the inequality $|\psi(x_2) - \tau| \le a\rho(\tau)$ fails to hold. Here, the segment $l$ touches two arcs $\sigma_1$ and $\sigma_2$ forming the outer boundary fragment. Obviously, the Lipschitz condition is only sufficient. In particular, Figure 3.2B represents a non-Lipschitz domain where the assertion of Lemma 3.2 remains valid.

Integrating (3.18) with respect to $x_2$, we find

$$\int_{\Omega_i} |\nabla v|^2\,dx \le K_{12} \int_{\widetilde{\Omega}_i} |\nabla v|^2\,dx + K_{12} \int_{\Omega_i} \rho^2(x) \left(\frac{\partial^2 v}{\partial x_k \partial x_1}\right)^2 dx, \tag{3.19}$$

where $i = 1, \ldots, N$. By Corollary 3.1,

$$\int_{\Omega'} |\nabla v|^2\,dx \le K \left( \int_{\Omega} |v|^2\,dx + \sum_{k,l=1}^{d} \int_{\widetilde{\Omega}\cup\Omega'} \left(\frac{\partial^2 v}{\partial x_k \partial x_l}\right)^2 dx \right). \tag{3.20}$$

Summing the inequalities (3.19) with respect to $i$ from 1 to $N$ and adding the inequality (3.20), we estimate the first integral on the right-hand side of (3.20) in accordance with Corollary 3.1 and obtain the inequality (3.10).

It remains to prove the elementary inequality (3.17). Integrating by parts, we find

$$\begin{aligned} \int_0^\tau f^2(t)\,dt &= f^2(\tau)\tau - 2\int_0^\tau t\, f(t) f'(t)\,dt \\ &\le f^2(\tau)\tau + \varepsilon \int_0^\tau f^2(t)\,dt + \frac{1}{\varepsilon}\int_0^\tau [f'(t)]^2 t^2\,dt. \end{aligned}$$

By the mean-value formula, we can choose $\tau$ such that

$$f^2(\tau) \le \frac{2}{b}\int_{\frac{b}{2}}^{b} f^2(t)\,dt, \quad \tau \in [b/2, b].$$

Then

$$\int_0^{\frac{b}{2}} f^2(t)\,dt \le 2\int_{\frac{b}{2}}^b f^2(t)\,dt + \varepsilon\int_0^b f^2(t)\,dt + \frac{1}{\varepsilon}\int_0^b [f'(t)]^2 t^2\,dt.$$

Adding $\int_{\frac{b}{2}}^b f^2(t)\,dt$ to both sides of the last inequality, we find

$$\int_0^b f^2(t)\,dt \le 3\int_{\frac{b}{2}}^b f^2(t)\,dt + \varepsilon\int_0^b f^2(t)\,dt + \frac{1}{\varepsilon}\int_0^b [f'(t)]^2 t^2\,dt.$$

Consequently,

$$(1-\varepsilon)\int_0^b f^2(t)\,dt \le 3\int_{\frac{b}{2}}^b f^2(t)\,dt + \frac{1}{\varepsilon}\int_0^b t^2[f'(t)]^2\,dt.$$

Setting, for example, $\varepsilon = 1/2$, we obtain the inequality (3.17) with $K_{11} = 6$. □

**3.4. The third important inequality.** The following assertion will be used in the proof of solvability of the main boundary value problems.

THEOREM 3.3. *Let $H$ be a closed subspace of $H^1(\Omega)$ containing no rigid displacements. Then for any $v \in H$ we have*

$$\|v\|^2_{H^1(\Omega)} \le C_3\|e(v)\|^2_{L_2(\Omega)},$$

*where the constant $C_3 > 0$ is independent of $v$.*

PROOF. Assume the contrary. Then there exists a sequence $v_i \in H$ such that $\|v_i\|^2_{H^1(\Omega)} = 1$, but $\|e(v_i)\|^2_{L_2(\Omega)} \to 0$. We extract a subsequence $v_{i_k}$ such that $v_{i_k} \to v_0$ in $L_2(\Omega)$. Since the sequence $v_{i_k}$ is uniformly bounded in $H^1(\Omega)$, we have $v_0 \in H$. Let us show that $v_0$ is different from zero. By the second Korn inequality,

$$\|\nabla v_{i_k}\|_{L_2(\Omega)} \le K(\|v_{i_k}\|_{L_2(\Omega)} + \|e(v_{i_k})\|_{L_2(\Omega)}).$$

By assumption, $\|e(v_{i_k})\|_{L_2(\Omega)} \to 0$ as $k \to \infty$. Thus, if $\|v_{i_k}\|_{L_2(\Omega)} \to 0$, then $\|v_{i_k}\|^2_{H^1(\Omega)} \to 0$, which contradicts the equality $\|v_i\|^2_{H^1(\Omega)} = 1$. Since the sequence $v_{i_k}$ strongly converges to $v_0$, it follows that $\|v_0\|_{L_2(\Omega)}$ is positive.

Further, $v_{i_k} \to v_0$ in $H^1(\Omega)$ since, in view of the second Korn inequality, we have

$$\|v_{i_k} - v_{i_n}\|^2 + \|\nabla(v_{i_k} - v_{i_n})\|^2 \le K_{13}(\|v_{i_k} - v_{i_n}\|^2 + \|e(v_{i_k} - v_{i_n})\|^2) \to 0$$

as $k \to \infty$. Consequently, $e(v_0) = 0$ and $v_0$ is a rigid displacement (cf. Exercise 3.3). We arrive at a contradiction with the assumption that $H$ does not contain rigid displacements, which proves the required assertion. □

**3.5. Solvability of the main problems for the elasticity system.** Using the Lax–Milgram theorem and Korn inequalities, we can establish the solvability of the boundary value problems for the elasticity system. We begin with the first boundary value problem.

Setting $w \equiv u - \varphi$ (cf. Definition 3.1), for $w$ we obtain the integral identity

$$\int_\Omega \left(A^{kl}(x)\frac{\partial w}{\partial x_l}, \frac{\partial v}{\partial x_k}\right)dx = -\int_\Omega (f,v)\,dx + \int_\Omega \left(A^{kl}(x)\frac{\partial \varphi}{\partial x_l}, \frac{\partial v}{\partial x_k}\right)dx \quad \forall v \in \overset{\circ}{H}{}^1(\Omega).$$

Denoting the right-hand side by $l(v)$, we reduce the question to the proof of the existence of $w \in \overset{\circ}{H}{}^1(\Omega)$ such that

$$\int_\Omega \left(A^{kl}(x)\frac{\partial w}{\partial x_l}, \frac{\partial v}{\partial x_k}\right) dx = l(v) \quad \forall v \in \overset{\circ}{H}{}^1(\Omega).$$

The existence of such $w$ follows from the Lax–Milgram theorem. To show that the bilinear form on the left-hand side of the last equality satisfies all the assumptions of the Lax–Milgram theorem, we use the first Korn inequality. Indeed,

$$\int_\Omega \left(A^{kl}(x)\frac{\partial w}{\partial x_l}, \frac{\partial w}{\partial x_k}\right) dx \geq 4\kappa_1 \int_\Omega |e(w)|^2 dx \geq K_{14} \int_\Omega (|u|^2 + |\nabla u|^2)\, dx,$$

where the last inequality holds by virtue of the first Korn inequality. The estimate

$$\left| \int_\Omega \left(A^{kl}(x)\frac{\partial w}{\partial x_l}, \frac{\partial v}{\partial x_k}\right) dx \right| \leq K_{15} \|w\|_{H^1(\Omega)} \|v\|_{H^1(\Omega)}$$

can be directly verified using the Cauchy–Schwarz–Bunyakovskii inequality.

EXERCISE 3.7. Prove that the solution $u(x)$ to the first boundary value problem satisfies the estimate

$$\|u\|_{\overset{\circ}{H}{}^1(\Omega)} \leq C(\Omega)(\|f\|_{L_2(\Omega)} + \|\varphi\|_{H^1(\Omega)}).$$

Express the constant $C(\Omega)$ in terms of $\kappa_1$, $\kappa_2$, and $a$, where $a$ is the edge of a cube containing the domain $\Omega$.

For the second and third (mixed) boundary value problems the solvability is established by the same scheme, but in the Lax–Milgram theorem we must take for $H$ the space of vector-valued functions in $H^1(\Omega)$ that are orthogonal to rigid displacements in the case of the second boundary value problem, and the space $H^1(\Omega, \Gamma)$ in the case of the third (mixed) boundary value problem. To estimate from below the integral

$$\int_\Omega \left(A^{kl}(x)\frac{\partial w}{\partial x_l}, \frac{\partial w}{\partial x_k}\right) dx,$$

we use the second Korn inequality formulated in Theorem 3.3. In addition, it is necessary to show that the space $H^1(\Omega, \Gamma)$ does not contain rigid displacements.

EXERCISE 3.8. Let the $(d-1)$-dimensional measure of $\Gamma$ be positive. Show that the space $H^1(\Omega, \Gamma)$ does not contain rigid displacements.

EXERCISE 3.9. Prove that the second boundary-value problem is solvable under the condition

$$\int_{\partial\Omega} (\varphi, \eta)\, ds - \int_\Omega (f, \eta) dx = 0 \quad \forall \eta(x) \in \mathcal{R},$$

where $\mathcal{R}$ is the space of rigid displacements. Show that the solution $u(x)$ satisfies the estimate

$$\|e(u)\|_{L_2(\Omega)} \leq C(\Omega)(\|f\|_{L_2(\Omega)} + \|\varphi\|_{L_2(\partial\Omega)}).$$

EXERCISE 3.10. Prove the solvability of the third boundary value problem. Show that the solution satisfies the estimate

$$\|u\|_{H^1(\Omega)} \leq C(\Omega)(\|f\|_{L_2(\Omega)} + \|\varphi\|_{L_2(S)} + \|\psi\|_{H^1(\Omega)}).$$

## 4. Stokes equations

**4.1. The main notions and statements of the problems.** The problem

$$
\begin{aligned}
&-\nabla p + \Delta \overline{v} = -\overline{f} \quad \text{in } \Omega,\\
&\operatorname{div} \overline{v} = 0 \quad \text{in } \Omega, \qquad (4.1)\\
&\overline{v} = 0 \quad \text{on } \Omega,
\end{aligned}
$$

describes the motion of a viscous incompressible fluid with small velocities. In the case under consideration, the kinematic viscosity coefficient $\mu$ at $\Delta\overline{v}$ is equal to 1. In the classical statement, the solution $(p, \overline{v})$ must satisfy the conditions $p \in C^1(\Omega)$ and $\overline{v} \in (C^2(\Omega) \cap C(\overline{\Omega}))^d$. However, as in the case of the elasticity system, it is reasonable to consider weak solutions to the problem (4.1), which require less smoothness of $p$ and $\overline{v}$. Before defining a weak solution to the problem (4.1), we introduce the necessary function spaces.

DEFINITION 4.1. The space $\overset{\circ}{H}{}^1(\operatorname{div}, \Omega)$ is the closure of smooth compactly supported vector-valued functions $\overline{\varphi}$ in $\Omega$ with zero divergence ($\operatorname{div}\overline{\varphi} = 0$) in the $H^1(\Omega)$-norm, i.e.,

$$
\|\varphi\|_{H^1(\Omega)} = \Big( \int_\Omega \big(|\overline{\varphi}|^2 + |\nabla\overline{\varphi}|^2\big)\, dx \Big)^{1/2}.
$$

As usual, the space of linear continuous functionals on $\overset{\circ}{H}{}^1(\Omega)$ is denoted by $H^{-1}(\Omega)$.

DEFINITION 4.2. A pair $(p(x), \overline{v}(x))$ with $p(x) \in L_2(\Omega)$ and $\overline{v}(x) \in \overset{\circ}{H}{}^1(\operatorname{div}, \Omega)$ is called a *weak solution* to the problem (4.1) with $f \in H^{-1}(\Omega)$ if

$$
-\nabla p + \Delta\overline{v} + \overline{f} = 0. \qquad (4.2)
$$

Equation (4.2) is understood as the equality of elements of $H^{-1}(\Omega)$ since each term on the left-hand side of (4.2) belongs to $H^{-1}(\Omega)$.

Another definition of a weak solution to the problem (4.1) is based on an integral identity.

DEFINITION 4.3. A vector-valued function $\overline{v}(x) \in \overset{\circ}{H}{}^1(\operatorname{div}, \Omega)$ is called a *weak solution* to the problem (4.1) *in the sense of integral identity* if

$$
\int_\Omega \sum_{i=1}^d (\nabla v_i, \nabla u_i)\, dx = -\int_\Omega (\overline{f}, \overline{u})\, dx \quad \forall\, \overline{u} \in \overset{\circ}{H}{}^1(\operatorname{div}, \Omega). \qquad (4.3)
$$

These definitions are equivalent under certain conditions on the domain $\Omega$. It is easy to check that (4.3) always follows from (4.2), but the converse implication is not elementary and is based on Theorem 4.1, which will be proved in Subsection 4.2 below. Consider the situation in more detail. From (4.3) it follows that

$$
\langle \Delta\overline{v} + \overline{f}, \overline{u} \rangle = 0 \quad \forall\, u \in \overset{\circ}{H}{}^1(\operatorname{div}, \Omega).
$$

By Theorem 4.1, under certain conditions on the domain $\Omega$, there exists $p \in L_2(\Omega)$ such that $\Delta\overline{v} + \overline{f} = \nabla p$. Consequently, the pair $(p(x), v(x))$ is a solution in the sense of Definition 4.2.

**4.2. Existence and uniqueness of a solution.** We establish the existence of a weak solution to the problem (4.1). As in the case of the elasticity system, the proof is based on the Lax–Milgram theorem. Consider the bilinear form

$$a(u, v) \equiv \sum_{i=1}^{d} \int_{\Omega} (\nabla \overline{v}_i, \nabla \overline{u}_i)\, dx$$

on $\overset{\circ}{H}{}^1(\operatorname{div}, \Omega) \times \overset{\circ}{H}{}^1(\operatorname{div}, \Omega)$. It is easy to verify (do it!) that $a(\overline{v}, \overline{u})$ satisfies all the assumptions of the Lax–Milgram theorem. Therefore, there exists $\overline{v} \in \overset{\circ}{H}{}^1(\operatorname{div}, \Omega)$ such that for all $\overline{u} \in \overset{\circ}{H}{}^1(\operatorname{div}, \Omega)$ we have $l(\overline{u}) = a(\overline{v}, \overline{u})$, where $l(\overline{u})$ is a linear functional on $\overset{\circ}{H}{}^1(\operatorname{div}, \Omega)$ defined by the formula $l(\overline{u}) \equiv \int_\Omega (\overline{f}, \overline{u})\, dx$. Then $\overline{v}(x)$ is a solution to the problem (4.1) in the sense of Definition 4.3. We regard $\overline{v}(x)$ as a "candidate" to the role of the second component of the weak solution $(p, \overline{v})$ in the sense of Definition 4.2. It remains to show that $\Delta \overline{v} + f \in H^{-1}(\Omega)$ is the gradient of some function $p(x) \in L_2(\Omega)$. This is the most difficult step in the proof of the existence of a weak solution to the problem (4.1). The following assertion holds (cf. [**56**]).

THEOREM 4.1. *Let $\Omega$ be a star-shaped domain with respect to a system of balls, and let $\overline{g} \in H^{-1}(\Omega)$ satisfy the condition $\langle \overline{g}, \overline{u} \rangle = 0$ for any $\overline{u} \in \overset{\circ}{H}{}^1(\operatorname{div}, \Omega)$. Then there exists $p(x) \in L_2(\Omega)$ such that $\overline{g} = \nabla p$.*

The existence of a solution to the problem (4.1) in the sense of Definition 4.2 immediately follows from Theorem 4.1.

PROOF OF THEOREM 4.1. Let $\overline{g}$ be a smooth vector-valued function. We show that the identity

$$\int_\Omega (\overline{g}, \overline{\varphi})\, dx = 0 \quad \forall\, \varphi \in C_0^\infty(\Omega),\ \operatorname{div} \overline{\varphi} = 0,$$

implies $\operatorname{curl} \overline{g} = 0$, where, by definition,

$$\operatorname{curl} \overline{g} := \begin{vmatrix} \overline{i} & \overline{j} & \overline{k} \\ \dfrac{\partial}{\partial x_1} & \dfrac{\partial}{\partial x_2} & \dfrac{\partial}{\partial x_3} \\ g_1 & g_2 & g_3 \end{vmatrix}$$

and $\overline{i}$, $\overline{j}$, $\overline{k}$ are the unit vectors of the $x_1$-, $x_2$-, $x_3$-axis respectively. Indeed,

$$\begin{aligned} &\int_\Omega (\operatorname{curl} \overline{g}, \overline{\varphi})\, dx \\ &= \int_\Omega \Big[ \Big( \frac{\partial g_3}{\partial x_2} - \frac{\partial g_2}{\partial x_3} \Big) \varphi_1 + \Big( \frac{\partial g_1}{\partial x_3} - \frac{\partial g_3}{\partial x_1} \Big) \varphi_2 + \Big( \frac{\partial g_2}{\partial x_1} - \frac{\partial g_1}{\partial x_2} \Big) \varphi_3 \Big]\, dx \\ &= - \int_\Omega \Big[ g_3 \Big( \frac{\partial \varphi_1}{\partial x_2} - \frac{\partial \varphi_2}{\partial x_1} \Big) + g_2 \Big( \frac{\partial \varphi_3}{\partial x_1} - \frac{\partial \varphi_1}{\partial x_3} \Big) + + g_1 \Big( \frac{\partial \varphi_2}{\partial x_3} - \frac{\partial \varphi_3}{\partial x_2} \Big) \Big]\, dx \\ &= - \int_\Omega (\overline{g}, \overline{\Psi})\, dx, \end{aligned}$$

where

$$\overline{\Psi} = \Big( \frac{\partial \varphi_2}{\partial x_3} - \frac{\partial \varphi_3}{\partial x_2}, \frac{\partial \varphi_3}{\partial x_1} - \frac{\partial \varphi_1}{\partial x_3}, \frac{\partial \varphi_1}{\partial x_2} - \frac{\partial \varphi_2}{\partial x_1} \Big)$$

and $\operatorname{div}\overline{\Psi} = 0$ because

$$\frac{\partial}{\partial x_3}\Big(\frac{\partial\varphi_1}{\partial x_2} - \frac{\partial\varphi_2}{\partial x_1}\Big) + \frac{\partial}{\partial x_2}\Big(\frac{\partial\varphi_3}{\partial x_1} - \frac{\partial\varphi_1}{\partial x_3}\Big) + \frac{\partial}{\partial x_1}\Big(\frac{\partial\varphi_2}{\partial x_3} - \frac{\partial\varphi_3}{\partial x_2}\Big) = 0.$$

Since $\overline{\varphi} \in C_0^\infty(\Omega)$ is arbitrary, we have $\operatorname{curl}\overline{g} = 0$ in $\Omega$. By the de Rham theorem, for a simply connected domain $\Omega$ and $g \in C^\infty(\Omega)$ such that $\operatorname{curl}\overline{g} = 0$ there exists a potential $w$ such that $\overline{g} = \nabla w$. Theorem 4.1 is valid not only for simply connected domains. We consider the case where the domain is an annulus with one hole (cf. Figure 4.1). In the general case, the proof is based on the same ideas, but it is rather difficult for presentation.

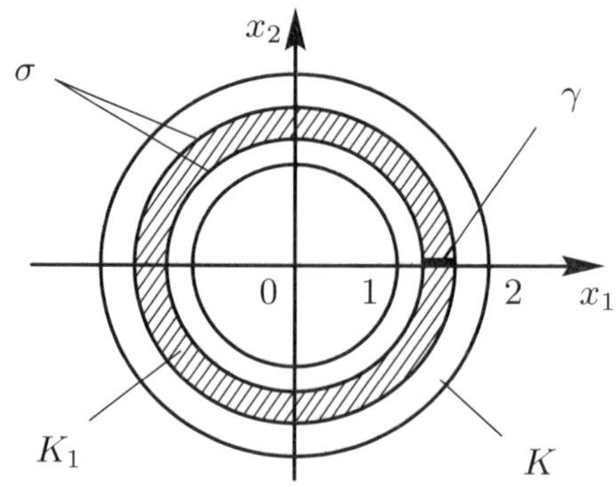


FIGURE 4.1

We cut the annulus $K = \{(r,\varphi) : 1 < r < 2\}$ along the segment $[1,2]$ on the $x_1$-axis. The domain $K\backslash[1,2]$ is simply connected. Since $\overline{g} = \operatorname{grad} w$, we have $\operatorname{curl}\overline{g} = 0$ in this domain, where

$$w(x) = \int_{(x_1^0,x_2^0)}^{(x_1,x_2)} (g_1\,dx_1 + g_2\,dx_2), \tag{4.4}$$

$x = (x_1, x_2)$, and $x^0 = (x_1^0, x_2^0)$ is a fixed point inside $K\backslash[1,2]$. Why can this formula fail for an uncut annulus $K$? The matter is that the function $w$ defined by (4.4) may be discontinuous on $[1,2]$, i.e., along the cut. We prove that it is continuous in $K$. We fix a segment $\gamma = [a,b]$ that is contained in $[1,2]$ and take the annulus $K_1 = \{(r,\varphi) : a < r < b\}$. Consider a continuous vector-valued function $\overline{a}(x_1,x_2)$ in $K$ that vanishes in $K\backslash K_1$, coincides with a smooth function $\overline{\alpha}(x_1)$ on $\gamma$, and satisfies the equality $\operatorname{div}\overline{a}(x_1,x_2) = 0$ in $K\backslash[1,2]$. The existence of such a vector-valued function for arbitrary $\overline{\alpha}(x_1)$ will be proved later. Then $\overline{a} \in \overset{\circ}{H}{}^1(\operatorname{div}, K)$, which implies

$$\int_K (\overline{g}, \overline{a})\,dx = 0.$$

On the other hand, taking into account that $g = \nabla w$ in the simply connected domain $K\backslash[1,2]$ and integrating by parts, we find

$$0 = \int_K (\overline{g}, \overline{a})\,dx = \int_{K\backslash[1,2]} (\overline{g}, \overline{a})\,dx = \int_{K\backslash[1,2]} (\nabla w, \overline{a})\,dx = \int_\gamma [w]\alpha_2(x_1)\,dx_1,$$

where $[w] \equiv w(x_1, x_2 + 0) - w(x_1, x_2 - 0)$ is the jump in $w$ across the segment $\gamma$. Since $\overline{\alpha}(x_1)$ is arbitrary, we see that $[w] = 0$, i.e., $w$ is continuous.

Thus, we have constructed a continuous scalar function $w$ in $K$ such that $\nabla w = \overline{g}$ in $K \setminus [1, 2]$. It is easy to show that the "potential" of $w$ is a smooth function in the entire annulus $K$ and the equality $\nabla w = \overline{g}$ holds at all points of $K$ (the proof is left to the reader).

The above proof can be used for domains of a more complicated structure (cf. Figure 4.2).

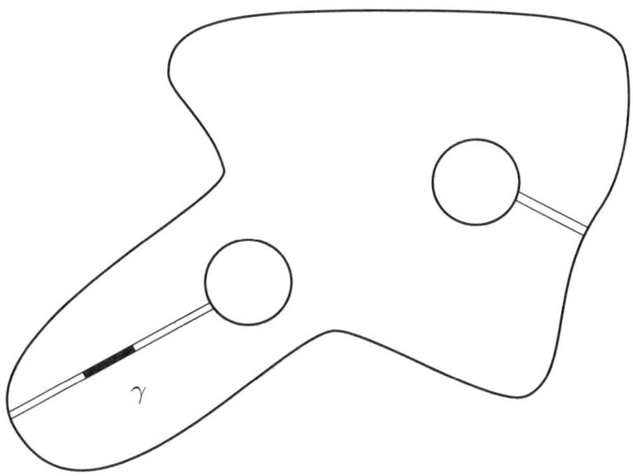


FIGURE 4.2

Making several cuts, we transform the initial domain to a simply connected domain. In the simply connected domain, we define $w$ by the equality (4.4) and prove the continuity of $w$ along the cuts arguing similarly to the case of an annulus.

We prove the existence of an auxiliary function $\overline{a}(x_1, x_2)$ with the above properties. For this purpose, we use the following assertion, which immediately implies the existence of $\overline{a}(x_1, x_2)$.

LEMMA 4.1. *Let $\Omega$ be a domain with smooth boundary, and let $\overline{\alpha}(x_1, x_2)$ be a vector field on the boundary of $\Omega$ such that $\int_{\partial\Omega} (\overline{\alpha}, \overline{\nu})\, d\sigma = 0$, where $\overline{\nu}$ is the outward normal to $\partial\Omega$. Then there exists a vector field $\overline{a}(x_1, x_2)$ in $\overline{\Omega}$ such that $\operatorname{div} \overline{a}(x_1, x_2) = 0$ and $\overline{a}(x_1, x_2) = \overline{\alpha}(x_1, x_2)$ for $(x_1, x_2) \in \partial\Omega$.*

PROOF. We look for a vector field $\overline{a}(x_1, x_2)$ in the form

$$\overline{a}(x_1, x_2) = \nabla\varphi + \operatorname{curl}\psi,$$

where $\varphi(x_1, x_2)$ and $\psi(x_1, x_2)$ are two unknown functions such that

$$\operatorname{curl}\psi \equiv \left(-\frac{\partial\psi}{\partial x_2}, \frac{\partial\psi}{\partial x_1}\right), \quad \nabla\varphi \equiv \left(\frac{\partial\varphi}{\partial x_1}, \frac{\partial\varphi}{\partial x_2}\right).$$

Let $\overline{\tau}$ be the unit tangent vector to $\partial\Omega$. We represent the vector field $\overline{\alpha}(x_1, x_2)$ on $\partial\Omega$ in the form

$$\overline{\alpha}(x_1, x_2) = \alpha_\nu(x_1, x_2)\overline{\nu}(x_1, x_2) + \alpha_\tau(x_1, x_2)\overline{\tau}(x_1, x_2)$$

and consider the solution $\varphi(x_1, x_2)$ to the Neumann problem

$$\Delta\varphi = 0 \quad \text{in } \Omega,$$
$$\left.\frac{\partial\varphi}{\partial\nu}\right|_{\partial\Omega} = \alpha_\nu,$$

which is solvable in view of the well-known compatibility condition for the Neumann problem and the condition $\int_{\partial\Omega} \alpha_\nu \, ds = 0$ of the lemma. It is easy to see that $\operatorname{div}(\nabla\varphi) = 0$ and $(\nabla\varphi, \nu) = \alpha_\nu$. Consequently, the vector-valued function $\operatorname{curl}\psi$ must have zero normal component on $\partial\Omega$ and a prescribed tangent component. For $\psi$ this assertion is equivalent to the conditions

$$\frac{\partial\psi}{\partial\overline{\tau}} = 0 \quad \text{on } \partial\Omega,$$
$$\frac{\partial\psi}{\partial\overline{\nu}} = \alpha_\tau - \frac{\partial\varphi}{\partial\overline{\tau}} \quad \text{on } \partial\Omega.$$

It is easy to construct a function vanishing on $\partial\Omega$ with a given normal derivative (of course, such a function is not unique). □

In the two-dimensional case, the above arguments allow us to represent the required function in terms of a solution to the Neumann problem for the Laplace equation. In the case of an arbitrary dimension, one can use Theorem 1.7 in Section 1 (details are left to the reader). For a three-dimensional domain, the proof of Lemma 4.1 can be found in [**75**].

EXERCISE 4.1. Let $\overline{\alpha}(x_1, x_2)$ be a vector field in the plane $\{x \in \mathbb{R}^3 : x_3 = 0\}$. Construct a vector-valued function $\overline{a}(x_1, x_2, x_3)$ in $\mathbb{R}^3$ such that $\operatorname{div}\overline{a} = 0$ and $\overline{a} = \overline{\alpha}$ for $x_3 = 0$.

EXERCISE 4.2. Let $\Omega \subset \mathbb{R}^2$ be a simply connected domain. Prove that it is possible to find a vector-valued function $\overline{a}(x_1, x_2)$ such that $\overline{a}(x_1, x_2) = \operatorname{curl}\psi(x_1, x_2)$.

To complete the proof of Theorem 4.1, we can approximate functions in $H^{-1}(\Omega)$ by smooth functions.

LEMMA 4.2. *Suppose that a domain $\Omega$ is star-shaped with respect to a system of balls, $p(x) \in L_2(\Omega)$, and $\int_\Omega p(x)\, dx = 0$. Then*

$$\|p(x)\|_{L_2(\Omega)} \le M \|\nabla p\|_{H^{-1}(\Omega)},$$

*where $M > 0$ is independent of $p(x)$.*

PROOF. By the definition of $H^{-1}(\Omega)$-norm, we have

$$\|\nabla p\|_{H^{-1}(\Omega)} = \sup_{\|u\|_{\overset{\circ}{H}{}^1(\Omega)} = 1} \left| \int_\Omega p(x) \; \operatorname{div}\overline{u}(x)\, dx \right|. \tag{4.5}$$

By Theorem 1.7, there is $\overline{U}(x) \in \overset{\circ}{H}{}^1(\Omega)$ such that

$$\operatorname{div}\overline{U}(x) = p(x), \quad \|\overline{U}\|_{\overset{\circ}{H}{}^1(\Omega)} \le C_0 \|p\|_{L_2(\Omega)},$$

where $C_0 > 0$ is independent of $p(x)$. Substituting $\overline{u}(x) = \overline{U}(x)/\|\overline{U}\|_{\overset{\circ}{H}{}^1(\Omega)}$ for a test function in (4.5), we find

$$\|\nabla p\|_{H^{-1}(\Omega)} \ge \frac{\|p\|^2_{L_2(\Omega)}}{\|\overline{U}\|_{\overset{\circ}{H}{}^1(\Omega)}} \ge \frac{\|p\|^2_{L_2(\Omega)}}{C_0 \|p\|_{L_2(\Omega)}} = C_0^{-1} \|p\|_{L_2(\Omega)},$$

which is what was to be proved. $\square$

Now, we prove Theorem 4.1 for $\overline{g} \in H^{-1}(\Omega)$. Suppose that $\Omega_k$ is a sequence of domains such that $\Omega_k \Subset \Omega$, $\bigcup_{k=1}^{\infty} \Omega_k = \Omega$, and the inequality in Lemma 4.2 holds with the same constant $C_0 > 0$ for all $\Omega_k$. Define smooth vector-valued functions $\overline{g}_k(x)$ on $\Omega_k$ by the formula

$$\overline{g}_k(x) = \int_{\mathbb{R}^d} \overline{g}(x)\delta_k^{-d} K\left(\frac{x-y}{\delta_k}\right) dy,$$

where $\delta_k = \rho(\Omega_k, \partial\Omega)$, and $K(y)$ is a symmetric nonnegative $C_0^\infty$-function such that $\int_{\mathbb{R}^d} K(y)\, dy = 1$.

EXERCISE 4.3. Verify that the vector-valued function $\overline{g}_k(x)$ satisfies the condition $\langle \overline{g}_k, \overline{u}\rangle = 0$ for any function $\overline{u}(x) \in \overset{\circ}{H}{}^1(\Omega_k)$ such that $\operatorname{div} \overline{u}(x) = 0$ in $\Omega_k$.

Since $\overline{g}_k(x)$ is smooth, the above arguments show that there exists a smooth function $w_k(x)$ in $\Omega_k$ such that $\nabla w_k = g_k$ in $\Omega_k$. We extend $w_k$ by zero to $\Omega$ and denote the extended function by the same symbol. We can assume that $\int_\Omega w_k\, dx = 0$, $k = 1, 2, \dots$ . By the construction of $\overline{g}_k$, we have $\langle \overline{g}_k, \overline{\varphi}\rangle \to \langle \overline{g}, \overline{\varphi}\rangle$ as $k \to \infty$, for any $\overline{\varphi}(x) \in \overset{\circ}{H}{}^1(\Omega)$ such that $\operatorname{div} \overline{\varphi}(x) = 0$ and $\operatorname{dist}(\operatorname{supp} \overline{\varphi}(x), \partial\Omega) > 0$. By the estimate in Lemma 4.2 and Exercise 4.3, we have

$$\|w_k\|_{L_2(\Omega)} = \|w_k\|_{L_2(\Omega_k)} \leq C(\Omega_k)\|\overline{g}_k\|_{H^{-1}(\Omega_k)} \leq C\|\overline{g}_k\|_{H^{-1}(\Omega_k)},$$

where the constant $C > 0$ is independent of $k$.

EXERCISE 4.4. Show that $\|\overline{g}_k\|_{H^{-1}(\Omega_k)} \leq \|\overline{g}\|_{H^{-1}(\Omega)}$.

Based on Exercise 4.4, we conclude that $\|w_k\|_{L_2(\Omega)} \leq C_1$, where $C_1$ is independent of $k$. Therefore, from $w_k$ we can chose a weakly convergent subsequence in $L_2(\Omega)$: $w_{k'} \rightharpoonup w^* \in L_2(\Omega)$ as $k' \to \infty$. In this case, $\nabla w_{k'} \rightharpoonup \nabla w^*$ weakly in $H^{-1}(\Omega)$ as $k' \to \infty$ and $\nabla w^* = g$. Indeed, suppose that $\overline{\varphi} \in \overset{\circ}{H}{}^1(\Omega)$, $\operatorname{supp} \varphi \Subset \Omega$, and $\operatorname{div} \overline{\varphi} = 0$. Then $\langle \nabla w_{k'}, \varphi\rangle = \langle g_{k'}, \varphi\rangle$ for sufficiently large $k'$. Passing to the limit as $k' \to \infty$, we find $\langle \nabla w^*, \varphi\rangle = \langle \overline{g}, \varphi\rangle$. Taking into account that the set of functions $\overline{\varphi}(x) \in \overset{\circ}{H}{}^1(\Omega)$ with $\operatorname{supp} \varphi(x) \Subset \Omega$ is dense in $\overset{\circ}{H}{}^1(\Omega)$, we conclude that $\nabla w^* = \overline{g}$. Thus, we have constructed a representation of functions in $L_2(\Omega)$. Theorem 4.1 is proved. $\square$

Using Lemma 4.2, one can prove the second Korn inequality. Although it was already proved in Subsection 3.3, the other proof based on Lemma 4.2 is of independent interest. As in Lemma 4.2, we assume that the domain $\Omega$ is star-shaped with respect to a system of balls. We use the equalities expressing the second order derivatives of an arbitrary vector-valued function $\overline{u} = (u_1, \dots, u_d)$ in terms of the first order derivatives of the components of the strain tensor $e_{ij}(\overline{u})$:

$$\frac{\partial^2 u_i}{\partial x_j^2} = 2\frac{\partial}{\partial x_j} e_{ij}(\overline{u}) - \frac{\partial}{\partial x_i} e_{jj}(\overline{u}),$$

$$\frac{\partial^2 u_i}{\partial x_k \partial x_l} = \frac{\partial}{\partial x_k} e_{il}(\overline{u}) + \frac{\partial}{\partial x_l} e_{ik}(\overline{u}) - \frac{\partial}{\partial x_i} e_{kl}(\overline{u}).$$

Using Lemma 4.2, it is easy to deduce the inequality

$$\|\nabla \overline{u}\|_{L_2(\Omega)}^2 \leq \frac{1}{|\Omega|}\Big(\int_\Omega \nabla \overline{u}(x) dx\Big)^2 + C\|e(\overline{u})\|_{L_2(\Omega)}^2. \tag{4.6}$$

The first integral on the right-hand side of (4.6) can be expressed in terms of the integrals along $\partial\Omega$. Then we can use the following estimates for the traces of functions in $H^1(\Omega)$:

$$\|u\|_{L_2(\partial\Omega)} \leq k\|\nabla u\|_{L_2(\Omega)} + K\|u\|_{L_2(\Omega)}. \tag{4.7}$$

Hence (4.6) and (4.7) imply the required inequality

$$\|\nabla u\|_{L_2(\Omega)} \leq C[\|u\|_{L_2(\Omega)} + \|e(u)\|_{L_2(\Omega)}].$$

It suffices to prove that the constant $k > 0$ in the inequality (4.7) can be taken small enough (moreover, the constant $K$ depends on $k$ and increases if $k$ decreases).

EXERCISE 4.5. Prove that the constant $k$ in the inequality (4.7) can be chosen as small as we like.

*Hint.* Begin by proving that the constant $k$ in the inequality

$$\|u\|_{H^\alpha(\Omega)} \leq k\|u\|_{H^\beta(\Omega)} + K\|u\|_{L_2(\Omega)} \quad (0 < \alpha < \beta)$$

can be arbitrarily small.

CHAPTER 2

# Homogenization Methods

Homogenization methods are used for studying the limit behavior of solutions to boundary value problems with rapidly oscillating coefficients. One of the main methods in homogenization theory is based on the compensated compactness lemma.

## 5. Compensated compactness

In asymptotic analysis, it is often required to pass to a limit in expressions of the form

$$(u_\varepsilon(x), v_\varepsilon(x)) = u_{\varepsilon,1}(x)v_{\varepsilon,1}(x) + \cdots + u_{\varepsilon,d}(x)v_{\varepsilon,d}(x),$$

where $u_\varepsilon$ and $v_\varepsilon$ are vector-valued functions converging weakly in a Hilbert space $H$ or in the dual Banach function spaces $B$ and $B'$. The convergence of the product is understood in some weak sense, say, in the sense of distributions. It is not difficult to pass to the limit if one of the factors strongly converges in $H$, but in general the factors must possess some additional properties in order to compensate for the absence of strong convergence. The compensated compactness method presented below deals with correct formulations of such additional conditions. An important example is the "div–curl" condition.

### 5.1. "Div–curl" condition.

LEMMA 5.1 (on compensated compactness). *Consider vector-valued functions $u_\varepsilon \in (L_2(\Omega))^d$ and $v_\varepsilon \in (L_2(\Omega))^d$ in a domain $\Omega \subset \mathbb{R}^d$ such that $u_\varepsilon \rightharpoonup u_0$ weakly in $[L_2(\Omega)]^d$, $v_\varepsilon \rightharpoonup v_0$ weakly in $[L_2(\Omega)]^d$, $\operatorname{curl} u_\varepsilon = 0$, and $\operatorname{div} v_\varepsilon \to f_0$ strongly in $H^{-1}(\Omega)$ as $\varepsilon \to 0$. Then*

$$(u_\varepsilon, v_\varepsilon) \overset{*}{\rightharpoonup} (u_0, v_0) \text{ weak* in } L_1(\Omega) \text{ as } \varepsilon \to 0. \tag{5.1}$$

PROOF. Without loss of generality, we can assume that $u_0 = 0$ and $v_0 = 0$. Then $f_0 = 0$. Indeed, it suffices to write $(u_\varepsilon, v_\varepsilon)$ as $(u_\varepsilon, v_\varepsilon) = (u_\varepsilon - u_0, v_\varepsilon - v_0) - (u_0, v_0) + (u_\varepsilon, v_0) + (u_0, v_\varepsilon)$ and note that the weak* limit of the sum of the last three terms is equal to $(u_0, v_0)$. The equality $f_0 = 0$ follows from Definition 1.4 of the weak divergence. The weak* convergence property in the space $L_1(\Omega)$ is local. Therefore, it suffices to prove that $(u_\varepsilon, v_\varepsilon)$ converges to zero in $L_1(Q)$ in an arbitrary ball $Q \Subset \Omega$. In $Q$, any curl-free field $u_\varepsilon$ (i.e., $\operatorname{curl} u_\varepsilon = 0$) is *potential*, i.e., there exists $U_\varepsilon$ such that $u_\varepsilon = \nabla_x U_\varepsilon$. The function $U_\varepsilon$ is unique up to an additive constant, which can be fixed by setting

$$\langle U_\varepsilon \rangle_Q \equiv \frac{1}{|Q|} \int_Q U_\varepsilon(x)\, dx = 0.$$

Since $\nabla_x U_\varepsilon \rightharpoonup 0$ weakly in $L_2(Q)$, we have $U_\varepsilon \rightharpoonup 0$ weakly in $H^1(Q)$. Indeed, since $\langle U_\varepsilon \rangle_Q = 0$, the Poincaré inequality implies that

$$\int_Q U_\varepsilon^2 dx \leq C \int_Q |\nabla U_\varepsilon|^2 dx,$$

i.e., $U_\varepsilon$ is bounded in $L_2(Q)$ and, consequently, $U_\varepsilon \rightharpoonup 0$ in $L_2(Q)$ and $U_\varepsilon \rightharpoonup 0$ in $H^1(Q)$. The weak convergence in $H^1(Q)$ implies the strong convergence in $L_2(Q)$ (the weak compactness of a sequence in $H^1(Q)$ implies the strong compactness of this sequence in $L_2(Q)$). Thus, $U_\varepsilon \to 0$ strongly in $L_2(Q)$ as $\varepsilon \to 0$.

For any $\varphi(x) \in C_0^\infty(Q)$ we have

$$\begin{aligned}\int_Q (u_\varepsilon, v_\varepsilon)\varphi\, dx &= \int_Q (\nabla_x U_\varepsilon, v_\varepsilon)\varphi\, dx = \int_Q (v_\varepsilon, \nabla_x(U_\varepsilon \varphi))\, dx - \int_Q (v_\varepsilon, \nabla_x \varphi)\, U_\varepsilon\, dx \\ &= -[\operatorname{div} v_\varepsilon, U_\varepsilon \varphi] - \int_Q (v_\varepsilon, \nabla_x \varphi) U_\varepsilon\, dx.\end{aligned}$$

Since $\operatorname{div} v_\varepsilon \to 0$ strongly in $H^{-1}(Q)$ and $U_\varepsilon \varphi$ is bounded in $\overset{\circ}{H}{}^1(Q)$, we have $[\operatorname{div} v_\varepsilon, U_\varepsilon \varphi] \to 0$ as $\varepsilon \to 0$. Similarly, since $U_\varepsilon \to 0$ strongly in $L_2(Q)$ and $(v_\varepsilon, \nabla_x \varphi)$ is bounded in $L_2(Q)$, we have

$$\int_Q (v_\varepsilon, \nabla_x \varphi)\, U_\varepsilon\, dx \to 0$$

as $\varepsilon \to 0$ and, consequently,

$$\int_Q (u_\varepsilon, v_\varepsilon)\varphi\, dx \to 0 \quad \forall\, \varphi \in C_0^\infty(Q).$$

The lemma is proved. □

REMARK 5.1. Lemma 5.1 provides us with only one of all possible conditions under which we can pass to the limit of the product of two weakly converging sequences. At present, this question is well investigated (cf., for example, [**49**]).

**5.2. Convergence of arbitrary solutions.** Consider a family of differential operators

$$L_\varepsilon = \frac{\partial}{\partial x_i}\Big(a_{ij}^\varepsilon(x) \frac{\partial}{\partial x_j}\Big)$$

with measurable bounded coefficients $a_{ij}^\varepsilon(x)$ satisfying the uniform ellipticity condition

$$\lambda |\xi|^2 \leq \sum_{i,j=1}^d a_{ij}^\varepsilon(x) \xi_i \xi_j \leq \Lambda |\xi|^2 \tag{5.2}$$

for any $\xi = (\xi_1, \dots, \xi_d) \in \mathbb{R}^d$ and $x \in \Omega$, where the constants $\lambda > 0$ and $\Lambda > 0$ are independent of the parameter $\varepsilon$. Such operators can be viewed as operators $L_\varepsilon : \overset{\circ}{H}{}^1(\Omega) \to H^{-1}(\Omega)$. It is easy to see that the uniform ellipticity condition (5.2) means that each operator $L_\varepsilon$ is coercive in the sense of Definition 1.9 and the constants $\kappa_1$ and $\kappa_2$ can be chosen to be the same for all operators of the family $\{L_\varepsilon\}$.

EXERCISE 5.1. Express $\kappa_1$ and $\kappa_2$ in terms of the ellipticity constants $\lambda$ and $\Lambda$.

An important example is a family $\{L_\varepsilon\}$ of operators with coefficients $a^\varepsilon_{ij}(x) = a_{ij}\left(\frac{x}{\varepsilon}\right)$, where $a_{ij}(y)$ are bounded, measurable, periodic with period 1 in all the variables $y_1, \dots, y_d$ and satisfy the ellipticity condition $\lambda|\xi|^2 \leq \sum_{i,j=1}^d a_{ij}(y)\xi_i\xi_j \leq \Lambda|\xi|^2$. By Theorem 1.2, we can choose a $G$-converging sequence of operators $L_{\varepsilon'}$ in $\{L_\varepsilon\}$ such that $L_{\varepsilon'} \xrightarrow{G} A_0$, where $A_0$ is an abstract coercive operator (possibly, with other constants $\kappa'_1$ and $\kappa'_2$). We can show that $A_0$ is a second order differential operator with constant coefficients.

This example illustrates the following important property of the $G$-convergence: the solutions to the boundary value problem for the operators $L_\varepsilon$ converge to the solution to the corresponding boundary value problem for the operator $A_0$ as $\varepsilon \to 0$ in spite of the fact that the coefficients of the operators $L_\varepsilon$ do not converge.

In applications, equations of the form $L_\varepsilon u = f$ describe various processes in a medium with rapidly oscillating spatial characteristics (microstructure), whereas the operator $A_0$ is related to a homogeneous medium.

Thus, a medium with rapidly oscillating spatial characteristics can be approximately described via homogeneous media.

The coefficients of the operator $A_0$ are usually referred to as *homogenized* or *effective*. The goal of *homogenization theory* is to construct homogenized characteristics and clarify whether the solutions to the boundary value problem for the operators $L_\varepsilon$ are close to the solution of the problem for the operator $A_0$. Sometimes, instead of the original operator, it is reasonable to study the $G$-limit operator because its structure is much simpler. In particular, this comes out when processes in an inhomogeneous medium are simulated with the help of computers. Indeed, in the numerical analysis of processes in a strongly inhomogeneous medium, in order to reach the desired accuracy of the difference method, the discretization step should be considerably less than the characteristic size of the inhomogeneity of the medium, which leads to a huge amount of calculations. But the coefficients of the $G$-limit operator usually oscillate slowly, which allows us to apply the difference method with a quite large step size.

Using Lemma 5.1, we establish a homogenization result for elliptic operators $L_\varepsilon$ with periodic coefficients $a^\varepsilon_{ij}(x) \equiv a_{ij}\left(\frac{x}{\varepsilon}\right)$. For this purpose, we introduce the notation

$$\Box \equiv [0,1] \times \cdots \times [0,1], \qquad A(y) \equiv \{a_{ij}(y)\}, \qquad \langle w \rangle_\Box \equiv \int_\Box w(y)\,dy$$

and for every constant vector $\zeta \in \mathbb{R}^d$ define a 1-periodic vector-valued function $\overline{w}_\zeta(y)$ with the following properties:

(1) $\operatorname{curl}_y \overline{w}_\zeta(y) = 0$ and $\overline{w}_\zeta(y) \in (L_{2,\mathrm{per}}(\Box))^d$;

(2) $\operatorname{div}_y(A(y)\overline{w}_\zeta(y)) = 0$;

(3) $\langle \overline{w}_\zeta(y) \rangle_Y = \zeta$.

EXERCISE 5.2. Prove that there exists a unique 1-periodic function $\overline{w}_\zeta(y)$ with properties (1)–(3).

*Hint.* Since $\operatorname{curl}_y \overline{w}_\zeta(y) = 0$, we can represent the unknown function $\overline{w}_\zeta(y)$ as $\overline{w}_\zeta(y) = \nabla U_\zeta(y) + \zeta$, where $U_\zeta(y)$ is a scalar 1-periodic function in $H^1(\Box)$. Then the problem for $\overline{w}_\zeta(y)$ is reduced to an elliptic equation for $U_\zeta(y)$ with periodic boundary conditions. The existence and uniqueness can be proved using the Fredholm theorem.

If $\xi$ is some coordinate vector $\overline{e}_k = (0, \dots, 1, \dots, 0)$ ("1" at the $k$th position), then $w_{\overline{e}_k}(y) \equiv w_k(y)$, $k = 1, \dots, d$, takes the form $w_k(y) = \nabla U_k + \overline{e}_k$, where $U_k(y)$ is 1-periodic with respect to $y_1, \dots, y_d$ and satisfies the equation

$$\sum_{i,j=1}^{d} \frac{\partial}{\partial y_i}\Big(a_{ij}(y)\frac{\partial U_k}{\partial y_j}\Big) = -\sum_{j=1}^{d} \frac{\partial a_{kj}}{\partial y_j}(y), \quad y \in \square.$$

Note that the solution is determined up to an additive constant.

The coefficient matrix $\widehat{A}$ of the homogenized equation is defined by the formula $\widehat{A}\zeta \equiv \langle A(y)\overline{w}_\zeta(y)\rangle_\square$ for all $\zeta \in \mathbb{R}^d$. It is well defined because $\overline{w}_\zeta(y)$ depends on $\zeta \in \mathbb{R}^d$ linearly (check it!).

EXERCISE 5.3. Prove that the entries $\widehat{a}_{ik}$ of the matrix $\widehat{A}$ can be expressed in terms of $U_k(y)$, $k = 1, \dots, d$, by the formula

$$\widehat{a}_{ik} \equiv \int_\square \Big[a_{ik}(y) + a_{ij}(y)\frac{\partial U_k(y)}{\partial y_j}\Big]\,dy, \quad i,k = 1, \dots, d. \tag{5.3}$$

EXERCISE 5.4. Prove that the entries $\widehat{a}_{ij}$ of the matrix $\widehat{A}$ satisfy the estimate

$$\lambda|\xi|^2 \le \sum_{i,j=1}^{d} \widehat{a}_{ij}\xi_i\xi_j \le \Lambda|\xi|^2.$$

THEOREM 5.1. *Suppose that $\Omega$ is a bounded domain and $f \in H^{-1}(\Omega)$. Then the solutions $u_\varepsilon$ to the Dirichlet problem*

$$\sum_{i,j=1}^{d} \frac{\partial}{\partial x_i}\Big(a_{ij}\Big(\frac{x}{\varepsilon}\Big)\frac{\partial u_\varepsilon}{\partial x_j}\Big) = f(x) \text{ in } \Omega, \quad u_\varepsilon(x) \in \overset{\circ}{H}{}^1(\Omega), \tag{5.4}$$

*converge to the solution $u_0(x)$ to the problem*

$$\sum_{i,j=1}^{d} \frac{\partial}{\partial x_i}\Big(\widehat{a}_{ij}\frac{\partial u_0}{\partial x_j}\Big) = f(x) \text{ in } \Omega, \quad u_0(x) \in \overset{\circ}{H}{}^1(\Omega), \tag{5.5}$$

*weakly in $\overset{\circ}{H}{}^1(\Omega)$ and strongly in $L_2(\Omega)$.*

PROOF. From $u_\varepsilon(x)$ we can extract a subsequence $u_{\varepsilon'}(x)$ such that $u_{\varepsilon'}(x) \rightharpoonup u_0(x)$ weakly in $\overset{\circ}{H}{}^1(\Omega)$ and strongly in $L_2(\Omega)$, $\nabla u_{\varepsilon'}(x) \rightharpoonup \nabla u_0(x)$ weakly in $L_2(\Omega)$, and $\overline{p}^{\,\varepsilon'}(x) \rightharpoonup \overline{p}^{\,0}(x)$ weakly in $L_2(\Omega)$, where $\overline{p}^{\,\varepsilon}(x) \equiv A(\frac{x}{\varepsilon})\nabla u_\varepsilon(x)$ is the flux of $u_\varepsilon$. It can be done in view of the weak compactness of the unit ball in $\overset{\circ}{H}{}^1(\Omega)$ in $L_2(\Omega)$ and the Rellich theorem.

Let $\overline{q}^{\,\varepsilon}(x) \equiv A(\frac{x}{\varepsilon})\overline{w}_\zeta(\frac{x}{\varepsilon})$, where the vector-valued function $\overline{w}_\zeta(y)$ was defined above. By definition, we have the equality $(\overline{p}^{\,\varepsilon}(x), \overline{w}_\zeta(\frac{x}{\varepsilon})) = (\nabla u_\varepsilon(x), \overline{q}^{\,\varepsilon}(x))$, which can be regarded as the equality of two inner products of vectors in $\mathbb{R}^d$ (for every $x \in \Omega$). Using Lemma 5.1, we can pass to the limit in this equality as $\varepsilon' \to 0$. Indeed, the vector-valued functions on the left-hand side satisfy the conditions $\operatorname{div}\overline{p}^{\,\varepsilon} = f$ and $\operatorname{curl}\overline{w}_\zeta = 0$, whereas the vector-valued functions on the right-hand side satisfy the equalities $\operatorname{div}\overline{q}^{\,\varepsilon} = 0$ and $\operatorname{curl}\nabla u_\varepsilon = 0$.

EXERCISE 5.5. Prove that $\overline{w}_\zeta(\frac{\cdot}{\varepsilon}) \rightharpoonup \zeta$ and $\overline{q}^{\,\varepsilon} \rightharpoonup \widehat{A}\zeta$ weakly in $L_2(\Omega)$ as $\varepsilon \to 0$.

Passing to the limit as $\varepsilon' \to 0$, we obtain the equality $(\overline{p}^0, \zeta) = (\nabla u_0, \widehat{A}\zeta)$, which holds almost everywhere in $\Omega$ and in $L_2(\Omega)$. Since $\zeta \in \mathbb{R}^d$ is arbitrary, $\overline{p}^0 = \widehat{A}\nabla u_0$. Since $\operatorname{div} \overline{p}^\varepsilon(x) = f(x)$, we have $\operatorname{div} \overline{p}^0(x) = f(x)$ in the sense of distributions. Therefore, $\operatorname{div}(\widehat{A}\nabla u_0) = f$. Recalling that $u_0(x) \in \overset{\circ}{H}{}^1(\Omega)$, we conclude that $u_0(x)$ is a solution to the problem (5.5). Since the Dirichlet problem has a unique solution, $u_0(x)$ is the limit of any converging subsequence $u_{\varepsilon'}(x)$ and, consequently, $u_\varepsilon(x)$ converge to $u_0(x)$. □

No we have the following important question: does the $G$-limit differential operator with constant coefficients depend on the choice of boundary conditions? Indeed, in addition to the Dirichlet boundary conditions, there are other natural boundary conditions. The answer is negative: the coefficients of the $G$-limit differential operator are determined by only the coefficients of the original operator and are independent of the choice of boundary conditions. In particular, the coefficient matrix of the homogenized problem (5.5) defines the homogenized equation also for all other boundary conditions.

To illustrate this fact, we consider the Neumann problem

$$(5.6)\qquad \begin{aligned} &\sum_{i,j=1}^{d} \frac{\partial}{\partial x_i}\Big(a_{ij}\Big(\frac{x}{\varepsilon}\Big)\frac{\partial v_\varepsilon}{\partial x_j}\Big) = \widetilde{f}(x) \quad \text{in } \Omega, \qquad \int_\Omega \widetilde{f}(x)\,dx = 0,\\ &\sum_{i,j=1}^{d} a_{ij}^\varepsilon \frac{\partial v_\varepsilon}{\partial x_j}\nu_i = 0 \quad \text{on } \partial\Omega, \qquad \int_\Omega v_\varepsilon(x)\,dx = 0 \end{aligned}$$

and the formally homogenized problem

$$(5.7)\qquad \begin{aligned} &\sum_{i,j=1}^{d} \widehat{a}_{ij}\frac{\partial^2 v_0}{\partial x_i \partial x_j} = \widetilde{f}(x) \text{ in } \Omega,\\ &\sum_{i,j=1}^{d} \widehat{a}_{ij}\frac{\partial v_0}{\partial x_j}\nu_i = 0 \text{ on } \partial\Omega, \qquad \int_\Omega v_0(x)\,dx = 0, \end{aligned}$$

where $\nu = (\nu_1, \dots, \nu_d)$ is the unit outward normal to $\partial\Omega$ and $\widetilde{f} \in L_2(\Omega)$.

THEOREM 5.2. *Suppose that $a_{ij}^\varepsilon$ are the same as in the problem (5.4) and $\widehat{a}_{ij}$ are the same as in the problem (5.5). Then $v_\varepsilon \rightharpoonup v_0$ weakly in $H^1(\Omega)$ as $\varepsilon \to 0$ and $a^\varepsilon \nabla v_\varepsilon \rightharpoonup a^0 \nabla v_0$ weakly in $(L_2(\Omega))^d$ as $\varepsilon \to 0$, where $v_\varepsilon$ is a solution to the problem (5.6) and $v_0$ is a solution to the problem (5.7).*

PROOF. It is easy to see that the $a^\varepsilon \nabla v_\varepsilon$ are uniformly bounded with respect to $\varepsilon$ in the space $L_2(\Omega)$, whereas the $v_\varepsilon$ are bounded in the space $H^1(\Omega)$. By the Rellich theorem, there exists a subsequence $\varepsilon' \to 0$ and functions $v_*$ and $q_0 \in (L_2(\Omega))^d$ such that $v_\varepsilon \rightharpoonup v_*$ weakly in $H^1(\Omega)$ as $\varepsilon' \to 0$, and $a^\varepsilon \nabla v_\varepsilon \rightharpoonup q_0$ weakly in $(L_2(\Omega))^d$ as $\varepsilon' \to 0$. Consequently, $\operatorname{div} q_0 = \widetilde{f}$.

Consider the identity $(\nabla v_\varepsilon, [a^\varepsilon \nabla u_\varepsilon]) = ([\nabla v_\varepsilon a^\varepsilon], \nabla u_\varepsilon)$, where $u_\varepsilon$ is a solution to the problem (5.4) with an arbitrary fixed $f$. Both pairs of sequences on the left- and right-hand sides of this equality satisfy the assumptions of Lemma 5.1. Passing to the weak* limit as $\varepsilon' \to 0$ and using Lemma 5.1, we find that

$$(5.8)\qquad (\nabla v_*, \widehat{a}\nabla u_0) = (q_0, \nabla u_0)$$

almost everywhere in $\Omega$, where $u_0$ is the solution to the problem (5.5). Since $f$ is arbitrary, (5.8) holds for an arbitrary function $u_0 \in \overset{\circ}{H}{}^1(\Omega)$. This means that $\widehat{a}\nabla v_* = q_0$ in $L_2(\Omega)$ and $\operatorname{div}(\widehat{a}\nabla v_*) = \widetilde{f}$ in the space of distributions in $\Omega$. Thus, $v_*$ satisfies the equation $\operatorname{div}(\widehat{a}\nabla v_*) = \widetilde{f}$ in $\Omega$.

The equality $\int_\Omega v_*\,dx = 0$ follows from the convergence $v_\varepsilon \to v_*$. It remains to verify the boundary condition $(\widehat{a}\nabla v_*, \overline{\nu}) = 0$ on $\partial\Omega$. To do this, we write the integral identity corresponding to the problem (5.6),

$$\int_\Omega a_{ij}\Big(\frac{x}{\varepsilon}\Big)\frac{\partial}{\partial x_j}v_\varepsilon \frac{\partial}{\partial x_i}\varphi\,dx = \int_\Omega \widetilde{f}\varphi\,dx \quad \forall\,\varphi \in C^\infty(\mathbb{R}^n),$$

and pass to the limit as $\varepsilon \to 0$. Since the $q_\varepsilon$ converge to $\widehat{a}\nabla v_*$, we have

$$\int_\Omega \widehat{a}_{ij}\frac{\partial}{\partial x_j}v_* \frac{\partial}{\partial x_i}\varphi\,dx = \int_\Omega \widetilde{f}\varphi\,dx \quad \forall\,\varphi \in C^\infty(\mathbb{R}^n),$$

i.e., $v_*$ satisfies the integral identity corresponding to the problem (5.7), which completes the proof. □

We emphasize the following important property of $q_\varepsilon$. If $q_\varepsilon$ and $\widetilde{f}$ are extended by zero to the entire space $\mathbb{R}^n$, then the extended functions $\widetilde{q}_\varepsilon$ and $\widetilde{\widetilde{f}}$ satisfy the equality $\operatorname{div}\widetilde{q}_\varepsilon = \widetilde{\widetilde{f}}(x)$ in $\mathbb{R}^d$ in the sense of distributions.

EXERCISE 5.6. Prove the last assertion.
*Hint.* Use the condition $(\overline{q}_\varepsilon, \nu) = 0$ on $\partial\Omega$.

We formulate a general result concerning the independence of the homogenized equation of the boundary conditions.

THEOREM 5.3 (convergence of arbitrary solutions). *Suppose that the $a^\varepsilon_{ij}(x)$ are the same as in the problem (5.4) and the $\widehat{a}_{ij}$ are defined by (5.3). Assume that for some $g \in (L_2(\Omega))^d$ and $f \in H^{-1}(\Omega)$ the functions $v_\varepsilon \in H^1(\Omega)$ satisfy the equation*

$$\sum_{i,j=1}^d \frac{\partial}{\partial x_i}\Big(a_{ij}\Big(\frac{x}{\varepsilon}\Big)\Big(\frac{\partial v_\varepsilon}{\partial x_j} + g_j\Big)\Big) = f(x) \quad \text{in } \Omega$$

*and $v_{\varepsilon'} \rightharpoonup v_*$ weakly in $H^1(\Omega)$ as $\varepsilon' \to 0$. Then $v_*$ satisfies the equation*

$$\operatorname{div}(\widehat{a}(\nabla v_* + g)) = f \quad \text{in } \Omega$$

*and $a^{\varepsilon'}(\nabla v_{\varepsilon'} + g) \rightharpoonup \widehat{a}(\nabla v_* + g)$ weakly in $L_2(\Omega)$ as $\varepsilon' \to 0$.*

**5.3. Structure of the $G$-limit operator.** The assertion that the class of uniformly elliptic operators in the divergence form is closed with respect to the $G$-convergence is the key result of $G$-convergence theory. We study the structure of the $G$-limit operator of an arbitrary sequence of elliptic differential operators.

Let $A_\varepsilon \xrightarrow{G} A_0$ as $\varepsilon \to 0$, where

$$A_\varepsilon \equiv \sum_{i,j=1}^d \frac{\partial}{\partial x_i}\Big(a^\varepsilon_{ij}(x)\frac{\partial}{\partial x_j}\Big), \quad \lambda|\xi|^2 \le \sum_{i,j=1}^d a^\varepsilon_{ij}(x)\xi_i\xi_j \le \Lambda|\xi|^2, \quad x \in \Omega_1, \tag{5.9}$$

and let $A_0 : \overset{\circ}{H}{}^1(\Omega) \to H^{-1}(\Omega)$ be an abstract coercive operator (cf. Theorem 1.2). In this case, $A_0$ is not a differential operator with constant coefficients, but it is still an elliptic differential operator in the divergence form. Let us prove this fact.

First, consider a domain $\Omega_1$ such that $\Omega \Subset \Omega_1$ and extend the operator $A_\varepsilon$ to $\Omega_1$ by setting $a_\varepsilon(x) = \Lambda\,\mathrm{Id}$ for $x \in \Omega_1 \setminus \Omega$. From $A_\varepsilon$ we choose a subsequence such that $A_{\varepsilon'} \xrightarrow{G} A_0$ in $\Omega_1$ and define $u_\varepsilon^\xi(x)$ as the solution to the boundary value problem

$$A_\varepsilon u_\varepsilon^\xi = A_0[\varphi(x)(x,\xi)], \quad u_\varepsilon^\xi \in \overset{\circ}{H}{}^1(\Omega_1), \quad (x,\xi) = \sum_{i=1}^d x_i\xi_i,$$

where $\varphi \in C_0^\infty(\Omega_1)$ is a cut-off function such that $\varphi = 1$ in $\Omega$, and $\xi$ is an arbitrary vector in $\mathbb{R}^d$. By the definition of $G$-convergence, we have $\nabla u_{\varepsilon'}^\xi \rightharpoonup \xi$ in $L_2(\Omega)$.

Let $q^\xi(x)$ be the weak limit of $a_{\varepsilon'}(x)\nabla u_{\varepsilon'}^\xi(x)$ in $L_2(\Omega)$ on some subsequence $\varepsilon'$ as $\varepsilon' \to 0$. Then, using "div–curl" condition for arbitrary vectors $\xi, \eta \in \mathbb{R}^d$, we find

$$(\nabla u_{\varepsilon'}^\eta, a_{\varepsilon'}(x)\nabla u_{\varepsilon'}^\xi) \overset{*}{\rightharpoonup} (\eta, q^\xi(x)), \quad (\nabla u_{\varepsilon'}^\xi, a_{\varepsilon'}(x)\nabla u_{\varepsilon'}^\eta) \overset{*}{\rightharpoonup} (\xi, q^\eta(x)) \quad \text{in } \Omega.$$

Since the matrix $a_\varepsilon(x)$ is symmetric, we have $(\eta, q^\xi(x)) = (\xi, q^\eta(x))$ for all $\xi, \eta \in \mathbb{R}^d$ in $L_2(\Omega)$. Therefore, $q^\xi(x)$ is linear with respect to $\xi$, i.e., $q^\xi(x) = \widehat{a}(x)\xi$ since $(\eta, q^{\alpha_1\xi_1+\alpha_2\xi_2}(x)) = (\alpha_1\xi_1 + \alpha_2\xi_2, q^\eta(x)) = \alpha_1(\xi_1, q^\eta(x)) + \alpha_2(\xi_2, q^\eta(x)) = (\eta, \alpha_1 q^{\xi_1}(x)) + (\eta, \alpha_2 q^{\xi_2}(x)) = (\eta, \alpha_1 q^{\xi_1}(x) + \alpha_2 q^{\xi_2}(x))$ for arbitrary $\eta, \xi_1, \xi_2 \in \mathbb{R}^d$, $\alpha_1, \alpha_2 \in \mathbb{R}$, and $\widehat{a}(x)$ is a symmetric matrix $((\eta, a_0(x)\xi) = (\xi, \widehat{a}(x)\eta)$ for any $\xi, \eta \in \mathbb{R}^d)$.

Let $v_\varepsilon(x)$ be a solution to the problem

$$A_\varepsilon v_\varepsilon = f(x), \quad f \in H^{-1}(\Omega), \quad v_\varepsilon \in \overset{\circ}{H}{}^1(\Omega).$$

By the "div–curl" condition, $(\nabla u_{\varepsilon''}^\xi, a_{\varepsilon''}(x)\nabla v_{\varepsilon''}) \overset{*}{\rightharpoonup} (\xi, p_0(x))$ in $\Omega$ on some subsequence $\varepsilon''$, where $p_0(x)$ is the weak limit of $a_{\varepsilon''}(x)\nabla v_{\varepsilon''}$ in $L_2(\Omega)$. On the other hand, using the "div–curl" condition and the symmetry property of the matrices $a_{\varepsilon''}(x)$, $\widehat{a}(x)$, we find

$$(\nabla u_{\varepsilon''}^\xi, a_{\varepsilon''}(x)\nabla v_{\varepsilon''}) = (\nabla v_{\varepsilon''}, a_{\varepsilon''}(x)\nabla u_{\varepsilon''}^\xi) \overset{*}{\rightharpoonup} (\nabla v_0(x), \widehat{a}(x)\xi) = (\xi, \widehat{a}(x)\nabla v_0(x)).$$

Consequently, $(\xi, p_0(x)) = (\xi, \widehat{a}(x)\nabla v_0(x))$ in $L_2(\Omega)$. Since $\xi \in \mathbb{R}^d$ is arbitrary, we have $p_0(x) = \widehat{a}(x)\nabla v_0(x)$, i.e., $a_\varepsilon(x)\nabla v_\varepsilon \rightharpoonup \widehat{a}(x)\nabla v_0(x)$ in $L_2(\Omega)$.

EXERCISE 5.7. Prove that the matrix $\widehat{a}(x)$ satisfies the estimate

$$\lambda|\xi|^2 \le (\widehat{a}(x)\xi, \xi) \le \Lambda|\xi|^2$$

with the same constants $\lambda$ and $\Lambda$ as in (5.9).

Since $\operatorname{div}(a_\varepsilon\nabla v_\varepsilon) = f$, we have $\operatorname{div} p_0 = f$. Therefore, $\operatorname{div}(\widehat{a}\nabla v_0) = f$ in $\Omega$. Consequently, $A_0$ is an elliptic differential operator in the divergence form. Thus, we have proved the following assertion.

THEOREM 5.4. *Let $A_\varepsilon \equiv \sum_{i,j=1}^d \frac{\partial}{\partial x_i}\big(a_{ij}^\varepsilon(x)\frac{\partial}{\partial x_j}\big)$, where $a_{ij}^\varepsilon$ are bounded measurable functions such that $\lambda|\xi|^2 \le \sum_{i,j=1}^d a_{ij}^\varepsilon\xi_i\xi_j \le \Lambda|\xi|^2$ almost everywhere in $\Omega$, and let $A_\varepsilon \xrightarrow{G} A_0$. Then $A_0$ is a differential operator of the form $A_0 \equiv \sum_{i,j=1}^d \frac{\partial}{\partial x_i}\big(\widehat{a}_{ij}(x)\frac{\partial}{\partial x_j}\big)$, where $\widehat{a}_{ij}$ are bounded measurable functions such that $\lambda|\xi|^2 \le \sum_{i,j=1}^d \widehat{a}_{ij}\xi_i\xi_j \le \Lambda|\xi|^2$.*

**5.4. Example of a homogenization of a nonlinear problem.** Lemma 5.1 can be used for analyzing solutions to nonlinear boundary value problems with rapidly oscillating coefficients.

Consider the system

$$-\operatorname{div}_x A\Big(\frac{x}{\varepsilon}, \nabla_x u_\varepsilon(x)\Big) = f(x) \quad \text{in } \Omega, \qquad u_\varepsilon = v \text{ on } \partial\Omega, \qquad v \in H^1(\Omega), \tag{5.10}$$

where the mapping $A : \mathbb{R}^d \times \mathbb{R}^d \to \mathbb{R}^d$ satisfies the following conditions:

(1) $A(\xi, \eta)$ is 1-periodic with respect to the first variable;

(2) $|A(\xi, \eta_2) - A(\xi, \eta_1)| \le \kappa_1 |\eta_2 - \eta_1|$ for all $\xi, \eta_1, \eta_2 \in \mathbb{R}^d$;

(3) $(A(\xi, \eta_2) - A(\xi, \eta_1), \eta_2 - \eta_1) \ge \kappa_2 |\eta_2 - \eta_1|^2$ for all $\xi, \eta_1, \eta_2 \in \mathbb{R}^d$.

Here, $0 < \kappa_2 < \kappa_1$.

Consider the formally homogenized system

$$-\operatorname{div}_x \widehat{A}(\nabla_x u_0(x))) = f(x) \text{ in } \Omega, \qquad u_0 = v \text{ on } \partial\Omega, \tag{5.11}$$

where $\widehat{A} : \mathbb{R}^d \to \mathbb{R}^d$ is defined by the formula

$$\widehat{A}(\eta) \equiv \int_{\square} A(\xi, \eta + \nabla_\xi N_\eta(\xi))\, d\xi \tag{5.12}$$

and the 1-periodic function $N_\eta(\xi)$ with respect to $\xi$ is found from the equation

$$\operatorname{div}_\xi A(\xi, \eta + \nabla_\xi N_\eta(\xi)) = 0 \quad \text{in } \square. \tag{5.13}$$

EXERCISE 5.8. Verify that $\widehat{A}(\eta)$ is a Lipschitz function such that

$$(\widehat{A}(\eta_1) - \widehat{A}(\eta_2), \eta_1 - \eta_2) \ge \kappa |\eta_1 - \eta_2|^2.$$

THEOREM 5.5. *$u_\varepsilon \rightharpoonup u_0$ weakly in $H^1(\Omega)$ as $\varepsilon \to 0$.*

PROOF. It is easy to show that properties (2) and (3) of $A(\xi, \eta)$ imply the uniform boundedness of $u_\varepsilon$ and $E_\varepsilon := A(x/\varepsilon, \nabla_x u_\varepsilon)$ in $H^1(\Omega)$ and $[L_2(\Omega)]^d$ respectively. Consequently, by the Rellich theorem, it is possible to choose weakly converging subsequences (denoted in the same way)

$$u_\varepsilon \rightharpoonup u_0 \quad \text{weakly in } H^1(\Omega), \qquad E_\varepsilon \rightharpoonup E_0 \quad \text{weakly in } [L_2(\Omega)]^d \tag{5.14}$$

as $\varepsilon \to 0$. It is easy to see that $E_0$ satisfies the equation

$$-\operatorname{div} E_0 = f \quad \text{in } \Omega. \tag{5.15}$$

To show that $E_0 = \widehat{A}(\nabla_x u_0(x))$, we introduce the functions $v_\varepsilon(x) = (\eta, x) + \varepsilon N_\eta\big(\frac{x}{\varepsilon}\big)$, where $N_\eta(\xi)$ was defined above. Since $A$ is monotone, we have, for all $\eta \in \mathbb{R}^d$ and nonnegative $\varphi \in C^\infty(\overline{\Omega})$,

$$\int_\Omega \Big(A\Big(\frac{x}{\varepsilon}, \nabla_x u_\varepsilon\Big) - A\Big(\frac{x}{\varepsilon}, \nabla_x v_\varepsilon\Big), \nabla_x u_\varepsilon - \nabla_x v_\varepsilon\Big)\varphi(x)\, dx \ge 0. \tag{5.16}$$

It is easy to see that

$$\operatorname{div}_x E_\varepsilon^1 \equiv \operatorname{div}_x A\Big(\frac{x}{\varepsilon}, \eta + \nabla_x N_\eta\Big(\frac{x}{\varepsilon}\Big)\Big) = 0,$$

$$A\Big(\frac{x}{\varepsilon}, \nabla_x v_\varepsilon\Big) = A\Big(\xi, \eta + \nabla_\xi N_\eta(\xi)\Big)\Big|_{\xi = x/\varepsilon} \rightharpoonup \int_\square A(\xi, \eta + \nabla_\xi N_\eta(\xi))\, d\xi \equiv A(\eta)$$

weakly in $[L_2(\Omega)]^d$, $\nabla u_\varepsilon \rightharpoonup \nabla u_0(x)$ weakly in $[L_2(\Omega)]^d$, and $\nabla v_\varepsilon \rightharpoonup \eta$ weakly in $[L_2(\Omega)]^d$ as $\varepsilon \to 0$. By Lemma 5.1, we can pass to the limit in (5.16) as $\varepsilon \to 0$, which yields the inequality

$$\int_\Omega (E_0(x) - \widehat{A}(\eta), \nabla_x u_0(x) - \eta)\varphi(x)\, dx \geq 0. \tag{5.17}$$

Since (5.17) holds for any nonnegative $\varphi \in C^\infty(\overline{\Omega})$, we have

$$(E_0(x) - \widehat{A}(\eta), \nabla u_0(x) - \eta) \geq 0 \quad \forall \eta \in \mathbb{R}^d \tag{5.18}$$

for almost all $x \in \Omega$. According to the theory of monotone operators (cf. [**78**]), (5.18) implies $E_0(x) - \widehat{A}(\nabla u_0(x)) = 0$. Let us prove this assertion. Assume the contrary: $E_0(x) - \widehat{A}(\nabla u(x)) \neq 0$. Then at least one of the components of the vector $E_0(x) - \widehat{A}(\nabla u(x))$ is different from zero, say, $[E_0(x) - \widehat{A}(\nabla u(x))]_i \neq 0$ in $L_2(\Omega)$. Let $x_0$ be a Lebesgue point of $E_0(x)$ and $\nabla u_0(x)$, and let $[E_0(x_0) - \widehat{A}(\nabla u_0(x_0))]_i \neq 0$. For definiteness, we assume that $[E_0(x_0) - \widehat{A}(\nabla u_0(x_0))]_i > 0$. We choose $\eta_\delta = \nabla u_0(x_0) + \delta e_i$, where $e_i$ is the $i$th unit coordinate vector in $\mathbb{R}^d$. Then for sufficiently small $\delta$ the inequality (5.18) fails on a set of positive measure (check it!). Similarly, in the case $[E_0(x_0) - \widehat{A}(\nabla u_0(x_0))]_i < 0$, setting $\eta_\delta = \nabla u_0(x_0) - \delta e_i$ and taking $\delta$ small enough, we arrive at a contradiction with (5.18) on a set of positive measure, which proves the equality $E_0(x) = \widehat{A}(\nabla u_0(x))$.

Since $E_0$ satisfies (5.15), we have $-\operatorname{div}_x \widehat{A}(\nabla_x u_0(x)) = f$ in $\Omega$. □

## 6. Homogenization of equations with random coefficients

Let $A(\omega)$ be a random $d \times d$-matrix with entries $a_{ij}(\omega) \in L_\infty(\Omega)$ satisfying the ellipticity condition for almost all $\omega \in \Omega$,

$$\kappa_1|\eta|^2 \leq a_{ij}(\omega)\eta_i\eta_j \leq \kappa_2|\eta|^2. \tag{6.1}$$

Consider the following problem: for any $\eta \in \mathbb{R}^d$,

$$\text{find } v_\eta \in \mathcal{V}_2^{\text{pot}}(\Omega) \text{ such that } \quad \int_\Omega \varphi_i a_{ij}(\eta_j + (v_\eta)_j)\, d\mu = 0 \quad \forall \varphi \in \mathcal{V}_2^{\text{pot}}(\Omega). \tag{6.2}$$

By Lemma 2.2, the problem (6.2) can be written as follows: for any $\eta \in \mathbb{R}^d$

$$\text{find } v_\eta \in \mathcal{V}_2^{\text{pot}}(\Omega) \text{ such that } A(\eta + v_\eta) \in L_2^{\text{sol}}(\Omega).$$

EXERCISE 6.1. Prove that the solution $v_\eta$ to the problem (6.2) exists and is unique.

On a typical realization, this problem becomes a usual elliptic equation. If $u(x)$ is the potential of $v(x) = v(T(x)\omega)$ and $a_{ij}(x) = a_{ij}(T(x)\omega)$, then

$$\operatorname{div}\left(A(T(x)\omega)\,(\eta + \nabla u)\right) = 0. \tag{6.3}$$

Conversely, if the potential of almost all realizations of $v \in \mathcal{V}_2^{\text{pot}}(\Omega)$ satisfies equation (6.3), then the vector $A(\omega)(\eta + v(\omega))$ is solenoidal and, consequently, (6.2) holds.

We also need the transposed problem: for any $\zeta \in \mathbb{R}^d$

$$\text{find } w_\zeta \in \mathcal{V}_2^{\text{pot}}(\Omega) \text{ such that } (w_\zeta + \zeta)^t A \in L_2^{\text{sol}}(\Omega). \tag{6.4}$$

Note that the ergodicity of the dynamical system $T(x)$ was not required. The invariant subsets of $\Omega$ with respect to $T(x)$ form a $\sigma$-algebra (check it!). If $T(x)$ is ergodic, the $\sigma$-algebra is trivial, i.e., contains only two subsets of $\Omega$: the empty set and $\Omega$ itself. If $T(x)$ is an arbitrary system, the $\sigma$-algebra is not necessarily trivial. For $u \in L^1(\Omega)$ denote by $\widetilde{E}(u)$ the conditional expectation of $u(\omega)$ with respect to the $\sigma$-algebra of invariant subsets of $\Omega$, i.e., $\widetilde{E}(u)$ is measurable with respect to the $\sigma$-algebra and $\int_B \widetilde{E}(u)\, d\mu = \int_B u\, d\mu$ for any invariant subset $B \subset \Omega$.

EXERCISE 6.2. Let $u \in \mathcal{V}_2^{\mathrm{pot}}(\Omega)$, $v \in L_2^{\mathrm{sol}}(\Omega)$. By definition, $E(u \cdot v) = 0$. Prove that $\widetilde{E}(u \cdot v) = 0$ for almost all $\omega \in \Omega$ (here, $\widetilde{E}(u \cdot v)$ takes values in $L_1(\Omega)$, but not in $\mathbb{R}^1$). If $\widetilde{E}(u \cdot v) = 0$ almost everywhere, we say that $u$ and $v$ are *conditionally orthogonal.*

Introduce the so-called homogenized matrix of a matrix $a_{ij}(\omega)$. It is easy to see that $\widetilde{E}(A(\eta + v_\eta))$ is linear with respect to $\eta \in \mathbb{R}^d$. Therefore, $\widetilde{E}(A(\eta + v_\eta)) = \widehat{A}\eta$ (check it!), where $\widehat{A} = \widehat{A}(\omega)$ is a matrix with measurable entries $\widehat{a}_{ij}$ with respect to the $\sigma$-algebra of invariant subsets. We say that $\widehat{A}(\omega)$ is the *homogenized matrix* of $A(\omega)$. If $T(x)$ is an ergodic system, the matrix $\widehat{a}_{ij}$ is not random.

LEMMA 6.1. *If a matrix $A(\omega)$ is symmetric, then the homogenized matrix $\widehat{A}$ is also symmetric. The matrix $\widehat{A}$ is positive definite and satisfies* (6.1).

PROOF. For arbitrary $\zeta$ and $\eta$ in $\mathbb{R}^d$ we have

$$\begin{aligned}\zeta^t \widehat{A}\eta &= \widetilde{E}(\zeta^t A(\eta + v_\eta)) = \widetilde{E}((\zeta + v_\zeta)^t A(\eta + v_\eta)) = \widetilde{E}((\eta + v_\eta)^t A(\zeta + v_\zeta)) \\ &= \widetilde{E}(\eta^t A(\zeta + v_\zeta)) = \eta^t \widehat{A}\zeta,\end{aligned}$$

i.e., $\widehat{A}$ is symmetric .

Since $A(\eta + v)$ and $v$ are conditionally orthogonal, we have

$$\begin{aligned}\eta\widehat{A}\eta &= \widetilde{E}(\eta A(\eta + v)) = \widetilde{E}((\eta + v)A(\eta + v)) \geq \kappa_1 \widetilde{E}((\eta + v)^2) \\ &\geq \kappa_1 (\widetilde{E}(\eta + v))^2 = \kappa_1 \eta^2.\end{aligned}$$

The fact that the upper ellipticity constant $\kappa_2$ is preserved for the matrix $\widehat{A}$ will be proved later using general theorems on $G$-convergence. □

Let $A(\omega)$ be a random matrix with entries $a_{ij} \in L_\infty(\Omega)$ satisfying the ellipticity condition, and let $A(x) = A(T(x)\omega)$ be a typical realization of $A(\omega)$. Consider the problem

$$u^\varepsilon \in \overset{\circ}{H}{}^1(D), \quad \operatorname{div}(A^\varepsilon \nabla u^\varepsilon) = f, \tag{6.5}$$

where $A^\varepsilon(x) = A\left(\frac{x}{\varepsilon}\right)$ and $D$ is a bounded domain in $\mathbb{R}^d$. Let $u^0$ be a solution to the problem

$$u^0 \in \overset{\circ}{H}{}^1(D), \quad \operatorname{div}(\widehat{A} \nabla u^0) = f. \tag{6.6}$$

THEOREM 6.1. *For almost all $\omega \in \Omega$*

$$u^\varepsilon \rightharpoonup u^0 \ \text{in}\ \overset{\circ}{H}{}^1(D), \quad A^\varepsilon \nabla u^\varepsilon \rightharpoonup \widehat{A}\nabla u^0 \ \text{in}\ L_2(D)\ \text{as}\ \varepsilon \to 0,$$

*where $u^\varepsilon$ and $u^0$ are solutions to the problems* (6.5) *and* (6.6) *respectively.*

*If $T(x)$ is an ergodic system, then the matrix $\widehat{A}(\omega)$ is independent of $\omega$, i.e., it is not random and, consequently, $u^0$ is a deterministic function.*

PROOF. The solutions $u^\varepsilon$ are bounded in $\overset{\circ}{H}{}^1(D)$ and $p^\varepsilon = A^\varepsilon \nabla u^\varepsilon$ are bounded in $L_2(D)$. Denote by $u^0$ and $p^0$ the weak limits of $u^\varepsilon$ and $p^\varepsilon$ respectively. Consider the problem (6.4) and set

$$w_\zeta(x) = w_\zeta(T(x)\omega), \quad q(x) = w_\zeta^t(x)A(x),$$
$$w_\zeta^\varepsilon(x) = w_\zeta\Big(\frac{x}{\varepsilon}\Big), \quad q^\varepsilon(x) = q\Big(\frac{x}{\varepsilon}\Big).$$

Note that $\operatorname{curl} w_\zeta^\varepsilon = 0$ and $\operatorname{div} q^\varepsilon = 0$. By the Birkhoff theorem and Exercise 2.1, $w_\zeta^\varepsilon \rightharpoonup \zeta$ and $q^\varepsilon \rightharpoonup \zeta^t \widehat{A}$ in $L_2(D)$ as $\varepsilon \to 0$. Moreover, $p^\varepsilon w_\zeta^\varepsilon = q^\varepsilon \nabla u^\varepsilon$. Passing to the limit as $\varepsilon \to 0$ and using Lemma 5.1, we find $p^0\zeta = \zeta^t \widehat{A} \nabla u^0$. Consequently, $p^0 = \widehat{A}\nabla u^0$. Since $\operatorname{div} p^0 = f$, we conclude that $u^0$ is a solution to the homogenized problem (6.6) and $\widehat{A}$ is the homogenized matrix. If the system $T(x)$ is ergodic, then the homogenized matrix is independent of $\omega$. □

## 7. Two-scale asymptotic expansion method

**7.1. One-dimensional case.** Consider a boundary value problem for an elliptic equation with rapidly oscillating periodic coefficients. The main idea of the two-scale asymptotic expansion method applied to this problem is as follows.

Let $\varepsilon$ be a small parameter determining the period. We look for a solution in the form

$$u_\varepsilon \sim \sum_{j=0}^{\infty} \varepsilon^j u_j(x,\xi)\Big|_{\xi = x/\varepsilon}, \tag{7.1}$$

where the (slow) variables $x$ describe the macroscopic properties and the (fast or local) variables $\xi$ describe the microscopic properties of the model. Assume that $u_j(x,\xi)$ are 1-periodic with respect to $\xi$. Substituting the asymptotic series, using repeatedly the formula

$$\frac{\partial}{\partial x_i} u\Big(x, \frac{x}{\varepsilon}\Big) = \frac{\partial}{\partial x_i} u(x,\xi) + \frac{1}{\varepsilon}\frac{\partial}{\partial \xi_i} u(x,\xi)\Big|_{\xi = x/\varepsilon}, \tag{7.2}$$

and collecting terms with equal power of $\varepsilon$, we obtain a chain of equations for $u_j(x,\xi)$.

Consider an example. Suppose we need to find a stationary heat field in a rod of length $l$ (the thickness of the rod is ignored) with rapidly oscillating heat conductivity, i.e., the rod consists of tiny alternating fragments of distinct materials with different properties. The zero temperature is set at the ends of the rod.

The temperature of the rod is simulated by the problem

$$\frac{d}{dx}\Big(a\Big(\frac{x}{\varepsilon}\Big)\frac{du}{dx}\Big) = f(x), \quad u(0) = u(l) = 0, \tag{7.3}$$

where $a(\xi) > 0$ is a 1-periodic function. We look for a solution in the form of asymptotic series (7.1). Substituting (7.1) in the equation, using (7.2), and collecting terms with equal power of $\varepsilon$, we find that

$$\Big[\varepsilon^{-2}\frac{\partial}{\partial\xi}\Big(a(\xi)\frac{\partial u_0(x,\xi)}{\partial\xi}\Big)+\varepsilon^{-1}\Big\{\frac{\partial}{\partial\xi}\Big(a(\xi)\frac{\partial u_1(x,\xi)}{\partial\xi}\Big)+\frac{\partial}{\partial\xi}\Big(a(\xi)\frac{\partial u_0(x,\xi)}{\partial x}\Big)+\frac{\partial}{\partial x}\Big(a(\xi)\frac{\partial u_0(x,\xi)}{\partial\xi}\Big)\Big\}+\varepsilon^{0}\Big\{\frac{\partial}{\partial\xi}\Big(a(\xi)\frac{\partial u_2(x,\xi)}{\partial\xi}\Big)+\frac{\partial}{\partial\xi}\Big(a(\xi)\frac{\partial u_1(x,\xi)}{\partial x}\Big)+\frac{\partial}{\partial x}\Big(a(\xi)\frac{\partial u_1(x,\xi)}{\partial\xi}\Big)+\frac{\partial}{\partial x}\Big(a(\xi)\frac{\partial u_0(x,\xi)}{\partial x}\Big)\Big\}+\cdots\Big]\Big|_{\xi=x/\varepsilon}\sim f(x).$$

Collecting terms with equal power of $\varepsilon$, we obtain

$$\frac{\partial}{\partial\xi}\Big(a(\xi)\frac{\partial u_0(x,\xi)}{\partial\xi}\Big)=0, \tag{7.4}$$

$$\frac{\partial}{\partial\xi}\Big(a(\xi)\frac{\partial u_1(x,\xi)}{\partial\xi}\Big)+\frac{\partial}{\partial\xi}\Big(a(\xi)\frac{\partial u_0(x,\xi)}{\partial x}\Big)+\frac{\partial}{\partial x}\Big(a(\xi)\frac{\partial u_0(x,\xi)}{\partial\xi}\Big)=0, \tag{7.5}$$

$$\frac{\partial}{\partial\xi}\Big(a(\xi)\frac{\partial u_2(x,\xi)}{\partial\xi}\Big)+\frac{\partial}{\partial\xi}\Big(a(\xi)\frac{\partial u_1(x,\xi)}{\partial x}\Big)+\frac{\partial}{\partial x}\Big(a(\xi)\frac{\partial u_1(x,\xi)}{\partial\xi}\Big)+\frac{\partial}{\partial x}\Big(a(\xi)\frac{\partial u_0(x,\xi)}{\partial x}\Big)=f(x). \tag{7.6}$$

The main assumption here is the independence of the variables $x$ and $\xi$. These equations are regarded as a recurrent sequence of differential equations, where $\xi$ is a variable and $x$ is a parameter.

EXERCISE 7.1. Prove that the set of periodic solutions to the equation

$$\frac{d}{d\xi}\Big(a(\xi)\frac{d}{d\xi}v(\xi)\Big)=0$$

consists of constants.

Since $x$ is a parameter in equation (7.4), the constants depend on $x$, i.e., $u_0(x,\xi)=u_0(x)$. Taking into account this fact, we write equation (7.5) in the form

$$\frac{\partial}{\partial\xi}\Big(a(\xi)\Big(\frac{\partial u_1(x,\xi)}{\partial\xi}+\frac{du_0(x)}{dx}\Big)\Big)=0. \tag{7.7}$$

Integrating (7.5) with respect to $\xi$ and dividing by $a(\xi)$, we find

$$\frac{\partial u_1(x,\xi)}{\partial\xi}+\frac{du_0(x)}{dx}=\frac{C_1(x)}{a(\xi)}. \tag{7.8}$$

Since $u_1(x,\xi)$ is periodic with respect to $\xi$, we have $\frac{du_0(x)}{dx}=C_1(x)\langle\frac{1}{a(\xi)}\rangle$, i.e., $C_1(x)=\frac{du_0(x)}{dx}\Big\langle\frac{1}{a(\xi)}\Big\rangle^{-1}$. Thus,

$$\frac{\partial u_1(x,\xi)}{\partial\xi}=\Big(\Big\langle\frac{1}{a(\xi)}\Big\rangle^{-1}\frac{1}{a(\xi)}-1\Big)\frac{du_0(x)}{dx}$$

and, consequently, $u_1(x,\xi)=N_1(\xi)\frac{du_0(x)}{dx}+M_1(x)$, where

$$N_1(\xi)=\int_0^{\xi}\Big(\Big\langle\frac{1}{a(\xi)}\Big\rangle^{-1}\frac{1}{a(\xi)}-1\Big)\,d\xi$$

and $M_1(x)$ is arbitrary, for example, the zero function.

Since equation (7.7) is linear, we can find $u_1(x,\xi)$. Indeed, the solution has the form

$$u_1(x,\xi) = N(\xi)\frac{du_0(x)}{dx} + C_2(x).$$

Substituting this expression in (7.7), we obtain the equation

$$\frac{d}{d\xi}\Big(a(\xi)\frac{d}{d\xi}\Big(N(\xi)+\xi\Big)\Big) = 0, \tag{7.9}$$

called the *cell problem*. One of the solutions to this problem is $N_1(\xi)$.

We write equation (7.6) in the form

$$\frac{\partial}{\partial\xi}\Big(a(\xi)\frac{\partial u_2(x,\xi)}{\partial\xi}\Big) - \Big\{\frac{d}{d\xi}(a(\xi)N_1(\xi)) - a(\xi)\Big(\frac{dN_1(\xi)}{d\xi}+1\Big)\Big\}\frac{d^2u_0(x)}{dx^2} + f(x). \tag{7.10}$$

The solvability condition in the space of periodic functions, saying that the mean value of the right-hand side over the period is to be equal to zero, takes the form

$$\Big\langle a(\xi)\Big(\frac{dN_1(\xi)}{d\xi}+1\Big)\Big\rangle\frac{d^2u_0(x)}{dx^2} = f(x).$$

Taking into account the formulas for $N_1(\xi)$ and the boundary conditions in (7.3), we find that

$$\Big\langle\frac{1}{a(\xi)}\Big\rangle^{-1}\frac{d^2u_0(x,\xi)}{dx^2} = f(x), \quad u_0(0) = u_0(l) = 0. \tag{7.11}$$

If $u_0(x)$ is a solution to equation (7.11), then the problem (7.6) is solvable in the class of 1-periodic functions with respect to $\xi$, and the first three terms in (7.1) formally satisfy the equation and the boundary condition up to a term of order $\varepsilon$. Therefore, for small $\varepsilon$ the function $u_0(x)$ well approximates $u_\varepsilon(x)$, i.e., the temperature distribution on an inhomogeneous rod with heat conductivity $a\big(\frac{x}{\varepsilon}\big)$ differs little from that on a homogeneous rod with coefficient $\big\langle\frac{1}{a(\xi)}\big\rangle^{-1}$.

We have presented the idea of the two-scale asymptotic expansion method for constructing an effective (homogenized) equation approximately describing a process in a strongly inhomogeneous medium. This method successively works in much more general situations. In particular, it is useful for analyzing the so-called perforated media.

**7.2. Perforated domains.** Denote by $\Omega$ a bounded domain in $\mathbb{R}^d$ and by $B$, a periodic open set in $\mathbb{R}^d$ with smooth boundary. We say that a set is *periodic* if it is invariant under shifts by integers along any unit basis vector $\overline{e}_1,\dots,\overline{e}_d$ in $\mathbb{R}^d$, i.e., $B = B + \overline{e}_j$, $j = 1,2,\dots,d$. The connectedness of $B$ is not assumed. On the other hand, we assume that $\mathbb{R}^d\setminus\overline{B}$ is connected.

Denote by $\varepsilon B$ the homothetic contraction of $B$ by a factor of $1/\varepsilon$, i.e., $\varepsilon B = \{x : x/\varepsilon \in B\}$. We set $\Omega_\varepsilon = \Omega\setminus\varepsilon\overline{B}$ (Figure 7.1). We require the set $\Omega_\varepsilon$ to be connected (note that this set is open by definition). Then it is possible to construct a periodic "lattice" $B$ such that its complement consists of disjoint closed components (Figure 7.2).

We forbid such cases: $\Omega_\varepsilon$ must be connected.

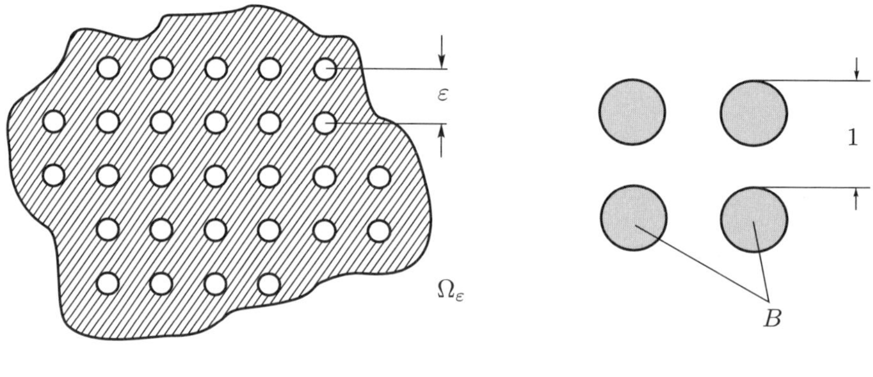


FIGURE 7.1

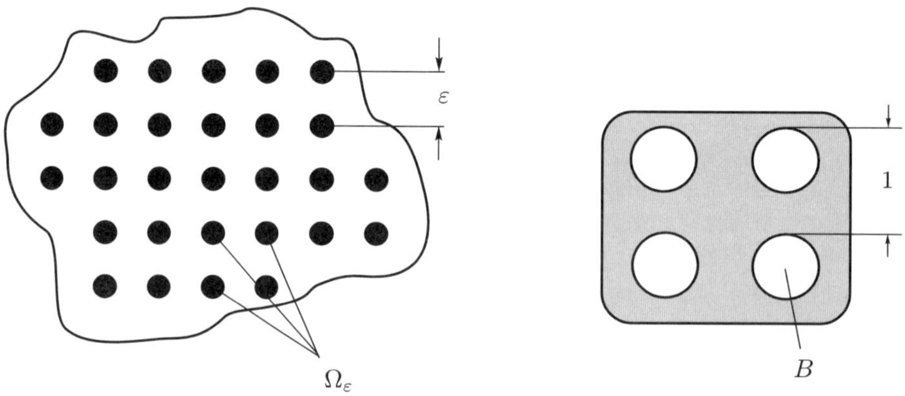


FIGURE 7.2

Denote $S_\varepsilon = \partial(\varepsilon B) \cap \Omega$ and $\Gamma_\varepsilon = \partial\Omega \setminus \varepsilon B$. It is obvious that $S_\varepsilon$ is the part of the boundary of $\Omega_\varepsilon$ lying on the boundary of $\varepsilon B$, and $\Gamma_\varepsilon$ is the part of the boundary of $\Omega_\varepsilon$ lying on the boundary of $\Omega$. Let $a_{ij}(y)$, $i, j = 1, 2, \dots, d$, be smooth 1-periodic functions such that $\lambda|\xi|^2 \le \sum_{i,j=1}^d a_{ij}(y)\xi_i\xi_j \le \Lambda|\xi|^2$, $\lambda > 0$. Let $u_\varepsilon$, $0 < \varepsilon < 1$, be a solution to the boundary value problem

$$
\begin{aligned}
&\sum_{i,j=1}^d \frac{\partial}{\partial x_i}\left(a_{ij}\left(\frac{x}{\varepsilon}\right)\frac{\partial u_\varepsilon}{\partial x_j}\right) = f(x) \quad \text{in } \Omega_\varepsilon, \quad f \in C^\infty(\overline{\Omega}),\\
&\sum_{i,j=1}^d a_{ij}\left(\frac{x}{\varepsilon}\right)\frac{\partial u_\varepsilon}{\partial x_j} n_i = 0 \quad \text{on } S_\varepsilon, \quad u_\varepsilon = 0 \quad \text{on } \Gamma_\varepsilon.
\end{aligned}
\tag{7.12}
$$

Our goal is to study the asymptotic behavior of $u_\varepsilon$ as $\varepsilon \to 0$.

We apply the asymptotic expansion method to this problem. As in the one-dimensional case, we look for an approximate solution in the form

$$\overline{u}_\varepsilon(x) = u_0(x) + \varepsilon \sum_{i=1}^{d} N_i\Big(\frac{x}{\varepsilon}\Big) \frac{\partial u_0}{\partial x_i}(x),$$

where $u_0(x)$ and $N_i(\xi)$, $i = 1, 2, \ldots, d$, are unknown. For this purpose, we substitute $\overline{u}_\varepsilon(x)$ in (7.12) and try to choose $u_0(x)$ and $N_i(\xi)$ in such a way that small "residuals" appear on the right-hand side of the equation and in boundary conditions. A computation shows that $N_i(\xi)$ must be a solution to the problem

$$\begin{aligned}
&\sum_{k,j=1}^{d} \frac{\partial}{\partial \xi_k}\left(a_{kj}(\xi)\frac{\partial N_i}{\partial \xi_j}\right) = -\sum_{k=1}^{d} \frac{\partial}{\partial \xi_k} a_{ki}(\xi) \quad \text{in } \mathbb{R}^d \setminus B,\\
&N_i(\xi) \text{ is 1-periodic}, \quad \int_Y N_i(\xi)\, d\xi = 0,\\
&\sum_{k,j=1}^{d} a_{kj}(\xi)\frac{\partial N_i}{\partial \xi_j}\nu_k = -\sum_{k=1}^{d} a_{ki}(\xi)\nu_k \quad \text{on } \partial B,
\end{aligned} \tag{7.13}$$

where $\nu = (\nu_1, \nu_2, \ldots, \nu_d)$ is the outward normal to the boundary of $B$, $Y = \square \setminus B$, and $\square = [0,1]^d$. Using the Stokes formula, it is easy to check that the compatibility condition for this problem is satisfied. Hence it has a periodic solution. The uniqueness of a solution follows from the maximum principle.

We define the coefficient matrix $\widehat{a}_{ij}$ by the formula

$$\widehat{a}_{ij} = \int_Y \left[a_{ij}(\xi) + a_{ik}\frac{\partial N_j}{\partial \xi_k}\right] d\xi.$$

Let $u_0(x)$ be a solution to the Dirichlet problem

$$\sum_{i,j=1}^{d} \frac{\partial}{\partial x_i}\left(\widehat{a}_{ij}\frac{\partial u_0}{\partial x_j}\right) = |Y| f(x) \quad \text{in } \Omega, \qquad u_0\big|_{\partial\Omega} = 0, \tag{7.14}$$

where $|Y|$ is the volume of a set $Y$. This problem is solvable if the matrix $\widehat{a}_{ij}$ is positive definite. In the case under consideration, there are explicit formulas for $\widehat{a}_{ij}$ and, consequently, the positive definiteness of $\widehat{a}_{ij}$ can be verified directly.

EXERCISE 7.2. Prove that the homogenized matrix $\widehat{a}_{ij}$ satisfies the inequalities

$$\widetilde{\lambda}|Y|\,|\xi|^2 \le \sum_{i,j=1}^{d} \widehat{a}_{ij}\xi_i\xi_j \le \Lambda|Y|\,|\xi|^2, \quad \widetilde{\lambda} > 0.$$

Note that the constant $\widetilde{\lambda}$ may be less than $\lambda$.

Thus, $u_0(x)$ and $N_i(\xi)$, $i = 1, 2, \ldots, d$, have been found. Substituting the difference between the exact solution and the approximation in (7.12), we find the residuals in the equation and boundary condition. For brevity, we denote by $L_\varepsilon$

the elliptic operator on the left-hand side of (7.12). A computation shows that

$$L_\varepsilon(u_\varepsilon - \overline{u}_\varepsilon) = -\varepsilon(a_{lk}(\xi)N_s(\xi))\big|_{\xi=x/\varepsilon}\frac{\partial^3 u_0(x)}{\partial x_l \partial x_k \partial x_s}$$

$$+\left[\widehat{a}_{pq} - a_{pq}(\xi) - a_{pj}(\xi)\frac{\partial N_q(\xi)}{\partial \xi_j} - \frac{\partial}{\partial \xi_j}(a_{lp}(\xi)N_q(\xi))\right]\Bigg|_{\xi=x/\varepsilon}\frac{\partial^2 u_0(x)}{\partial x_p \partial x_q} \quad \text{in } \Omega_\varepsilon,$$

$$(u_\varepsilon - \overline{u}_\varepsilon)\big|_{\Gamma_\varepsilon} = -\varepsilon\sum_{i=1}^d N_i\Big(\frac{x}{\varepsilon}\Big)\frac{\partial u_0}{\partial x_i}(x),$$

$$a_{kj}\Big(\frac{x}{\varepsilon}\Big)\frac{\partial(u_\varepsilon - \overline{u}_\varepsilon)}{\partial x_j}\nu_k\bigg|_{S_\varepsilon} = -\varepsilon a_{kj}\Big(\frac{x}{\varepsilon}\Big)N_i\Big(\frac{x}{\varepsilon}\Big)\frac{\partial^2 u_0(x)}{\partial x_i \partial x_i}\nu_k.$$

The expression on the right-hand side of the equation contains a term of order $\varepsilon$, whereas the remaining term does not contain the factor $\varepsilon$ and is a finite sum of functions of type $h\big(\frac{x}{\varepsilon}\big)w(x)$, where $h(\xi)$ is a periodic function with zero average over $\square$ (after extending it by zero from $\square \cap B$ to $\square$) and $w(x)$ is a sufficiently smooth function.

In spite of the fact that the residual in the boundary condition on $S_\varepsilon$ contains $\varepsilon$, the integral norm of the residual is not small since the area of the surface $S_\varepsilon$ increases like $\frac{1}{\varepsilon}$ as $\varepsilon \to 0$.

Thus, to prove that $(u_\varepsilon - \overline{u}_\varepsilon)$ is small, we use more delicate arguments based on the following auxiliary assertion.

LEMMA 7.1. *The solution $v_\varepsilon(x)$ to the boundary value problem*

$$L_\varepsilon v_\varepsilon(x) = f_0(x) + \sum_{i=1}^d \frac{\partial f_i}{\partial x_i}(x) \quad \text{in } \Omega_\varepsilon, \qquad f_0 \in L_2(\Omega),\ f_i \in H^1(\Omega),$$

$$v_\varepsilon\big|_{\Gamma_\varepsilon} = \psi\big|_{\Gamma_\varepsilon}, \quad \psi \in H^1(\Omega_\varepsilon), \qquad a_{ij}\Big(\frac{x}{\varepsilon}\Big)\frac{\partial v_\varepsilon}{\partial x_i}\nu_j\bigg|_{S_\varepsilon} = \nu_i f_i\big|_{S_\varepsilon},$$

*satisfies the estimate*

$$\|v_\varepsilon\|_{H^1(\Omega_\varepsilon)} \le C\Big[\|f_0\|_{L_2(\Omega)} + \sum_{i=1}^d \|f_i\|_{L_2(\Omega)} + \|\psi\|_{H^1(\Omega_\varepsilon)}\Big],$$

*where the constant $C$ is independent of $\varepsilon$.*

We show how to estimate the norm $\|u_\varepsilon - \overline{u}_\varepsilon\|_{H^1(\Omega_\varepsilon)}$ using this lemma. Consider the boundary value problem

$$\frac{\partial}{\partial \xi_k}\left(a_{kj}(\xi)\frac{\partial N_{pq}}{\partial \xi_j}(\xi)\right)$$

$$= -\frac{\partial}{\partial \xi_k}(a_{kp}(\xi)N_q(\xi)) - a_{pj}(\xi)\frac{\partial N_q}{\partial \xi_j}(\xi) - a_{pq}(\xi) + \widehat{a}_{pq} \quad \text{in } \mathbb{R}^d \setminus B,$$

$$N_{pq} \text{ is 1-periodic with respect to } \xi, \qquad \int_Y N_{pq}(\xi)\,d\xi = 0,$$

$$a_{ij}(\xi)\frac{\partial N_{pq}}{\partial \xi_j}(\xi)\nu_i = -\nu_k a_{kp}(\xi)N_q(\xi) \quad \text{on } \partial B.$$

This problem is solvable because of the condition

$$\int_Y \left( \frac{\partial}{\partial \xi_j}(a_{pj}(\xi)N_q(\xi)) + a_{pj}(\xi)\frac{\partial N_q}{\partial \xi_j}(\xi) + a_{pq}(\xi) - \widehat{a}_{pq} \right) d\xi + \int_{\partial B \cap \square} \nu_j(\xi) a_{pj}(\xi) N_q(\xi)\, d\sigma(\xi) = \int_Y \left( a_{pj}(\xi)\frac{\partial N_q}{\partial \xi_j}(\xi) + a_{pq}(\xi) - \widehat{a}_{pq} \right) d\xi = 0,$$

which follows from the definition of $\widehat{a}_{pq}$. Using the functions $N_{pq}(\xi)$, we can write the boundary value problem for $(u_\varepsilon - \overline{u}_\varepsilon)$ in the form

$$L_\varepsilon(u_\varepsilon - \overline{u}_\varepsilon) = \varepsilon f_0^\varepsilon(x) + \varepsilon \sum_{i=1}^d \frac{\partial f_i^\varepsilon}{\partial x_i}(x) \quad \text{in } \Omega_\varepsilon,$$

$$(u_\varepsilon - \overline{u}_\varepsilon) = -\varepsilon N_s\left(\frac{x}{\varepsilon}\right)\frac{\partial u_0}{\partial x_s}(x) \quad \text{on } \Gamma_\varepsilon,$$

$$a_{ij}\left(\frac{x}{\varepsilon}\right)\frac{\partial (u_\varepsilon - \overline{u}_\varepsilon)}{\partial x_j}\nu_i = \varepsilon \nu_j f_j^\varepsilon \quad \text{on } S_\varepsilon,$$

where

$$f_0^\varepsilon(x) = -a_{kh}\left(\frac{x}{\varepsilon}\right)N_s\left(\frac{x}{\varepsilon}\right)\frac{\partial^3 u_0(x)}{\partial x_h \partial x_k \partial x_s} - a_{kj}\left(\frac{x}{\varepsilon}\right)\frac{\partial N_{pq}}{\partial \xi_j}\left(\frac{x}{\varepsilon}\right)\frac{\partial^3 u_0(x)}{\partial x_p \partial x_q \partial x_k},$$

$$f_k^\varepsilon(x) = a_{kj}\left(\frac{x}{\varepsilon}\right)\frac{\partial N_{pq}}{\partial \xi_j}\left(\frac{x}{\varepsilon}\right)\frac{\partial^2 u_0(x)}{\partial x_p \partial x_q}.$$

Now we can use the lemma since $\|f_0^\varepsilon\|_{L_2(\Omega_\varepsilon)} + \sum_{i=1}^d \|f_i^\varepsilon\|_{L_2(\Omega_\varepsilon)} \le C$, where $C > 0$ is independent of $\varepsilon$. The expression $\left\|\varepsilon N_s\left(\frac{x}{\varepsilon}\right)\frac{\partial u_0}{\partial x_s}\right\|_{H^{1/2}(\Gamma_\varepsilon)}$ is bounded above by $C_1\sqrt{\varepsilon}$ (cf. Exercise 13.5 in the next chapter). Consequently, $\|u_\varepsilon - \overline{u}_\varepsilon\|_{H^1(\Omega_\varepsilon)} \le C_2\sqrt{\varepsilon}$, where the constant $C_2$ is independent of $\varepsilon$. Thus, we have estimated the rate of convergence of $\overline{u}_\varepsilon$ to $u_\varepsilon$ as $\varepsilon \to 0$.

PROOF OF LEMMA 7.1. We set $w_\varepsilon = v_\varepsilon - \psi$. The function $w_\varepsilon$ is a solution to the boundary value problem

$$L_\varepsilon w_\varepsilon = f(x) + \sum_{i=1}^d \frac{\partial f_i}{\partial x_i} - L_\varepsilon \psi \quad \text{in } \Omega_\varepsilon, \qquad w_\varepsilon\big|_{\Gamma_\varepsilon} = 0,$$

$$a_{ij}\left(\frac{x}{\varepsilon}\right)\frac{\partial w_\varepsilon}{\partial x_i}\nu_j\Big|_{S_\varepsilon} = -\nu_i f_i - a_{ij}\left(\frac{x}{\varepsilon}\right)\frac{\partial \psi}{\partial x_j}\nu_i.$$

Multiplying by $w_\varepsilon$ and integrating by parts, we find that

$$-\int_{\Omega_\varepsilon} a_{ij}\left(\frac{x}{\varepsilon}\right)\frac{\partial w_\varepsilon}{\partial x_i}\frac{\partial w_\varepsilon}{\partial x_j}dx = \int_{\Omega_\varepsilon} f(x)w_\varepsilon(x)\,dx - \int_{\Omega_\varepsilon} f_i(x)\frac{\partial w_\varepsilon}{\partial x_i}(x)\,dx + \int_{\Omega_\varepsilon} a_{ij}\left(\frac{x}{\varepsilon}\right)\frac{\partial \psi}{\partial x_j}\frac{\partial w_\varepsilon}{\partial x_i}dx, \tag{7.15}$$

where the boundary integrals cancel in pairs by virtue of the boundary conditions. The last equality leads to an estimate for $\|w_\varepsilon\|_{H^1(\Omega_\varepsilon)}$. Denote by $H^1(\Omega_\varepsilon, \Gamma_\varepsilon)$ the closure of the set of functions in $C^\infty(\Omega_\varepsilon)$ vanishing in a neighborhood of $\Gamma_\varepsilon$ in the $H^1(\Omega_\varepsilon)$-norm. Define the space $H^1(\Omega, \Gamma_\varepsilon)$ in a similar way. The norm $\|w_\varepsilon\|_{H^1(\Omega_\varepsilon)}$

can be estimated by using the following Friedrichs inequality in the perforated domain $\Omega_\varepsilon$: for any $v_\varepsilon \in H^1(\Omega_\varepsilon, \Gamma_\varepsilon)$,

$$\|v_\varepsilon\|^2_{H^1(\Omega_\varepsilon)} \leq C \int_{\Omega_\varepsilon} a_{ij}\Big(\frac{x}{\varepsilon}\Big) \frac{\partial v_\varepsilon}{\partial x_i} \frac{\partial v_\varepsilon}{\partial x_j} dx, \tag{7.16}$$

where the constant $C > 0$ is independent of $\varepsilon$. The inequality (7.16) is similar to the standard Friedrichs inequality, but the constant $C$ is independent of $\varepsilon$. The proof of (7.16) is based on the known fact that there exists an extension operator $P_\varepsilon : H^1(\Omega_\varepsilon, \Gamma_\varepsilon) \to H^1(\Omega)$ of $u \in H^1(\Omega_\varepsilon, \Gamma_\varepsilon)$ to the domain $\Omega$ (without perforation) that satisfies the following conditions:

(1) $P_\varepsilon u\big|_{\Omega_\varepsilon} = u$;
(2) $\|P_\varepsilon u\|^2_{L_2(\Omega)} \leq C\|u\|^2_{L_2(\Omega_\varepsilon)}$;
(3) $\|\nabla(P_\varepsilon u)\|^2_{L_2(\Omega)} \leq C\|\nabla u\|^2_{L_2(\Omega_\varepsilon)}$.

The existence of $P_\varepsilon$ with properties (1)–(3) is proved, for example, in [**102, 103, 1**].

Taking into account the properties of the extension operator $P_\varepsilon$ based on the Friedrichs inequality for functions in $H^1(\Omega, \Gamma_\varepsilon)$ with the constant independent of $\varepsilon$ (cf., for example, [**37, 38**]), we have

$$\begin{aligned} \|v_\varepsilon\|^2_{H^1(\Omega_\varepsilon)} &\leq \|P_\varepsilon v_\varepsilon\|^2_{H^1(\Omega)} \leq C_1 \int_\Omega |\nabla(P_\varepsilon v_\varepsilon)|^2 dx \\ &\leq C_2 \int_{\Omega_\varepsilon} |\nabla v_\varepsilon|^2 dx \leq C_3 \int_{\Omega_\varepsilon} a_{ij}\Big(\frac{x}{\varepsilon}\Big) \frac{\partial v_\varepsilon}{\partial x_i} \frac{\partial v_\varepsilon}{\partial x_j} dx, \end{aligned}$$

which proves the inequality (7.16). Now (7.15) implies

$$\begin{aligned} \frac{1}{C}\|w_\varepsilon\|^2_{H^1(\Omega_\varepsilon)} &\leq \frac{1}{\alpha}\int_{\Omega_\varepsilon} w_\varepsilon^2(x)\,dx + \alpha \int_{\Omega_\varepsilon} f_0^2(x)\,dx + \alpha \int_{\Omega_\varepsilon} \sum_{i=1}^d f_i^2(x)\,dx \\ &+ \frac{1}{\alpha}\int_{\Omega_\varepsilon} \sum_{i=1}^d \Big|\frac{\partial w_\varepsilon}{\partial x_i}\Big|^2 dx + \Lambda\alpha \int_{\Omega_\varepsilon} \sum_{i=1}^d \Big|\frac{\partial \psi}{\partial x_i}\Big|^2 dx + \frac{\Lambda}{\alpha}\int_{\Omega_\varepsilon} \sum_{i=1}^d \Big|\frac{\partial w_\varepsilon}{\partial x_i}\Big|^2 dx, \end{aligned}$$

where $\alpha = \text{const} > 0$. Taking sufficiently large $\alpha$ and applying obvious transformations, we find that $\|w_\varepsilon\|^2_{H^1(\Omega_\varepsilon)} \leq K\{\|f_0\|^2_{L_2(\Omega)} + \sum_{i=1}^d \|f_i\|^2_{L_2(\Omega)} + \|\psi\|^2_{H^1(\Omega_\varepsilon)}\}$, as required. □

Using the asymptotic expansion method, we can establish a homogenization result and estimate the rate of convergence of the solutions of the original problem to the solution of the homogenized problem. This is one of the advantages of the asymptotic expansion method. However, this method requires the high smoothness of the data. The smoothness conditions can be essentially weakened if we use other methods for proving the homogenization result, for example, the compensated compactness method.

Note that the homogenization result is still valid even if there are no extension operators with the above properties. In fact, the condition that the periodically perforated set is open and connected is sufficient (cf. [**132**]).

Analyzing the behavior of solutions to boundary value problems in perforated domains, we observe the following interesting phenomenon. In order for a process in a perforated medium to have effective (homogenized) characteristics (i.e., for the existence of average macroscopic characteristics), the conectedness of the perforated set is not necessary. Sometimes, such a process (for example, diffusion or electric current) may proceed in an entire set consisting of many disconnected components. However, the intersection of the closures of these components must be nonempty and certain conditions must be satisfied. Consider two examples.

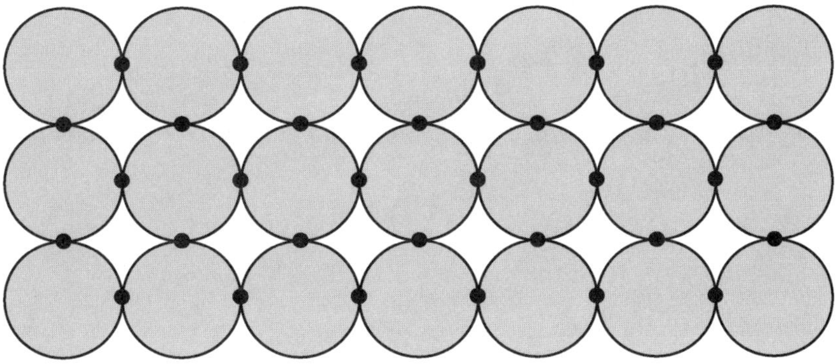

FIGURE 7.3

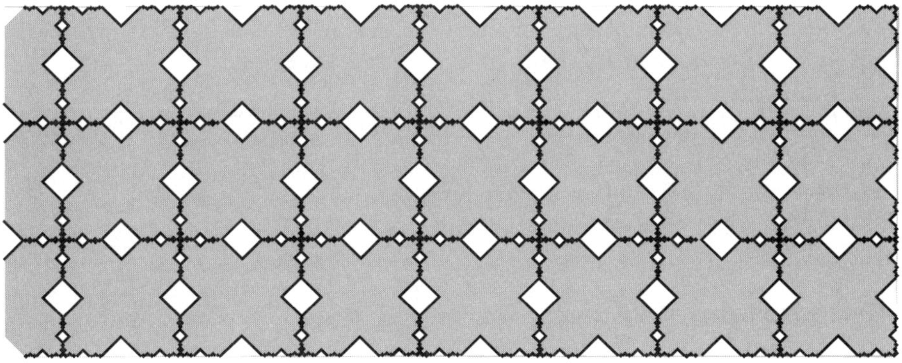

FIGURE 7.4

Figure 7.3 represents the system of disjoint tangent open balls that are located periodically. The closures of the balls meet at the tangent points. Figure 7.4 represents the system of periodically repeated disjoint open sets. The intersection of their closures consists of the Cantor sets located on each line segment of the standard rectangular lattice.

In the first case, the macroscopic process does not propagate and homogenization does not hold. The structure splits into disconnected elements. In the second case, the process proceeds in the entire structure. What is the reason?

It turns out that, in the first case, the electrostatic capacity of the intersection of the closures of open balls vanishes, whereas, in the second case, the intersection of closures is a Cantor set on the plane and its electrostatic capacity is positive. In Section 8, we present the $p$-connectedness technique, which allows us to formulate and prove an assertion on homogenization of periodic structures in the form of the union of disconnected components.

## 8. $p$-Connectedness technique (in the sense of V. V. Zhikov)

**8.1. $p$-Connectedness on a periodicity torus.** Denote by $\square = [0,1)^d$ a periodicity cell (or torus) and by $F$ an open periodic set. Consider periodic functions in $L_{2,\mathrm{per}}(F)$. The relation $a = \operatorname{div} b$, where $a \in L_{2,\mathrm{per}}(F)$, $b \in (L_{2,\mathrm{per}}(F))^d$, is understood in the weak sense, i.e.,

$$\int_{F\cap\square} a\varphi\, dx = -\int_{F\cap\square} (b, \nabla\varphi)\, dx, \quad \varphi \in C^\infty_{\mathrm{per}}(\square). \tag{8.1}$$

We assume that the functions in $L_{2,\mathrm{per}}(F)$ are extended by zero to the entire space $\mathbb{R}^d$. Then

$$\int_\square a\varphi\, dx = -\int_\square (b, \nabla\varphi)\, dx, \quad \varphi \in C^\infty_{\mathrm{per}}(\square).$$

In particular, if $a = 0$, then the vector $b$ is said to be *solenoidal* or *$F$-solenoidal*, i.e.,

$$\int_{F\cap\square} (b, \nabla\varphi)\, dx = \int_\square (b, \nabla\varphi)\, dx = 0 \quad \forall\varphi \in C^\infty_{\mathrm{per}}(\square).$$

EXERCISE 8.1. Consider a domain $\Omega \subset \mathbb{R}^3$ with a ball $B$ eliminated. A smooth solenoidal vector $b(x)$ is defined in the spherical layer, i.e., $\frac{\partial b_1}{\partial x_1} + \frac{\partial b_2}{\partial x_2} + \frac{\partial b_3}{\partial x_3} = 0$. Extend the vector $b$ by zero to the eliminated ball. When does it remain solenoidal?

*Answer.* The necessary and sufficient condition is $(b,n)\big|_{\partial B} = 0$, where $n$ is a normal to the boundary of the ball.

We introduce the Sobolev space $W^1_{p,\mathrm{per}}(F)$ as the closure of $C^\infty_{\mathrm{per}}(\square)$ in the norm

$$\left(\int_{F\cap\square} (u^p + |\nabla u|^p)\, dx\right)^{1/p}.$$

In the case $p = 2$, we also use the notation $H^1_{\mathrm{per}}(F) = W^1_{2,\mathrm{per}}(F)$.

DEFINITION 8.1. An open periodic set $F$ is said to be *$p$-connected* if $u \in W^1_{p,\mathrm{per}}(F)$ and if $\nabla u = 0$ implies $u \equiv \mathrm{const}$ almost everywhere in $F$.

PROPOSITION 8.1. *If $F$ is connected in a periodicity torus, then it is $p$-connected for any $p \geq 1$.*

EXERCISE 8.2. Prove Proposition 8.1.

**8.2. 2-Connected sets.**

EXAMPLE 8.1. Consider the periodic set $F$ introduced at the end of Subsection 7.2. Recall that $F$ consists of disconnected components that intersect in the Cantor set (cf. Figure 8.1). We prove that $F$ is 2-connected. Assume the contrary. Then there exists a sequence $u_n \in C^\infty(\overline{F})$ such that $u_n \to u$ in $H^1(F)$, $u \in H^1(F)$, $\nabla u_n \to 0$ in $L_2(F)$, and $u \not\equiv \mathrm{const}$ in $F$. It is obvious that $u = \mathcal{C}_1$ in $F_0$ and $u = \mathcal{C}_2$ in $F_1$, where $F_0 = \left((0, \frac{1}{2}) \times (0,1)\right) \cap F$ and $F_1 = ((1/2, 1) \times (0,1)) \cap F$. Let $\mathcal{C}_1 \neq \mathcal{C}_2$. Without loss of generality, we may assume that $u \equiv 0$ in $F_0$ and $u \equiv 1$ in $F_1$.

To arrive at a contradiction, we note that the Cantor set $K$ is a part of the boundary of $F_0$ and, simultaneously, a part of the boundary of $F_1$. Let $G(x)$ be the "Cantor ladder," i.e., a continuous function taking the value $\frac{2k-1}{n}$ on each $(n,k)$th

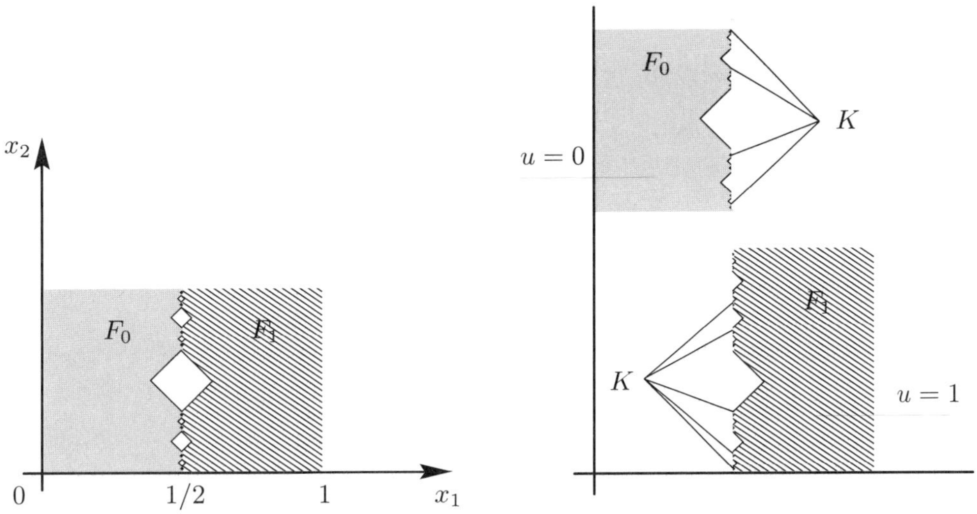


FIGURE 8.1

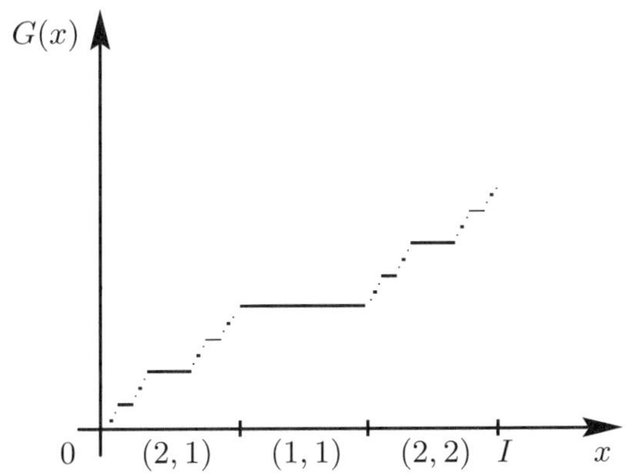


FIGURE 8.2

interval of the complement of the Cantor set (cf. Figure 8.2). We introduce measure $d\mu$ supported in $K$, by the formula

$$\int_{[0,1]} f(x)\, d\mu = \int_{[0,1]} f(x)\, dG,$$

where the left-hand side contains the Lebesgue–Stieltjes integral, i.e., $\mu(0, \alpha) = G(\alpha)$.

LEMMA 8.1. *Let $w(x) \in C^\infty(\overline{F}_0)$. Then there exists a constant $C > 0$ independent of $w(x)$ such that*

$$\int_K w^2(x)\, d\mu \le C\|w\|^2_{H^1(F_0)}.$$

A similar assertion holds for $F_1$.

Using Lemma 8.1, it is easy to arrive at a contradiction with the assumption that $u$ is not a constant. On one hand,

$$\int_K u_n^2(x)\,d\mu \le C\int_{F_0}(u_n^2(x)+|\nabla u_n|^2)\,dx \to 0$$

as $n\to\infty$. On the other hand,

$$\int_K [u_n(x)-1]^2\,d\mu \le C\int_{F_1}\left([u_n(x)-1]^2+|\nabla u_n|^2\right)\,dx \to 0$$

as $n\to\infty$. Hence $\int_K 1\,d\mu = 0$, and we obtain a contradiction.

PROOF OF LEMMA 8.1. Let $G(x)$ be the Cantor ladder, and let $P : H^1(F_0) \to H^1(\square^-)$ be the extension operator for functions in $H^1(F_0)$ to the domain $\square^- = (0,1/2)\times(0,1)$ such that

(1) $\|Pu\|_{L_2(\square^-)} \le C\|u\|_{L_2(F_0)}$ for all $u\in H^1(F_0)$;

(2) $\|\nabla(Pu)\|_{L_2(\square^-)} \le C\|\nabla u\|_{L_2(F_0)}$ for all $u\in H^1(F_0)$;

(3) $u\in C(\overline{F}_0)$ implies $\|Pu\|_{C(\square^-)} \le C\|u\|_{C(\overline{F}_0)}$,

where $C>0$ is independent of $u\in H^1(F_0)$. Such an operator exists because the boundary of $F_0$ is Lipschitz (cf., for example, [**88**]).

EXERCISE 8.3. Prove the existence of the extension operator $P$.

From the properties of the extension operator it follows that

$$\int_K w(x)\,d\mu = \int_0^1 Pw(x)\,dG \quad \forall w\in C^\infty(\overline{F}_0).$$

EXERCISE 8.4. Prove that any Hölder continuous function on $[0,1]$ with exponent $\alpha>1/2$ belongs to $H^{1/2}[0,1]$.

*Hint.* Use the definition of the $H^{1/2}$-norm in Subsection 1.1.

EXERCISE 8.5. Prove that the "Cantor ladder" $G(x)$ is a Hölder continuous function with exponent $\alpha = \frac{\ln 2}{\ln 3} > \frac{1}{2}$, and consequently it belongs to $H^{1/2}[0,1]$.

EXERCISE 8.6. Prove that for any $\varphi(x)\in H^{1/2}[0,1]$,

$$\int_0^1 \varphi^2\,dG \le C\|\varphi\|^2_{H^{1/2}[0,1]},$$

where the constant $C>0$ is independent of $\varphi(x)\in H^{1/2}[0,1]$.

*Hint.* Use Exercise 8.5.

We continue the proof of Lemma 8.1. Using the above exercises, the trace theorem, and the properties of the extension operator, we find that

$$\left(\int_K w^2(x)\,d\mu\right)^{1/2} = \left(\int_0^1 [Pw(x)]^2\,dG\right)^{1/2} \le C\|Pw\|_{H^{1/2}[0,1]} \le C\|Pw\|_{H^1(\square^-)}$$
$$\le C_1\|w\|_{H^1(F_0)},$$

as required. □

The following approximation lemma yields an important characteristic of 2-connected sets.

Denote by $\langle f\rangle_F$ the mean value $\int_{F\cap\square} f(x)\,dx$ of a function $f(x)$.

LEMMA 8.2 (on approximation). *Suppose that $F$ is a 2-connected set and the mean value of $a \in L_{2,\mathrm{per}}(F)$ vanishes: $\langle a\rangle_F = 0$. Then for any $\delta > 0$ there are $a_0, b_1, \dots, b_d \in L_{2,\mathrm{per}}(F)$ such that $\|a_0\|_{L_{2,\mathrm{per}}(F)} \le \delta$, $a - a_0 = \operatorname{div} b$ in $\square$, where $b = (b_1, \dots, b_d)$.*

EXERCISE 8.7. Let $F$ be a connected periodic set with Lipschitz boundary. Is it true that $a_0$ can be taken equal to zero?

PROOF OF LEMMA 8.2. Consider the set $S$ of all $\varsigma \in L_{2,\mathrm{per}}(F)$ admitting the representation $\varsigma = \operatorname{div}\vartheta$, $\vartheta \in (L_{2,\mathrm{per}}(F))^d$. It is required to prove that $S$ is dense in the subspace $X = \{a \in L_{2,\mathrm{per}}(F), \langle a\rangle_F = 0\}$. Suppose that $\alpha \in L_{2,\mathrm{per}}(F)$, $\langle\alpha\rangle_F = 0$, and $\langle\alpha\varsigma\rangle_F = 0$ for any $\varsigma \in S$. If we prove that $\alpha = 0$ in $L_{2,\mathrm{per}}(F)$, then the required assertion is true.

Consider the periodic problem

$$-\Delta u + u = \alpha \quad \text{in } F$$

with homogeneous Neumann boundary condition on $\partial F$. By definition, $u \in H^1_{\mathrm{per}}(F)$ is a solution to this equation provided that

$$\int_{F\cap\square} ((\nabla u, \nabla\varphi) + u\varphi)\, dx = \int_{F\cap\square} \alpha\varphi\, dx \quad \forall\varphi \in C^\infty_{\mathrm{per}}(\overline{F}). \tag{8.2}$$

By the Riesz theorem, this problem is uniquely solvable. From (8.2) it follows that $\alpha - u = -\Delta u$. Therefore, $\alpha - u \in S$ and, consequently, $\langle\alpha(\alpha - u)\rangle_F = 0$. Hence

$$\langle\alpha^2\rangle_F = \langle u\alpha\rangle_F \le \langle u^2\rangle_F^{1/2}\langle\alpha^2\rangle_F^{1/2}, \tag{8.3}$$

i.e., $\langle\alpha^2\rangle_F \le \langle u^2\rangle_F$. On the other hand, from equation (8.2) it follows that

$$\int_{F\cap\square} (|\nabla u|^2 + u^2)\, dx = \int_{F\cap\square} \alpha u\, dx \le \langle\alpha^2\rangle_F^{\frac12}\langle u^2\rangle_F^{1/2},$$

which implies $\langle u^2\rangle_F \le \langle\alpha^2\rangle_F$. Consequently, $\langle u^2\rangle_F = \langle\alpha^2\rangle_F$ and $\int_{F\cap\square} |\nabla u|^2\, dx = 0$. By the definition of the 2-connectedness, $u \equiv \text{const}$. Using (8.3), we find $\langle\alpha^2\rangle_F = \langle u\alpha\rangle_F = u\langle\alpha\rangle_F$. Since $\langle\alpha\rangle_F = 0$, we have $\alpha = 0$ in $L_{2,\mathrm{per}}(F)$. □

We return to arbitrary sets $F$.

REMARK 8.1. It is easy to prove Lemma 8.2 if $F$ is connected on a periodicity torus. It suffices to check that the set $\left\{\sum_{i=1}^d \frac{\partial\varphi_i}{\partial x_i} : \varphi_i \in C^\infty_{\mathrm{per}}(F),\ \operatorname{supp}\varphi_i \subset F\right\}$ is dense in the space $X$. The proof of this assertion is left to the reader.

Lemma 8.2 implies the following assertion. Let $\Omega$ be a bounded domain in $\mathbb{R}^d$. Denote $F^\varepsilon := \varepsilon F$ and define the perforated domain $\Omega^\varepsilon = \Omega \cap F^\varepsilon$, as above.

LEMMA 8.3. *Let $F$ be a 2-connected set, and let $a(\xi) \in L_{2,\mathrm{per}}(F)$ be such that $\langle a\rangle_F = 0$. Assume that the functions $w_\varepsilon(x) \in C_0^\infty(\Omega)$ possess the following property:*

$$\overline{\lim_{\varepsilon\to 0}} \int_{\Omega^\varepsilon} (|w_\varepsilon|^2 + |\nabla w_\varepsilon|^2)\, dx = A < \infty.$$

*Then*

$$\lim_{\varepsilon\to 0} \int_{\Omega^\varepsilon} a\Big(\frac{x}{\varepsilon}\Big) w_\varepsilon(x)\, dx = 0. \tag{8.4}$$

PROOF. By Lemma 8.2,

$$\int_{\Omega^\varepsilon} a\Big(\frac{x}{\varepsilon}\Big) w_\varepsilon(x)\,dx = \int_{\Omega^\varepsilon} a_0\Big(\frac{x}{\varepsilon}\Big) w_\varepsilon(x)\,dx - \sum_{i=1}^{d} \varepsilon \int_{\Omega^\varepsilon} b_i\Big(\frac{x}{\varepsilon}\Big)\frac{\partial w_\varepsilon}{\partial x_i}\,dx.$$

By the properties of the mean value,

$$\lim_{\varepsilon\to 0}\int_{\Omega^\varepsilon}\Big|b_i\Big(\frac{x}{\varepsilon}\Big)\Big|^2 dx = |\Omega|\langle b_i^2\rangle_F, \quad \lim_{\varepsilon\to 0}\int_{\Omega^\varepsilon}\Big|a_0\Big(\frac{x}{\varepsilon}\Big)\Big|^2 dx = |\Omega|\langle a_0^2\rangle_F \le |\Omega|\delta^2,$$

which implies the required assertion. □

**8.3. Properties of $p$-connected microstructures.** Consider $u_\varepsilon \in C^\infty(\overline{\Omega})$ such that

$$\int_{\Omega^\varepsilon} (|u_\varepsilon|^2 + |\nabla u_\varepsilon|^2)\,dx \le C < \infty. \tag{8.5}$$

We study the asymptotic behavior of $\chi_\varepsilon(x)u_\varepsilon(x)$ as $\varepsilon \to 0$, where $\chi_\varepsilon(x) = \chi\big(\frac{x}{\varepsilon}\big)$ and $\chi(\xi)$ is the characteristic function of $F$:

$$\chi(\xi) = \begin{cases} 1, & \xi \in F, \\ 0, & \xi \notin F. \end{cases}$$

The sequence $u_\varepsilon$ is bounded in $L_2(\Omega)$. Without loss of generality, we can assume that

$$\chi_\varepsilon u_\varepsilon \rightharpoonup \theta u \quad \text{weakly in } L_2(\Omega), \quad \theta = |\square \cap F|. \tag{8.6}$$

Now we have the following important question: Does the limit function $u$ belong to the Sobolev space $H^1(\Omega)$? This fact does not follow directly from the estimate (8.5) for $\nabla u_\varepsilon$. The answer depends on how rich is the set of $F$-solenoidal vectors.

EXERCISE 8.8. Give an example of a periodic set $F$ and a sequence $u_\varepsilon \in H^1(\Omega^\varepsilon)$ such that the limit function $u$ does not belong to the space $H^1(\Omega)$.

The following lemma plays the key role.

LEMMA 8.4. *Suppose that $F$ is a 2-connected set, $z$ is a solenoidal vector, and $\zeta = \langle z\rangle_F$ is its average. Let $z_\varepsilon(x) = z\big(\frac{x}{\varepsilon}\big)$. Then for $u_\varepsilon$ and $u$ the following assertions hold:*

(i) $(\nabla u, \zeta) \in L_2(\Omega)$;

(ii) *for any* $\varphi \in C^\infty(\overline{\Omega})$,

$$\lim_{\varepsilon\to 0}\int_{\Omega^\varepsilon} \varphi(x)(\nabla u_\varepsilon(x), z_\varepsilon(x))\,dx = \int_\Omega \varphi(x)(\nabla u(x),\zeta)\,dx.$$

*Furthermore, if $u_\varepsilon(x) \in C_0^\infty(\Omega)$, then $u\big|_{\partial\Omega} = 0$ in the following sense:*

$$\int_\Omega u(\nabla\varphi,\zeta)\,dx = -\int_\Omega \varphi(\nabla u,\zeta)\,dx, \quad \varphi \in C^\infty(\overline{\Omega}). \tag{8.7}$$

PROOF. By the properties of the mean value, we have

$$\int_\Omega \Big|z\Big(\frac{x}{\varepsilon}\Big)\Big|^2 dx \le 2|\Omega|\langle z^2\rangle_F$$

for sufficiently small $\varepsilon$. Consequently,

$$\int_{\Omega^\varepsilon} |(\nabla u_\varepsilon, z_\varepsilon)|\,dx \le \Big(\int_{\Omega^\varepsilon}|\nabla u_\varepsilon|^2 dx\Big)^{1/2}\Big(\int_\Omega |z_\varepsilon|^2 dx\Big)^{1/2} \le C_1 < \infty.$$

Therefore, the sequence of functionals

$$l_\varepsilon(\varphi) = \int_{\Omega^\varepsilon} \varphi(\nabla u_\varepsilon, z_\varepsilon)\, dx$$

on $C(\overline{\Omega})$ is bounded in the norm. Without loss of generality, we can assume that this sequence converges weakly so that

$$\lim_{\varepsilon\to 0} \int_{\Omega^\varepsilon} \varphi(\nabla u_\varepsilon, z_\varepsilon)\, dx = l(\varphi).$$

On the other hand,

$$|l_\varepsilon(\varphi)| \le \Big(\int_{\Omega^\varepsilon} |\nabla u_\varepsilon|^2\, dx\Big)^{1/2} \Big(\int_{\Omega^\varepsilon} |\varphi|^2 |z_\varepsilon|^2\, dx\Big)^{1/2} \le C\Big(\int_{\Omega^\varepsilon} |\varphi|^2 |z_\varepsilon|^2\, dx\Big)^{1/2}.$$

By the properties of the mean value,

$$\lim_{\varepsilon\to 0} \int_{\Omega} |\varphi|^2 |z_\varepsilon|^2\, dx = \langle z^2\rangle_F \int_\Omega |\varphi|^2\, dx.$$

Hence

$$|l(\varphi)| \le C_2\Big(\int_\Omega |\varphi|^2\, dx\Big)^{1/2},$$

i.e., $l(\varphi)$ is a continuous functional on $L_2(\Omega)$. Therefore, there is $v_0 \in L_2(\Omega)$ such that

$$\lim_{\varepsilon\to 0} \int_{\Omega^\varepsilon} \varphi(\nabla u_\varepsilon, z_\varepsilon)\, dx = \int_\Omega \varphi v_0\, dx, \quad \varphi \in C^\infty(\overline{\Omega}).$$

It remains to verify that $v_0 = (\nabla u, \zeta)$. For this purpose, we show that

$$\lim_{\varepsilon\to 0} \int_{\Omega^\varepsilon} \varphi(\nabla u_\varepsilon, z_\varepsilon)\, dx = -\int_\Omega u(\nabla\varphi, \zeta)\, dx$$

for all $\varphi \in C_0^\infty(\Omega)$. To obtain (8.7) in the case $u_\varepsilon \in C_0^\infty(\Omega)$, it suffices to prove the last relation for any $\varphi \in C^\infty(\overline{\Omega})$.

We have

$$\begin{aligned} \int_{\Omega^\varepsilon} \varphi(\nabla u_\varepsilon, z_\varepsilon)\, dx &= \int_{\Omega^\varepsilon} (\nabla(\varphi u_\varepsilon), z_\varepsilon)\, dx - \int_{\Omega^\varepsilon} u_\varepsilon(\nabla\varphi, z_\varepsilon)\, dx \\ &= -\int_{\Omega^\varepsilon} u_\varepsilon(\nabla\varphi, z_\varepsilon)\, dx. \end{aligned} \tag{8.8}$$

Indeed, the first integral vanishes since $z_\varepsilon$ is solenoidal and $\varphi u_\varepsilon$ is compactly supported. Further,

$$\int_{\Omega^\varepsilon} u_\varepsilon(\nabla\varphi, z_\varepsilon)\, dx = \frac{1}{\theta}\int_\Omega \chi_\varepsilon u_\varepsilon(\nabla\varphi, \zeta)\, dx + \int_{\Omega^\varepsilon} u_\varepsilon\Big(\nabla\varphi, \Big(z_\varepsilon - \frac{\zeta\chi_\varepsilon}{\theta}\Big)\Big)\, dx.$$

The first term on the right-hand side of the last equality converges to the expression

$$\int_\Omega u(\nabla\varphi, \zeta)\, dx$$

in view of (8.6). The second term converges to zero by Lemma 8.3 with $a = z_j - \frac{\chi\zeta_j}{\theta}$ and $w_\varepsilon = u_\varepsilon \frac{\partial\varphi}{\partial x_j}$. □

The following assertion says that there are sufficiently many solenoidal vectors.

PROPOSITION 8.2. *If $F$ is connected in $\mathbb{R}^d$, then any vector in $\mathbb{R}^d$ is the average of some $F$-solenoidal vector.*

COROLLARY 8.1. *If $F$ is connected in $\mathbb{R}^d$, then the limit function $u$ belongs to $H^1(\Omega)$. If, in addition, $u_\varepsilon \in C_0^\infty(\Omega)$ and $\partial\Omega$ is Lipschitz, then $u \in \overset{\circ}{H}{}^1(\Omega)$.*

PROOF. It suffices to prove the last assertion. From (8.7) we have

$$\int_\Omega u\nabla\varphi\, dx = -\int_\Omega \varphi\nabla u\, dx \quad \forall\, \varphi \in C^\infty(\overline{\Omega}).$$

This means that $u \in H^1(\mathbb{R}^d)$ and $u(x) = 0$ outside the domain $\Omega$. Since the boundary of $\Omega$ is Lipschitz, we have $u \in \overset{\circ}{H}{}^1(\Omega)$. □

**8.4. Definition and properties of homogenized matrices.** Let $A(x)$ be a measurable symmetric periodic matrix on $F$ satisfying the ellipticity condition $\alpha E \le A(x) \le \frac{1}{\alpha}E$, $\alpha > 0$, $x \in F$. Denote by $V_{\text{pot}}$ the closure of the set $\{\nabla\varphi : \varphi \in C^\infty_{\text{per}}(F)\}$ in the space $(L_{2,\text{per}}(F))^d$. Vector-valued functions in $V_{\text{pot}}$ are said to be *potential*. Unlike solenoidal ones, they are defined only on $F$.

For arbitrary $\eta \in \mathbb{R}^d$ we consider the periodic problem

$$\text{find } v_\eta \in V_{\text{pot}} \text{ such that } \operatorname{div}(A(\eta + v_\eta)) = 0$$

or, in detail,

$$\text{find } v_\eta \in V_{\text{pot}} \text{ such that } \int_{\square\cap F} (A(\eta + v_\eta), \nabla\varphi)\, dx = 0 \quad \forall\, \varphi \in C^\infty_{\text{per}}. \tag{8.9}$$

We show that a solution to this problem exists and is unique.

In $V_{\text{pot}}$, we set

$$[u, v] := \int_{\square\cap F} (A(x)u, v)\, dx.$$

It is the inner product because the matrix $A(x)$ is positive definite and symmetric. The problem (8.9) can be written as follows:

$$\text{find } v_\eta(x) \in V_{\text{pot}} \text{ such that } \quad [v_\eta, \nabla\varphi] = -\int_{\square\cap F} (A(x)\eta, \nabla\varphi)\, dx \quad \forall\, \varphi \in C^\infty_{\text{per}}(F).$$

Since the expression on the right-hand side defines a linear continuous functional on $V_{\text{pot}}$, $v_\eta(x) \in V_{\text{pot}}$ exists and is unique by the Riesz theorem.

We note that the solution $v_\eta$ linearly depends on $\eta$; in addition, the vector $A(\eta + v_\eta)$ is solenoidal. Define $A^{\text{hom}}$ by the formula

$$A^{\text{hom}}\eta = \int_{\square\cap F} A(\eta + v_\eta)\, dx \tag{8.10}$$

and verify that it is symmetric. Taking another copy of the problem (8.9), where $\eta$ is replaced with $\zeta$ and $v_\eta$ is replaced with $w_\zeta$, we obtain the relation

$$\begin{aligned} \zeta A^{\text{hom}}\eta &= \left(\zeta, \int_{\square\cap F} A(\eta + v_\eta)\, dx\right) = \int_{\square\cap F} (A(\zeta + w_\zeta), (\eta + v_\eta))\, dx \\ &= \int_{\square\cap F} (A(\zeta + w_\zeta), \eta)\, dx = \eta A^{\text{hom}}\zeta. \end{aligned}$$

Thus, the symmetry property holds. We have the variational representation

$$\eta A^{\text{hom}}\eta = \min_{v\in V_{\text{pot}}} \int_{\square\cap F} (\eta + v_\eta)A(\eta + v_\eta)\, dx$$

since (8.9) is the Euler equation corresponding to the variational problem.

The following natural question arises: When is the matrix $A^{\text{hom}}$ singular? Obviously, it may happen only if $\eta + v_\eta = 0$, i.e., if a constant vector is potential.

EXERCISE 8.9. Give an example of a periodic set $F$ and a nonzero constant potential vector on $F$ (the answer is shown in Figure 8.3).

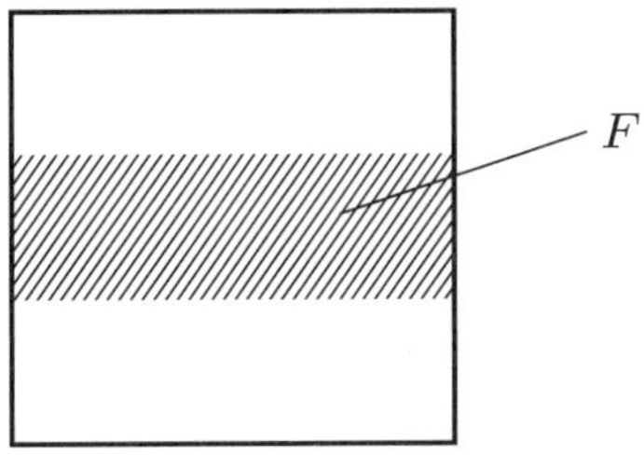


FIGURE 8.3

If the matrix $A^{\text{hom}}$ is positive definite, then every vector $\zeta \in \mathbb{R}^d$ is the average of some solenoidal vector-valued function. This fact follows from (8.10).

LEMMA 8.5. *If $F$ is connected in $\mathbb{R}^d$, then the homogenized matrix is positive definite.*

PROOF. From the variational representation we have

$$\eta A^{\text{hom}} \eta \geq \alpha \min_{v \in V_{\text{pot}}} \int_{\square \cap F} |\eta + v|^2 \, dx = \alpha \inf_{\varphi \in C^\infty_{\text{per}}(\square)} \int_{\square \cap F} |\eta + \nabla \varphi|^2 \, dx.$$

Let $F_1$ be an open connected periodic subset of $F$ with smooth boundary.

EXERCISE 8.10. Prove the existence of $F_1$.

By the obvious inequality

$$\inf_{\varphi \in C^\infty_{\text{per}}(\square)} \int_{\square \cap F} |\eta + \nabla \varphi|^2 \, dx \geq \inf_{\varphi \in C^\infty_{\text{per}}(\square)} \int_{\square \cap F_1} |\eta + \nabla \varphi|^2 \, dx$$

and the fact that the infimum on the right-hand side is attained, we find

$$\eta A^{\text{hom}} \eta \geq \alpha \min_{\varphi \in C^\infty_{\text{per}}(\square)} \int_{\square \cap F_1} |\eta + \nabla \varphi|^2 \, dx.$$

The quadratic form on the right-hand side of the last inequality defines the homogenized matrix of the Laplace operator on the smooth periodic structure $F_1$. Then the positive definiteness of this matrix follows from Exercise 7.2. □

**8.5. Flux convergence.** Now we have all necessary mathematical tools for studying homogenization in the perforated domain $\Omega^\varepsilon = \Omega \cap \varepsilon F$, where $F$ is an arbitrary 2-connected periodic set. As has been mentioned before, $F$ is not necessarily connected from the geometric point of view. Moreover, no smoothness conditions are imposed on the boundary.

Consider the boundary value problem

$$\begin{aligned} &- \operatorname{div}\Big(A\Big(\frac{x}{\varepsilon}\Big)\nabla u_\varepsilon\Big) + u_\varepsilon = f(x) \quad \text{in } \Omega^\varepsilon, \\ &a_{ij}(x)\frac{\partial u_\varepsilon}{\partial x_i}\nu_j = 0 \quad \text{on } S^\varepsilon, \qquad u_\varepsilon\big|_{\Gamma_\varepsilon} = 0, \end{aligned} \tag{8.11}$$

which is understood in the weak sense, i.e., a function $u_\varepsilon$ in $H^1(\Omega^\varepsilon, \Gamma_\varepsilon)$ is a solution to the problem (8.11) if

$$\int_{\Omega^\varepsilon} \Big[\Big(A\Big(\frac{x}{\varepsilon}\Big)\nabla u_\varepsilon, \nabla\psi\Big) + (u_\varepsilon, \psi)\Big]\,dx = \int_{\Omega^\varepsilon} f(x)\psi(x)\,dx \tag{8.12}$$

for any $\psi \in H^1(\Omega^\varepsilon, \Gamma_\varepsilon)$. The existence and uniqueness of a solution to this problem easily follows from the Lax–Milgram theorem. Furthermore, setting $\psi(x) \equiv u_\varepsilon(x)$, it is easy to derive the estimate $\|u_\varepsilon\|_{H^1(\Omega^\varepsilon)} \le C$, where the constant $C > 0$ is independent of $\varepsilon$.

Setting $\psi(x) = \varepsilon w\big(\frac{x}{\varepsilon}\big)\varphi(x)$, where $\varphi(x) \in C_0^\infty(\Omega)$, $w(\xi) \in C^\infty_{\rm per}(\square)$, in (8.12), we have

$$\begin{aligned}\int_{\Omega^\varepsilon} \Big(A\Big(\frac{x}{\varepsilon}\Big)\nabla u_\varepsilon, \nabla_\xi w\Big(\frac{x}{\varepsilon}\Big)\Big)\,\varphi(x)\,dx + \varepsilon\int_{\Omega^\varepsilon} \Big(A\Big(\frac{x}{\varepsilon}\Big)\nabla u_\varepsilon, \nabla\varphi(x)\Big)\, w\Big(\frac{x}{\varepsilon}\Big)\,dx \\ + \varepsilon\int_{\Omega^\varepsilon} u_\varepsilon w\Big(\frac{x}{\varepsilon}\Big)\varphi(x)\,dx = \varepsilon\int_{\Omega^\varepsilon} f(x) w\Big(\frac{x}{\varepsilon}\Big)\varphi(x)\,dx.\end{aligned}$$

Consequently,

$$\int_{\Omega^\varepsilon} \Big(A\Big(\frac{x}{\varepsilon}\Big)\nabla u_\varepsilon, \nabla_\xi w\Big(\frac{x}{\varepsilon}\Big)\Big)\,\varphi(x)\,dx \to 0 \quad \text{as } \varepsilon \to 0 \tag{8.13}$$

for any $\varphi(x) \in C_0^\infty(\Omega)$ and $w(\xi) \in C^\infty_{\rm per}(\square)$. Denote by $u(x)$ the weak limit of the subsequence $\chi_{\varepsilon'} u_{\varepsilon'}$ in $L_2(\Omega)$, where $\chi_\varepsilon$ is the characteristic function of $\Omega^\varepsilon$. What equation and boundary condition are satisfied by this function? The following lemma answers this question.

Denote by $\chi_\varepsilon$ the characteristic function of $\Omega^\varepsilon$.

LEMMA 8.6. *Let $u_\varepsilon \in H^1(\Omega^\varepsilon)$ be such that $\|u_\varepsilon\|_{H^1(\Omega^\varepsilon)} \le C$ and $\chi_\varepsilon u_\varepsilon \rightharpoonup u$ weakly in $L_2(\Omega)$ as $\varepsilon \to 0$. Let*

$$\lim_{\varepsilon\to 0}\int_{\Omega^\varepsilon} \Big(A\Big(\frac{x}{\varepsilon}\Big)\nabla u_\varepsilon, \nabla_\xi w\Big(\frac{x}{\varepsilon}\Big)\Big)\varphi(x)\,dx = 0 \quad \forall\, w \in C^\infty_{\rm per}(\square), \varphi \in C_0^\infty(\Omega). \tag{8.14}$$

*Then $\chi_\varepsilon A_\varepsilon \nabla u_\varepsilon \rightharpoonup A^{\rm hom}\nabla u$ weakly in $L_2(\Omega)$ as $\varepsilon \to 0$, i.e.,*

$$\int_{\Omega^\varepsilon} \varphi(x) A\Big(\frac{x}{\varepsilon}\Big)\nabla u_\varepsilon(x)\,dx \to \int_\Omega \varphi(x) A^{\rm hom}\nabla u\,dx \quad \forall\, \varphi \in L_2(\Omega).$$

As usual, we denote by $(\cdot,\cdot)$ the Euclidean inner product of two $d$-dimensional vectors.

PROOF OF LEMMA 8.6. Let $v_\eta \in V_{\rm pot}$ be a solution to the problem (8.9). Then the vector $z(\xi) = \chi(\xi)A(\xi)(\eta + v_\eta)$ is solenoidal, i.e., $\operatorname{div}(\chi_\varepsilon A_\varepsilon(\eta + v)) = 0$ in $\square$ in the weak sense, and $\langle z\rangle_F = A^{\rm hom}\eta$. We set $z_\varepsilon(x) = z\big(\frac{x}{\varepsilon}\big)$. By Lemma 8.4, we have

$$\int_\Omega \varphi\nabla u_\varepsilon z_\varepsilon\,dx \to \int_\Omega \varphi\nabla u A^{\rm hom}\eta\,dx \quad \text{as } \varepsilon \to 0. \tag{8.15}$$

REMARK 8.2. Lemma 5.1 is not suitable for proving the limit relation (8.15) since, in general, $u_\varepsilon$ does not admit an extension to the space $H^1(\square)$.

It is easy to verify that the equality (8.14) is preserved if we take $v_\eta\left(\frac{x}{\varepsilon}\right)$ instead of $\nabla_\xi w\left(\frac{x}{\varepsilon}\right)$ (check it!), i.e.,

$$\int_\Omega \chi_\varepsilon(x)\left(A\left(\frac{x}{\varepsilon}\right)\nabla u_\varepsilon(x), v_\eta\left(\frac{x}{\varepsilon}\right)\right)\varphi(x)\,dx \to 0 \quad \text{as } \varepsilon \to 0.$$

Therefore,

$$\int_{\Omega^\varepsilon} \varphi A_\varepsilon \nabla u_\varepsilon \eta\,dx \to \int_\Omega \varphi \nabla u A^{\text{hom}} \eta\,dx.$$

Since $\eta \in \mathbb{R}^d$ is arbitrary, we obtain the desired flux convergence. □

We see that any weak limit $u(x)$ of solutions to the problem (8.11) satisfies the equation $\operatorname{div}\left(A^{\text{hom}}\nabla u\right) + u = f$ in the sense of distributions. However, $u(x)$ does not necessarily belong to $H^1(\Omega)$ and the matrix $A^{\text{hom}}$ is not necessarily positive definite. Therefore, the equation obtained for $u(x)$ is not necessarily elliptic. We cannot assert that $u\big|_{\partial\Omega} = 0$ either. To determine whether the limit equation is elliptic, we need to study the set of solenoidal vector-valued functions on $F$ and clarify what constant vectors in $\mathbb{R}^d$ can serve as the average of solenoidal vector fields on $F$. If this set coincides with the entire space $\mathbb{R}^d$, then the limit problem is elliptic; otherwise, $A^{\text{hom}}$ is a singular matrix and $u$ has first order weak derivatives only along certain directions (cf. Lemma 8.4 and the equality (8.10)).

## 9. Two-scale convergence method

**9.1. Definitions and statements.** The two-scale convergence method was first proposed by Nguetseng [**100**] and was further developed by Allaire [**3**] and others.

Let $\Omega \subset \mathbb{R}^d$ be a bounded domain.

DEFINITION 9.1. A sequence $u_\varepsilon(x) \in L_2(\Omega)$ such that

$$\|u_\varepsilon\|_{L^2(\Omega)} \le C$$

*two-scale converges* to a function $u(x,\xi) \in L_2(\Omega \times \square)$ if for any $\varphi(x) \in C_0^\infty(\Omega)$ and $\Phi(\xi) \in C^\infty_{\text{per}}(\square)$,

$$\int_\Omega u_\varepsilon(x)\varphi(x)\Phi\left(\frac{x}{\varepsilon}\right)dx \to \int_{\Omega\times\square} u(x,\xi)\varphi(x)\Phi(\xi)\,dx d\xi \quad \text{as } \varepsilon \to 0.$$

In this case, we write

$$u_\varepsilon(x) \xrightarrow{2} u(x,\xi) \text{ as } \varepsilon \to 0.$$

If $u_\varepsilon(x) \to u(x)$ strongly in $L_2(\Omega)$ as $\varepsilon \to 0$, then $u_\varepsilon(x) \xrightarrow{2} u(x)$ as $\varepsilon \to 0$ (check it!). Note that if $F(\xi)$ is a 1-periodic function, then $F\left(\frac{x}{\varepsilon}\right) \rightharpoonup \langle F\rangle_\square$ weakly in $\Omega$ and $F\left(\frac{x}{\varepsilon}\right) \xrightarrow{2} F(\xi)$. The two-scale convergence generalizes the notion of weak convergence.

EXERCISE 9.1. Let $u(x,\xi)$ be periodic with respect to $\xi$ and continuous in $\overline{\Omega \times \square}$. Prove that $u\left(x, \frac{x}{\varepsilon}\right) \xrightarrow{2} u(x,\xi)$.

EXERCISE 9.2. Let $f(\xi) \in L_2(\square)$ be periodic. Does the sequence $f(x/\varepsilon^2)$ have a two-scale limit? If an answer is "yes," what is this limit?

We formulate (without proof) the main assertions about the two-scale convergence of functions. These results were first established by Nguetseng.

THEOREM 9.1. *Suppose that*

$$\int_\Omega |u_\varepsilon(x)|^2\,dx < C,$$

*where $C > 0$ is independent of $\varepsilon > 0$. Then from $u_\varepsilon(x)$ we can extract a two-scale converging subsequence.*

THEOREM 9.2. *Assume that $u_\varepsilon(x) \in H^1(\Omega)$ and*

$$\int_\Omega (|u_\varepsilon(x)|^2 + |\nabla_x u_\varepsilon(x)|^2)\,dx < C,$$

*where $C > 0$ is independent of $\varepsilon > 0$. Then from $u_\varepsilon(x)$ we can extract a subsequence $u_{\varepsilon'}(x)$ such that $u_{\varepsilon'} \overset{2}{\to} u(x)$ as $\varepsilon' \to 0$ and $\nabla_x u_{\varepsilon'} \overset{2}{\to} \nabla_x u(x) + \nabla_\xi u^{(1)}(x,\xi)$ as $\varepsilon' \to 0$, where $u(x) \in H^1(\Omega)$ and $u^{(1)}(x,\xi)$ is a function of variables $x \in \Omega$ and $\xi \in \square$ such that $u^{(1)} \in L_2(\Omega, H^1_{\mathrm{per}}(\square))$.*

THEOREM 9.3. *Assume that $u_\varepsilon(x) \in H^1(\Omega)$,*

$$\int_\Omega |u_\varepsilon(x)|^2\,dx < C, \quad \varepsilon^2 \int_\Omega |\nabla_x u_\varepsilon(x)|^2\,dx < C,$$

*where $C > 0$ is independent of $\varepsilon > 0$. Then from $u_\varepsilon(x)$ we can extract a subsequence such that $u_{\varepsilon'} \overset{2}{\to} u(x,\xi)$ as $\varepsilon' \to 0$ and $\varepsilon\nabla_x u_{\varepsilon'} \overset{2}{\to} \nabla_\xi u(x,\xi)$ as $\varepsilon' \to 0$, where $u(x,\xi) \in L_2(\Omega, H^1_{\mathrm{per}}(\square))$.*

We set $\Omega^h_\varepsilon := \Omega \cap \varepsilon E$, $\Omega^s_\varepsilon := \Omega \backslash \Omega^h_\varepsilon$, where $E$ is an open connected periodic set in $\mathbb{R}^d$, $\square^h := \square \cap E$, and $\square^s := \square \backslash \square^h$.

THEOREM 9.4. *Suppose that $u_\varepsilon(x) \in H^1(\Omega)$,*

$$\int_{\Omega^h_\varepsilon} (|u_\varepsilon(x)|^2 + |\nabla_x u_\varepsilon(x)|^2)\,dx < C, \quad \varepsilon^2 \int_\Omega |\nabla u_\varepsilon(x)|^2\,dx + \int_\Omega u_\varepsilon^2\,dx < C,$$

*where $C > 0$ is independent of $\varepsilon > 0$. Then from $u_\varepsilon(x)$ we can extract a subsequence such that $u_{\varepsilon'} \overset{2}{\to} u(x,\xi)$ as $\varepsilon' \to 0$, $u(x,\xi) \in L_2(\Omega, H^1_{\mathrm{per}}(\square))$; moreover, $u(x,\xi) = u(x)$ for $\xi \in \square^h$, $u(x) \in H^1(\Omega)$, and $\chi_{\Omega^h_{\varepsilon'}} \nabla_x u_{\varepsilon'} \overset{2}{\to} \chi_{\square^h}(\nabla_x u(x) + \nabla_\xi u^{(1)}(x,\xi))$ as $\varepsilon' \to 0$, where $u^{(1)}(x,\xi)$ is a function of variables $x \in \Omega$ and $\xi \in \square$, $u^{(1)}(x,\xi) \in L_2(\Omega, H^1_{\mathrm{per}}(\square^h))$. In addition, $\varepsilon' \chi_{\Omega^s_{\varepsilon'}} \nabla_x u_{\varepsilon'} \overset{2}{\to} \nabla_\xi u(x,\xi)$ as $\varepsilon' \to 0$.*

Theorems 9.1–9.4 are convenient and effective tools for treating homogenization problems. The proof of these theorems will be given later. Now, we show how to use Theorems 9.1–9.4 for the homogenization of a second order differential equation of elliptic type.

**9.2. Two examples.** 1. Let $u_\varepsilon(x) \in H^1(\Omega)$ be a solution to the Dirichlet problem

$$L_\varepsilon u_\varepsilon \equiv \sum_{i,j=1}^d \frac{\partial}{\partial x_i}\left(a_{ij}\left(\frac{x}{\varepsilon}\right)\frac{\partial u_\varepsilon}{\partial x_j}\right) = f(x) \quad \text{in } \Omega, \tag{9.1}$$

where $u_\varepsilon \in \overset{\circ}{H}{}^1(\Omega)$ and $f(x) \in L_2(\Omega)$. As usual, a solution to the problem (9.1) is understood in the sense of the integral identity

$$-\int_\Omega a_{ij}\Big(\frac{x}{\varepsilon}\Big)\frac{\partial u_\varepsilon}{\partial x_j}\frac{\partial v}{\partial x_i}\,dx = \int_\Omega f(x)v(x)\,dx \tag{9.2}$$

for any test function $v \in \overset{\circ}{H}{}^1(\Omega)$. Using Theorems 9.1–9.4, we can clarify the behavior of $u_\varepsilon(x)$. Note that we have already considered this problem and obtained its homogenization by other methods. Here, we establish a homogenization result by the two-scale convergence technique.

For a test function in the integral identity (9.2) we take $\varepsilon\varphi(x)\Phi\big(\frac{x}{\varepsilon}\big)$, where $\varphi \in C_0^\infty(\Omega)$ and $\Phi(\xi)$ is smooth and 1-periodic with respect to $\xi_1,\dots,\xi_d$. Then

$$\begin{aligned}-\int_\Omega a_{ij}\Big(\frac{x}{\varepsilon}\Big)\frac{\partial u_\varepsilon}{\partial x_j}\frac{\partial \Phi}{\partial \xi_i}\varphi(x)\,dx - \varepsilon\int_\Omega a_{ij}\Big(\frac{x}{\varepsilon}\Big)\frac{\partial u_\varepsilon}{\partial x_j}\Phi\Big(\frac{x}{\varepsilon}\Big)\frac{\partial\varphi}{\partial x_i}\,dx\\ = \varepsilon\int_\Omega f(x)\varphi(x)\Phi\Big(\frac{x}{\varepsilon}\Big)\,dx.\end{aligned} \tag{9.3}$$

Since the solution $u_\varepsilon(x)$ to the problem (9.1) satisfies the estimate $\|u_\varepsilon\|_{H^1(\Omega)} \leq C$ with a constant $C > 0$ independent of $\varepsilon$, we can choose a subsequence $u_{\varepsilon'}$ such that (cf. Theorem 9.2)

$$u_{\varepsilon'} \xrightarrow{2} u(x),\ \nabla_x u_{\varepsilon'} \xrightarrow{2} \nabla_x u(x) + \nabla_\xi u^{(1)}(x,\xi) \quad \text{as } \varepsilon' \to 0, \tag{9.4}$$

where $u^{(1)}(x,\xi)$ is a function on $\Omega\times\square$ of class $L_2(\Omega, H^1_{\text{per}}(\square))$. Passing to the limit in the integral identity (9.3) as $\varepsilon' \to 0$, we find that

$$\int_{\Omega\times\square} a_{ij}(\xi)\Big[\frac{\partial}{\partial x_i}u(x) + \frac{\partial}{\partial \xi_i}u^{(1)}(x,\xi)\Big]\frac{\partial\Phi}{\partial\xi_i}\varphi(x)\,dx\,d\xi = 0 \tag{9.5}$$

for any $\Phi$ and $\varphi$. Since $\varphi(x)$ is arbitrary, from (9.5) it follows that

$$\int_\square a_{ij}(\xi)\frac{\partial u^{(1)}(x,\xi)}{\partial\xi_i}\frac{\partial\Phi(\xi)}{\partial\xi_i}\,d\xi = -\int_\square a_{ij}(\xi)\frac{\partial u(x)}{\partial x_i}\frac{\partial\Phi(\xi)}{\partial\xi_i}\,d\xi$$

for any $\Phi \in C^\infty_{\text{per}}(\square)$, where $x \in \Omega$ is regarded as a parameter. Let $N^p(\xi) \in H^1_{\text{per}}(\square)$ be a solution to the following problem: find $N^p(\xi) \in H^1_{\text{per}}(\square)$ such that

$$\int_\square a_{ij}(\xi)\frac{\partial N^p(\xi)}{\partial\xi_i}\frac{\partial\Phi(\xi)}{\partial\xi_i}\,d\xi = -\int_\square a_{ip}(\xi)\frac{\partial\Phi(\xi)}{\partial\xi_i}\,d\xi$$

for any $\Phi \in C^\infty_{\text{per}}(\square)$. The existence of a solution follows from the Lax–Milgram theorem. The function $u^{(1)}(x,\xi)$ can be represented in the form

$$u^{(1)}(x,\xi) = \sum_{p=1}^d N^p(\xi)\frac{\partial u(x)}{\partial x_p}. \tag{9.6}$$

In the integral identity (9.2), we take a test function $v(x) \in \overset{\circ}{H}{}^1(\Omega)$ independent of $\varepsilon$ and $\xi$. Then we pass to the limit as $\varepsilon' \to 0$ and use Theorem 9.2 again. We have

$$-\int_{\Omega\times\square} a_{ij}(\xi)\Big(\frac{\partial u(x)}{\partial x_j} + \frac{\partial u^{(1)}(x,\xi)}{\partial\xi_j}\Big)\frac{\partial v(x)}{\partial x_i}\,dx\,d\xi = \int_\Omega f(x)v(x)\,dx.$$

Substituting the representation (9.6) for $u^{(1)}(x,\xi)$ in the last equality and changing the integration order (integrate first with respect to $\xi$ and then with respect to $x$), we find

$$-\int_\Omega \Big\{ \int_\square \Big( a_{ij}(\xi) + \frac{\partial N^j}{\partial \xi_k} a_{ik}(\xi) \Big)\, d\xi \Big\} \frac{\partial u(x)}{\partial x_j} \frac{\partial v(x)}{\partial x_i}\, dx = \int_\Omega f(x) v(x)\, dx.$$

Denoting

$$\widehat{a}_{ij} \equiv \int_\square \Big( a_{ij}(\xi) + \frac{\partial N^j}{\partial \xi_k} a_{ik}(\xi) \Big)\, d\xi,$$

we conclude that $u(x)$ satisfies the homogenized boundary value problem in the sense of the integral identity

$$(9.7) \qquad -\int_\Omega \widehat{a}_{ij} \frac{\partial u(x)}{\partial x_j} \frac{\partial v(x)}{\partial x_i}\, dx = \int_\Omega f(x) v(x)\, dx \quad \forall\, v(x) \in \overset{\circ}{H}{}^1(\Omega).$$

EXERCISE 9.3. Prove that the function $u(x) \in H^1(\Omega)$ in (9.4) belongs to the space $\overset{\circ}{H}{}^1(\Omega)$.

We have found the values of the coefficients of the homogenized equation. They are the same as those obtained by other methods in the previous sections. As has already been mentioned, $(\widehat{a}_{ij})$ is a positive definite matrix.

Since a solution to the problem (9.7) is unique, not only the subsequence $u_{\varepsilon'}$, but also the entire sequence $u_\varepsilon$ converges to $u(x)$. Thus, we have proved the following assertion.

THEOREM 9.5. *The solutions $u_\varepsilon(x)$ of the problem* (9.1) *two-scale converge to the solution $u(x) \in \overset{\circ}{H}{}^1(\Omega)$ of the problem* (9.7), *and*

$$\nabla_x u_\varepsilon \overset{2}{\to} \nabla_x u(x) + \sum_{p=1}^{d} \nabla_\xi N^p(\xi) \frac{\partial u(x)}{\partial x_p}.$$

Note that only Theorem 9.2 was used here, whereas the remaining theorems did not work yet. However, there is a large class of problems where Theorems 9.1, 9.3, and 9.4 are actually used.

2. Here, we use the results of an unpublished manuscript by V. V. Zhikov, which he kindly allowed us to use (cf. also [**134**]).

Let $F_i$, $i = 1, 2, \dots, k$, be open connected periodic sets with Lipschitz boundaries such that $\overline{F}_i \cap \overline{F}_j = \varnothing$. Denote $F = F_1 \cup \dots \cup F_k$ and $S = \mathbb{R}^d \setminus F$. The domains $F_i$ are called *hard phases*, and $S$ is referred to as a *soft phase*. We set $F^\varepsilon = \varepsilon F$, $F_j^\varepsilon = \varepsilon F_j$, and $S^\varepsilon = \varepsilon S$. Let

$$a_\varepsilon(x) = \begin{cases} 1 & \text{in } F^\varepsilon, \\ \varepsilon^2 & \text{in } S^\varepsilon. \end{cases}$$

Consider the Dirichlet problem in an arbitrary domain $\Omega$:

$$-\operatorname{div}(a_\varepsilon \nabla u_\varepsilon) + \lambda u_\varepsilon = f \quad \text{in } \Omega, \quad u_\varepsilon \in \overset{\circ}{H}{}^1(\Omega), \quad f \in C^\infty(\overline{\Omega}),$$

where $\lambda$ is a positive parameter. An equivalent statement can be written as the integral identity

$$(9.8) \qquad \int_{\Omega_\varepsilon^h} \nabla u_\varepsilon \nabla \psi\, dx + \varepsilon^2 \int_{\Omega_\varepsilon^s} \nabla u_\varepsilon \nabla \psi\, dx + \lambda \int_\Omega u_\varepsilon \psi\, dx = \int_\Omega f \psi\, dx$$

for any $\psi \in \overset{\circ}{H}{}^1(\Omega)$, where $\Omega_\varepsilon^h = \Omega \cap F^\varepsilon$ and $\Omega_\varepsilon^s = \Omega \cap S^\varepsilon$. For every fixed $\varepsilon > 0$ the problem (9.8) is uniquely solvable. However, the estimate $\|u_\varepsilon\|_{H^1(\Omega)} < C$ with a constant $C > 0$ independent of $\varepsilon$ is not valid because the lower ellipticity constant for the coefficients $a_\varepsilon(x)$ degenerates as $\varepsilon \to 0$.

The energy estimate for this problem takes the form (check it!)

$$\int_{\Omega_\varepsilon^h} |\nabla u_\varepsilon|^2\,dx + \varepsilon^2 \int_{\Omega_\varepsilon^s} |\nabla u_\varepsilon|^2\,dx + \lambda \int_\Omega (u_\varepsilon)^2\,dx \le C(\lambda) \int_\Omega f^2(x)\,dx.$$

Let $\Phi_0(\xi), \Phi_1(\xi), \dots, \Phi_k(\xi) \in H^1_{\mathrm{per}}(\square \cap S)$ be a solution to the problem

$$\begin{aligned} -\Delta_\xi \Phi_0(\xi) + \lambda \Phi_0(\xi) &= 1 \quad \text{in } S, \quad \Phi_0(\xi)\big|_{\partial S} = 0, \\ -\Delta_\xi \Phi_i(\xi) + \lambda \Phi_i(\xi) &= 0 \quad \text{in } S, \quad \Phi_i(\xi)\big|_{\partial F_j} = \delta_{ij}. \end{aligned}$$

On $F_j$, we define 1-periodic functions $N_j^p(\xi)$, $p = 1, \dots, d$, $j = 1, \dots, k$, that belong to the space $H^1_{\mathrm{per}}(F_j)$ and satisfy the integral identity

$$\int_{\square \cap F_j} \left(\bar{e}_k + \nabla_\xi N_j^k(\xi)\right) \nabla_\xi \Psi(\xi)\,d\xi = 0 \quad \forall\, \Psi(\xi) \in H^1_{\mathrm{per}}(\square \cap F_j). \tag{9.9}$$

We also need auxiliary functions $E_j(\xi) \in H^1_{\mathrm{per}}(\square \cap S)$, $j = 1, \dots, k$, defined as solutions to the problems

$$\Delta_\xi E_j(\xi) = 0 \quad \text{in } S, \quad E_j(\xi)\big|_{\partial F_i} = \delta_{ij},$$

where $\delta_{ij}$ is the Kronecker symbol. We extend $\Phi_j$ and $E_j$ to $\mathbb{R}^d_\xi$ in such a way that $\Phi_0 \equiv 0$ in $\square \setminus S$, $\Phi_j \equiv 1$ in $F_j$, $\Phi_j \equiv 0$ in $F_i$ for $j \ne i$, $E_j \equiv 1$ in $F_j$, and $E_j \equiv 0$ in $E_i$ for $j \ne i$. It is easy to see that $E_1 + E_2 + \dots + E_k \equiv 1$ in $\mathbb{R}^d$ (check it!).

EXERCISE 9.4. Using the weak statement of the boundary value problems and the Lax–Milgram theorem, prove the existence and uniqueness of $\Phi_0(\xi)$, $\Phi_i(\xi)$, $E_j(\xi)$.

We denote by $\widehat{A}_j$ the matrix with entries $a^j_{il} \equiv \left\langle \delta_{il} + \frac{\partial N_j^i}{\partial x_l} \right\rangle_{\square \cap F_j}$ and formulate the main result.

THEOREM 9.6. *The solution $u_\varepsilon(x)$ to the problem (9.8) satisfies the conditions*

$$\lim_{\varepsilon \to 0} \int_{\Omega \cap F_j^\varepsilon} \left(u_\varepsilon(x) - u_j(x)\right)^2 dx = 0, \quad j = 1, \dots, k,$$

$$\lim_{\varepsilon \to 0} \int_{\Omega \cap S^\varepsilon} \left(u_\varepsilon(x) - u_0(x, \frac{x}{\varepsilon})\right)^2 dx = 0,$$

$$\lim_{\varepsilon \to 0} \int_{\Omega \cap F_j^\varepsilon} \left(\nabla_x u_\varepsilon(x) - \nabla_x u_j(x) - \sum_{k=1}^d \nabla_\xi N_j^k\left(\frac{x}{\varepsilon}\right) \frac{\partial u_j}{\partial x_k}(x)\right)^2 dx = 0$$

*(there is no summation with respect to $j = 1, \dots, k$), and*

$$\lim_{\varepsilon \to 0} \varepsilon^2 \int_{\Omega \cap S^\varepsilon} \left(\nabla_x u_\varepsilon(x) - \nabla_x u_0(x, \xi)\big|_{\xi = \frac{x}{\varepsilon}}\right)^2 dx = 0,$$

*where* $u_j(x)$, $j = 1, \ldots, k$, *are solutions to the system of differential equations*

$$(9.10)\qquad \begin{aligned} &-\operatorname{div}(\widehat{A}_j \nabla_x u_j) + \sum_{i=1}^{k} c_{ji} u_i(x) + \lambda \theta_j u_j(x) + \lambda \sum_{i=1}^{k} b_{ji} u_i(x) \\ &\quad = f(x)\Big\{ \langle E_j \rangle - \lambda \int_{S\cap\square} E_j(\xi)\Phi_0(\xi)\, d\xi \Big\}, \quad j = 1, \ldots, k, \end{aligned}$$

*and the boundary conditions*

$$(9.11)\qquad u_j = 0 \quad \text{on } \partial\Omega, \quad j = 1, \ldots, k.$$

*Here,*

$$\begin{aligned} c_{ji} &= \int_{S\cap\square} (\nabla_\xi E_j(\xi), \nabla_\xi \Phi_i(\xi))\, d\xi \ \text{ is the connection matrix,} \\ b_{ji} &= \int_{S\cap\square} E_j(\xi)\Phi_i(\xi)\, d\xi \ \text{ is the memory matrix,} \\ \theta_j &= \langle 1 \rangle_{F_j\cap\square} \ \text{ is the porosity coefficient,} \end{aligned}$$

*and*

$$u_0(x,\xi) \equiv \sum_{j=1}^{k} \Phi_j(\xi) u_j(x) + \Phi_0(\xi) f(x).$$

EXERCISE 9.5. Prove that the entries $c_{ji}$ of the connection matrix satisfy the equality $c_{ij} = c_{ji} = \langle \nabla_\xi E_j, \nabla_\xi E_i \rangle$, and consequently $c_{ij}$ are independent of $\lambda$.

REMARK 9.1. It is possible to determine the coefficients of the limit system (9.10) from only the functions $\Phi_j$ without using $E_j$, $j = 1, \ldots, k$. In this case, the limit problem takes the form

$$(9.12)\qquad \begin{aligned} &-\operatorname{div}(\widehat{A}_j \nabla_x u_j) + \sum_{i=1}^{k} \langle \nabla_\xi \Phi_i, \nabla_\xi \Phi_j \rangle u_i(x) + \lambda \theta_j u_j(x) + \lambda \sum_{i=1}^{k} \langle \Phi_i, \Phi_j \rangle u_i(x) \\ &\quad = f(x) \int_{S\cap\square} \Phi_j(\xi) d\xi, \quad j = 1, \ldots, k, \end{aligned}$$

with the boundary conditions (9.11).

Before proving Theorem 9.6, let us discuss the formulation. Consider a homogenization of a family of Dirichlet problems, where the lower ellipticity constant is not bounded away from zero as $\varepsilon \to 0$. It degenerates as $\varepsilon^2$ on some set whose geometry depends on $\varepsilon$. The abandonment of the uniform ellipticity condition leads to the following new phenomenon of interest: unlike the previous example, the knowledge of the limit function $u_0(x)$ is not sufficient for describing the limit of solutions as $\varepsilon \to 0$. In this case, we need to solve a system of differential equations for functions $u_1(x), \ldots, u_k(x)$, each describing the limit behavior of solutions only in some part $\widetilde{F}_j^\varepsilon = F_j^\varepsilon \cap \Omega$ depending on $\varepsilon$ of the domain $\Omega$. It is obvious that $\widetilde{F}_j^\varepsilon \cap \widetilde{F}_i^\varepsilon = \varnothing$, $i \neq j$, but $\bigcup_{i=1}^k \widetilde{F}_i^\varepsilon \neq \Omega$, i.e., in the domain $\widetilde{S}_\varepsilon = \Omega \setminus \bigcup_{i=1}^k \widetilde{F}_i^\varepsilon$, none of the functions $u_i(x)$, $i = 1, \ldots, k$, approximates $u_\varepsilon(x)$.

In $\widetilde{S}_\varepsilon$, the sequence $u_\varepsilon(x)$ rapidly oscillates. However, due to Theorem 9.6, we can find the magnitude of these oscillations and study their character since the theorem explicitly indicates the function $u_0\big(x, \frac{x}{\varepsilon}\big)$ that is approached by $u_\varepsilon(x)$ on $\widetilde{S}_\varepsilon$ as $\varepsilon \to 0$ (cf. Figure 9.1).

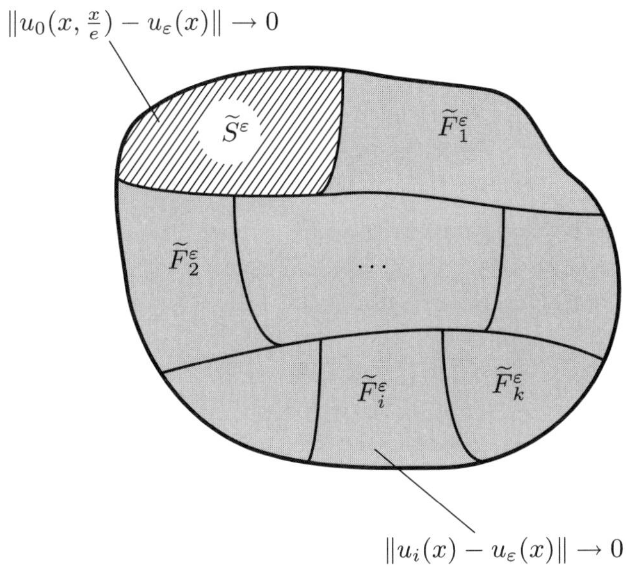


FIGURE 9.1

It is important to note that the parameter $\lambda$ occurs in the limit problem in a nonlinear way (the memory matrix depends on $\lambda$). Therefore, if $\lambda$ is a spectral parameter appearing after the Laplace transform of the evolution problem, then the appearance of integral operators in the limit problem is caused by the inverse Laplace transform. These operators correspond to a medium with memory. This important phenomenon was studied in detail by many authors. It is of great interest because of a number of applied problems arising, in particular, in oil production (cf., for example, [**5, 68, 29, 27, 28, 29**], and also [**67**]).

PROOF OF THEOREM 9.6. Substituting $\psi(x) = E_j\left(\frac{x}{\varepsilon}\right)\varphi(x)$, where $\varphi(x) \in C_0^\infty(\mathbb{R}^d)$, in (9.8), we have

$$\begin{aligned}\int_{\widetilde{F}_j^\varepsilon} \nabla u_\varepsilon \nabla\varphi\, dx + \varepsilon \int_{\widetilde{S}_\varepsilon} \nabla u_\varepsilon \nabla_\xi E_j \varphi\, dx + \varepsilon^2 \int_{\widetilde{S}_\varepsilon} \nabla u_\varepsilon \nabla \varphi E_j\, dx \\ + \lambda \int_{\widetilde{F}_j^\varepsilon} u_\varepsilon \varphi\, dx + \lambda \int_{\widetilde{S}_\varepsilon} u_\varepsilon \varphi E_j\, dx = \int_\Omega f\varphi E_j\, dx.\end{aligned} \tag{9.13}$$

Substituting $\varphi = \varepsilon\Psi\left(\frac{x}{\varepsilon}\right)v(x)$, where $\Psi(\xi)$ is a smooth periodic function and $v(x) \in C_0^\infty(\Omega)$, passing to the limit on the two-scale converging subsequence $u_\varepsilon$, and using Theorem 9.4, we obtain the equality

$$\int_{\Omega\times(\square\cap F_j)} [\nabla_x u_j(x) + \nabla_\xi u_j^{(1)}(x,\xi)]\nabla_\xi\Psi(\xi)v(x)\, dx\, d\xi = 0.$$

Since $v(x)$ is arbitrary, we have

$$\int_{\square\cap F_j} \nabla_\xi u_j^{(1)}(x,\xi)\nabla_\xi\Psi(\xi)\, d\xi = -\nabla_x u_j(x) \int_{\square\cap F_j} \nabla_\xi\Psi(\xi)\, d\xi$$

for almost all $x \in \Omega$. Consequently,

$$u_j^{(1)}(x,\xi) = \sum_{p=1}^{d} \frac{\partial u_j}{\partial x_p} N_j^p(\xi), \quad \xi \in \square \cap F_j, \tag{9.14}$$

where $N_j^p(\xi)$ were introduced as solutions to the problem (9.9).

For a test function in the integral identity (9.13) we take $\varphi = \Psi\left(\frac{x}{\varepsilon}\right) v(x)$, where $\Psi(\xi)$ is 1-periodic with respect to $\xi_1, \dots, \xi_d$ and $\operatorname{supp} \Psi(\xi) \subset \square \cap S$. Passing to the limit of a two-scale converging subsequence $u_{\varepsilon'}$ and using Theorem 9.3, we find

$$\begin{aligned} &\int_{\Omega \times (\square \cap S)} \nabla_\xi u(x,\xi) \nabla_\xi E_j(\xi) \Psi(\xi) v(x)\, dx\, d\xi \\ &\qquad + \int_{\Omega \times (\square \cap S)} \nabla_\xi u(x,\xi) E_j(\xi) \nabla_\xi \Psi(\xi) v(x)\, dx\, d\xi \\ &\qquad + \lambda \int_{\Omega \times (\square \cap S)} u(x,\xi) E_j(\xi) \Psi(\xi) v(x)\, dx\, d\xi \\ &\quad = \int_{\Omega \times (\square \cap S)} E_j f(x) \Psi(\xi) v(x)\, dx\, d\xi. \end{aligned}$$

Since $v(x) \in C_0^\infty(\Omega)$ is arbitrary, we conclude that

$$\begin{aligned} &\int_{\square \cap S} \nabla_\xi u(x,\xi) \nabla_\xi E_j(\xi) \Psi(\xi)\, d\xi + \int_{\square \cap S} \nabla_\xi u(x,\xi) E_j(\xi) \nabla_\xi \Psi(\xi)\, d\xi \\ &\qquad + \lambda \int_{(\square \cap S)} u(x,\xi) E_j(\xi) \Psi(\xi)\, d\xi = \left( \int_{(\square \cap S)} E_j(\xi) \Psi(\xi)\, d\xi \right) f(x) \end{aligned}$$

for almost all $x \in \Omega$ or

$$\begin{aligned} \int_{\square \cap S} \nabla_\xi u(x,\xi) \nabla_\xi \left(E_j(\xi)\Psi(\xi)\right) d\xi + \lambda \int_{(\square \cap S)} u(x,\xi) \left(E_j(\xi)\Psi(\xi)\right) d\xi& \\ = \left( \int_{(\square \cap S)} E_j(\xi) \Psi(\xi)\, d\xi \right) f(x).& \end{aligned}$$

The last integral identity is equivalent to the equation

$$-\Delta_\xi u(x,\xi) + \lambda u(x,\xi) = f(x), \quad \xi \in \square \cap S,$$

since $E_j(\xi)\Psi(\xi)$ is arbitrary in $\square \cap S$ (prove that $E_j > 0$ in $S$!).

Furthermore, since $u(x,\cdot) \in H^1_{\text{per}}(\square)$ for almost all $x \in \Omega$, the boundary condition $u(x,\xi)\big|_{\partial F_j} = u_j(x)$ holds. Hence

$$u(x,\xi) = u_0(x,\xi) = \sum_{j=1}^{k} \Phi_j(\xi) u_j(x) + f(x) \Phi_0(\xi), \quad \xi \in \square \cap S, \tag{9.15}$$

where $\Phi_j(\xi)$ and $\Phi_0(\xi)$ have already been found.

For a test function in the integral identity (9.13) we take $\varphi = \varphi(x)$ and pass to the limit as $\varepsilon' \to 0$ on a two-scale converging subsequence. Then

$$\int_{\Omega\times(\square\cap F_j)} [\nabla_x u_j(x) + \nabla_\xi u_j^{(1)}(x,\xi)]\nabla_x\varphi(x)\,dx\,d\xi$$
$$+ \int_{\Omega\times(\square\cap S)} \nabla_\xi u(x,\xi)\varphi(x)\nabla_\xi E_j(\xi)\,dx\,d\xi + \lambda\int_{\Omega\times(\square\cap F_j)} u_j(x)\varphi(x)\,dx\,d\xi$$
$$+ \lambda\int_{\Omega\times(\square\cap S)} u(x,\xi)\varphi(x)E_j(\xi)\,dx\,d\xi = \lambda\int_{\Omega\times\square} f(x)\varphi(x)E_j(\xi)\,dx\,d\xi.$$

Substituting (9.14) and (9.15) in this equation, we obtain the equality

$$-\operatorname{div}\int_{\square\cap F_j}\Big(\nabla_x u_j(x) + \frac{\partial u_j}{\partial x_p}\nabla_\xi N_j^p(\xi)\Big)\,d\xi + \sum_{i=1}^k\Big(u_i(x)\int_{\square\cap S}\nabla_\xi\Phi_i(\xi)\nabla_\xi E_j(\xi)\,d\xi\Big)$$
$$+ f(x)\int_{\square\cap S}\nabla_\xi\Phi_0(\xi)\nabla_\xi E_j(\xi)\,d\xi + \lambda\theta_j u_j(x) + \lambda\sum_{i=1}^k u_i(x)\int_{\square\cap S}\Phi_i(\xi)E_j(\xi)\,d\xi$$
$$+ \lambda\Big(\int_{S\cap\square}\Phi_0(\xi)E_j(\xi)\,d\xi\Big)f(x) = f(x)\int_\square E_j(\xi)\,d\xi,$$

which is equivalent to the system (9.10).

In the above arguments leading to the system (9.10), we can take $\Phi_j$ instead of $E_j$. Then the limit system (9.10) takes the form (9.12).

Thus, we have proved the two-scale convergence of $u_\varepsilon(x)$. To prove the convergence in the $L_2(\Omega)$-norm, we note that for test functions in the definition of two-scale convergence we can take not only functions of the form $\varphi(x)\psi(\xi)$, but also functions in larger classes, for example, $\Psi(x,\xi) \in C(\Omega\times\square)$ (check it!). On the other hand, an arbitrary function in $L_2(\Omega\times\square)$ cannot be taken for a test function because, in particular, the trace $v\big(x, \frac{x}{\varepsilon}\big)$ of an arbitrary function $v(x,\xi) \in L_2(\Omega,\square)$ is not necessarily defined. Moreover, there are examples of sequences $u_\varepsilon(x)$ that two-scale converge to $u_0 \in L_2(\Omega\times\square)$ and of test functions $v(x,\xi)$ in $L_2(\Omega\times\square)$ such that

$$\lim_{\varepsilon\to0}\int_\Omega u_\varepsilon(x)v\Big(x,\frac{x}{\varepsilon}\Big)\,dx \neq \int_{\Omega\times\square} u_0(x,\xi)v(x,\xi)\,dx\,d\xi.$$

EXERCISE 9.6. Let $u_\varepsilon(x)$ satisfy the estimate $\|u_\varepsilon\|_{L_2(\Omega)} \le C$, where $C > 0$ is independent of $\varepsilon$, and let $u_\varepsilon(x)$ two-scale converge to $u(x,\xi) \in L_2(\Omega\times\square)$. Then

$$\varliminf_{\varepsilon\to0}\|u_\varepsilon(x)\|^2_{L_2(\Omega)} \ge \int_{\Omega\times\square} u^2(x,\xi)\,dx\,d\xi.$$

*Hint.* Consider a "smoothed" function $u^\delta(x,\xi)$ that is close to $u(x,\xi)$ in the $L_2(\Omega\times\square)$-norm. Use the inequality

$$\int_\Omega\Big(u_\varepsilon(x) - u^\delta\Big(x,\frac{x}{\varepsilon}\Big)\Big)^2\,dx \ge 0.$$

EXERCISE 9.7. Suppose that a uniformly bounded sequence $u_\varepsilon(x)$ in $L_2(\Omega)$ two-scale converges to $u(x,\xi) \in C(\overline{\Omega}\times\square)$. Let

$$\lim_{\varepsilon\to0}\|u_\varepsilon(x)\|^2_{L_2(\Omega)} = \int_{\Omega\times\square} u^2(x,\xi)\,dx\,d\xi. \tag{9.16}$$

Show that

$$\lim_{\varepsilon\to 0}\Big\|u_\varepsilon(x)-u\Big(x,\frac{x}{\varepsilon}\Big)\Big\|_{L_2(\Omega)}=0.$$

*Hint.* Pass to the limit in the integral $\int_\Omega \Big(u_\varepsilon(x)-u\Big(x,\frac{x}{\varepsilon}\Big)\Big)^2 dx$ as $\varepsilon\to 0$ and use the properties of two-scale convergence.

REMARK 9.2. A two-scale converging sequence $u_\varepsilon(x)$ satisfying (9.16) is called a *strongly two-scale converging sequence*.

By Exercise 9.7, if the strong two-scale limit $u(x,\xi)$ satisfies some additional regularity conditions (the continuity on $\Omega\times\square$), then $\Big\|u_\varepsilon(x)-u\Big(x,\frac{x}{\varepsilon}\Big)\Big\|_{L_2(\Omega)}\to 0$ as $\varepsilon\to 0$, i.e., the strong two-scale convergence condition can be expressed in terms of convergence in the $L_2(\Omega)$-norm. This fact is important in applications since the two-scale convergence, as well as the weak convergence, does not provide us with a sufficiently complete information concerning the asymptotic structure of $u_\varepsilon$, whereas such an information is often useful for describing real physical processes in inhomogeneous media.

We can weaken the continuity condition on $u(x,\xi)$ in $\Omega\times\square$. In fact, we need conditions on $u(x,\xi)$ that would guarantee that $u\Big(x,\frac{x}{\varepsilon}\Big)$ belongs to $L_2(\Omega)$ and

$$\lim_{\varepsilon\to 0}\int_\Omega u^2\Big(x,\frac{x}{\varepsilon}\Big)\,dx=\int_{\Omega\times\square}u^2(x,\xi)\,dx\,d\xi.$$

For this purpose, the *Carathéodory condition* $u(x,\xi)\in L_2(\Omega;C_{\mathrm{per}}(\square))$ is often used.

We now return to the proof of Theorem 9.6. We extend $N_j^p$ from $F_j$ to the entire periodicity cell $\square$ in such a way that the extended function belongs to the space $H^1_{\mathrm{per}}(\square)$ and its support lies in $\overline{F}_j\cup S$. In the integral identity (9.8), we set

$$\psi\equiv\psi_\varepsilon(x)\equiv u\Big(x,\frac{x}{\varepsilon}\Big)+\varepsilon\sum_{j=1}^k u_j^{(1)}\Big(x,\frac{x}{\varepsilon}\Big)+\delta_\varepsilon^{(1)}(x)+\delta_\varepsilon^{(2)}(x),$$

where

$$u(x,\xi)=\begin{cases}u_0(x,\xi), & \xi\in S,\\ u_j(x), & \xi\in F_j,\ j=1,\dots,k,\end{cases}$$

$u_j^{(1)}$ are defined by formula (9.14), and $\delta_\varepsilon^{(1)}$ and $\delta_\varepsilon^{(2)}$ are boundary layer functions in a neighborhood of $\partial\Omega$, such that $\psi_\varepsilon(x)\in \overset{\circ}{H}{}^1(\Omega)$; moreover, $\|\delta_\varepsilon^{(1)}\|_{H^1(\Omega)}\to 0$ as $\varepsilon\to 0$ and $\|\delta_\varepsilon^{(2)}\|_{L_2(\Omega)}\to 0$, $\varepsilon\|\nabla\delta_\varepsilon^{(2)}\|\to 0$ as $\varepsilon\to 0$, $\operatorname{supp}\delta_\varepsilon^{(2)}\subset\Omega_\varepsilon^s$. These functions will be written explicitly.

Under the above assumptions, it is easy to verify that $u\Big(x,\frac{x}{\varepsilon}\Big)$ and $u_j^{(1)}\Big(x,\frac{x}{\varepsilon}\Big)$, $j=1,\dots,k$, belong to the space $H^1(\Omega)$. Therefore, we can take $\psi_\varepsilon$ for a test function in (9.8).

The functions $\delta_\varepsilon^{(1)}$ and $\delta_\varepsilon^{(2)}$ "correct" nonzero values of the functions $\varepsilon\sum_{j=1}^k u_j^{(1)}$ and $f\Phi_0$ on $\partial\Omega$ so that the sum vanishes on $\partial\Omega$. For $\delta_\varepsilon^{(1)}(x)$ and $\delta_\varepsilon^{(2)}(x)$ we can take

$$\delta_\varepsilon^{(1)}(x)=-\varepsilon\Big[\sum_{j=1}^k u_j^{(1)}(x,x/\varepsilon)\Big]\gamma\Big(\frac{\rho(x)}{\varepsilon}\Big),\quad \delta_\varepsilon^{(2)}(x)=-f(x)\Phi_0\Big(\frac{x}{\varepsilon}\Big)\gamma\Big(\frac{\rho(x)}{\varepsilon}\Big),$$

where $\rho(x)$ is the distance from a point $x$ to $\partial\Omega$ and $\gamma(s) \geq 0$ belongs to $C_0^\infty(\mathbb{R})$ and is equal to 1 in a neighborhood of the origin.

EXERCISE 9.8. Verify that $\delta_\varepsilon^{(1)}$ and $\delta_\varepsilon^{(2)}$ satisfy the above-mentioned conditions.

In the integral identity (9.8), we take $\psi_\varepsilon(x)$ for a test function and pass to the limit as $\varepsilon \to 0$. Using the equality

$$\nabla_x \psi_\varepsilon(x) = \nabla_x u_j(x) + \nabla_\xi u_j^{(1)}\Big(x, \frac{x}{\varepsilon}\Big) + \text{“additional terms”} \quad (x \in F_\varepsilon^j)$$

and the limit relations

$$u_\varepsilon(x) \xrightarrow{2} u(x,\xi), \quad \chi_{\Omega\cap S^\varepsilon}(x)\varepsilon\nabla u_\varepsilon(x) \xrightarrow{2} \nabla_\xi u(x,\xi),$$
$$\chi_{\Omega\cap F_\varepsilon^j}(x)(\nabla u_\varepsilon(x)) \xrightarrow{2} (\nabla_x u_j(x) + \nabla_\xi u_j^{(1)}(x,\xi))\chi_{F_j}(\xi),$$

and taking into account that $u(x,\xi)$ and $u_j^{(1)}(x,\xi)$ are continuous and piecewise smooth, we find that

(9.17)
$$\int_{\Omega\times(\square\cap S)} |\nabla_\xi u_0(x,\xi)|^2\,dx\,d\xi + \sum_{j=1}^k \int_{\Omega\times(\square\cap F_j)} |\nabla_x u_j(x) + \nabla_\xi u_j^{(1)}(x,\xi)|^2\,dx\,d\xi$$
$$+ \lambda\int_{\Omega\times\square} u^2(x,\xi)\,dx\,d\xi = \int_{\Omega\times\square} f(x)u(x,\xi)\,dx\,d\xi.$$

On the other hand, passing to the limit on an arbitrary converging subsequence in the equality

$$\varepsilon^2\int_{\Omega_\varepsilon^s}|\nabla u_\varepsilon|^2 dx + \int_{\Omega_\varepsilon^h}|\nabla u_\varepsilon|^2 dx + \lambda\int_\Omega (u_\varepsilon)^2 dx = \int_\Omega f u_\varepsilon dx$$

(which follows from (9.8) if $\psi = u_\varepsilon$), we find

$$\begin{aligned}\varepsilon^2\lim_{\varepsilon\to0}\int_{\Omega_\varepsilon^s}|\nabla u_\varepsilon(x)|^2\,dx + \lim_{\varepsilon\to0}\int_{\Omega_\varepsilon^h}|\nabla u_\varepsilon(x)|^2\,dx + \lambda\lim_{\varepsilon\to0}\int_\Omega (u_\varepsilon(x))^2\,dx \\ = \int_{\Omega\times\square} f(x)u_0(x,\xi)\,dx\,d\xi.\end{aligned} \tag{9.18}$$

Since none of the terms on the left-hand side of (9.17) exceeds the corresponding term in (9.18), we have

$$\varepsilon^2\lim_{\varepsilon\to0}\int_{\Omega_\varepsilon^s}|\nabla u_\varepsilon(x)|^2\,dx = \int_{\Omega\times(\square\cap S)}|\nabla u_0(x,\xi)|^2\,dx\,d\xi,$$
$$\lim_{\varepsilon\to0}\int_{\Omega_\varepsilon^h}|\nabla u_\varepsilon(x)|^2\,dx = \sum_{j=1}^k\int_{\Omega\times(\square\cap F_j)}|\nabla_x u_j(x) + \nabla_\xi u_j^{(1)}(x,\xi)|^2\,dx\,d\xi,$$
$$\lim_{\varepsilon\to0}\int_\Omega (u_\varepsilon(x))^2\,dx = \int_{\Omega\times\square}(u(x,\xi))^2\,dx\,d\xi.$$

By Exercise 9.7 (note that for $f \in C^\infty(\overline{\Omega})$ the functions $u(x,\xi)$ and $u_j^{(1)}(x,\xi)$ are continuous), we obtain the required assertion. $\square$

### 9.3. Proof of the Nguetseng theorems.

PROOF OF THEOREM 9.1. By assumption, $\|u_\varepsilon\|_{L_2(\Omega)} \le C$. For any $\psi(x,\xi) \in C_{\rm per}(\overline{\Omega}, \square)$, where $C_{\rm per}(\overline{\Omega}, \square)$ is the space of continuous 1-periodic functions with respect to $\xi$ on $\overline{\Omega} \times \square$, we have

$$\left|\int_\Omega u_\varepsilon(x)\psi\Big(x, \frac{x}{\varepsilon}\Big)\,dx\right| \le C\Big\|\psi\Big(x, \frac{x}{\varepsilon}\Big)\Big\|_{L_2(\Omega)} \le C\|\psi(x,\xi)\|_{C_{\rm per}(\overline{\Omega},\square)}.$$

Here, $C_{\rm per}(\overline{\Omega}, \square)$ is a separable Banach space equipped with the norm $\|\psi\| = \sup|\psi(x,\xi)|$, where the supremum is taken over all $x \in \Omega$, $\xi \in \square$. Thus, for fixed $\varepsilon$ the integral on the left-hand side is a linear continuous functional, denoted by $\mu_\varepsilon$, on the space $C_{\rm per}(\overline{\Omega}, \square)$ :

$$\langle \mu_\varepsilon, \psi\rangle = \int_\Omega u_\varepsilon(x)\psi\Big(x, \frac{x}{\varepsilon}\Big)\,dx.$$

Consequently, from $\mu_\varepsilon$ we can extract a subsequence $\mu_{\varepsilon'}$ such that $\mu_{\varepsilon'} \to \mu_0$ as $\varepsilon' \to 0$ in the weak topology of the space $\big(C_{\rm per}(\overline{\Omega}, \square)\big)^*$, i.e., for any $\psi \in C_{\rm per}(\overline{\Omega}, \square)$ we have $\langle \mu_{\varepsilon'}, \psi\rangle \to \langle \mu_0, \psi\rangle$ or

$$\lim_{\varepsilon\to 0}\int_\Omega u_\varepsilon(x)\psi\Big(x, \frac{x}{\varepsilon}\Big)\,dx = \langle \mu_0, \psi\rangle.$$

EXERCISE 9.9. Let $\psi(x,\xi)$ be continuous on $\overline{\Omega} \times \square$ and 1-periodic with respect to $\xi$. Prove that

$$\lim_{\varepsilon\to 0}\Big\|\psi\Big(x, \frac{x}{\varepsilon}\Big)\Big\|_{L_2(\Omega)} = \|\psi(x,\xi)\|_{L_2(\Omega\times\square)}.$$

By Exercise 9.9,

$$\langle \mu_0, \psi\rangle \le C\lim_{\varepsilon\to 0}\Big\|\psi\Big(x, \frac{x}{\varepsilon}\Big)\Big\|_{L_2(\Omega)} = C\|\psi(x,\xi)\|_{L_2(\Omega\times\square)}.$$

Since $C_{\rm per}(\overline{\Omega}, \square)$ is dense in $L_2(\Omega\times\square)$, this estimate allows us to extend $\mu_0$ to $L_2(\Omega\times\square)$ as a continuous functional. By the Riesz theorem, there exists $u(x,\xi) \in L_2(\Omega\times\square)$ such that

$$\langle \mu_0, \psi\rangle = \int_{\Omega\times\square} u(x,\xi)\psi(x,\xi)\,dx\,d\xi, \quad u(x,\xi) \in L_2(\Omega\times\square).$$

Theorem 9.1 is proved. □

PROOF OF THEOREM 9.2. By Theorem 9.1 and the assumption $\|\nabla u_\varepsilon\|_{L_2(\Omega)} \le C$, there exists a vector-valued function $\chi_0(x,\xi)$ such that

$$\lim_{\varepsilon'\to 0}\int_\Omega \Big(\nabla u_{\varepsilon'}(x), \psi\Big(x, \frac{x}{\varepsilon'}\Big)\Big)dx = \int_{\Omega\times\square}\chi_0(x,\xi)\psi(x,\xi)\,dx\,d\xi \tag{9.19}$$

as $\varepsilon' \to 0$ for any vector-valued functions $\psi(x,\xi) \in (C_0^\infty(\Omega, C_{\rm per}^\infty(\square)))^d$ (hereinafter, $C_0^\infty(\Omega, C_{\rm per}^\infty(\square))$ denotes the space of functions that are infinitely smooth in $\Omega\times\square$, 1-periodic with respect to $\xi$, and compactly supported in $\Omega$). Integrating by parts, we find

$$\varepsilon\int_\Omega \Big(\nabla u_\varepsilon(x), \psi\Big(x, \frac{x}{\varepsilon}\Big)\Big)dx = -\int_\Omega u_\varepsilon(x)\Big[{\rm div}_\xi\psi\Big(x, \frac{x}{\varepsilon}\Big) + \varepsilon\,{\rm div}_x\psi\Big(x, \frac{x}{\varepsilon}\Big)\Big]dx. \tag{9.20}$$

Since the left-hand side and the second term on the right-hand side of (9.20) converge to zero as $\varepsilon \to 0$, we have

$$\lim_{\varepsilon\to 0} \int_\Omega u_\varepsilon(x)\, \mathrm{div}_\xi\, \psi\Big(x, \frac{x}{\varepsilon}\Big)\, dx = 0,$$

and consequently

$$\int_{\Omega\times\square} u_0(x,\xi)\, \mathrm{div}_\xi\, \psi(x,\xi)\, dx\, d\xi = 0,$$

where $u_0(x,\xi)$ is the two-scale limit of $u_\varepsilon$ on some subsequence $\varepsilon' \to 0$. The limit exists in view of Theorem 9.1. Since $\psi(x,\xi)$ is arbitrary, the last equality implies

$$\int_\square u_0(x,\xi)\, \mathrm{div}\, \Psi(\xi)\, d\xi = 0 \quad \text{for almost all } x \in \Omega$$

for any $\Psi(\xi) \in (C^\infty_{\mathrm{per}}(\square))^d$. To show that $u_0$ is independent of $\xi$, we introduce a distribution on $C^\infty_{\mathrm{per}}(\square)$ by the formula

$$\langle \overline{F}, \Phi\rangle = \int_\square u_0(x,\xi)\Phi(\xi)\, d\xi.$$

Then

$$\Big(\frac{\partial}{\partial \xi_k}\overline{F}, \Phi\Big) = -\int_\square u_0(x,\xi)\Big(\frac{\partial}{\partial\xi_k}\Phi(\xi)\Big)\, d\xi = 0.$$

Consequently, the weak derivatives of $u_0(x,\xi)$ with respect to $\xi_1,\dots,\xi_d$ vanish. Therefore, $u_0(x,\xi) = u(x) \in L_2(\Omega)$. Let $\psi(x,\xi)$ be a test vector-valued function such that $\mathrm{div}_\xi\, \psi(x,\xi) = 0$. Then

$$\int_\Omega u_\varepsilon(x)\, \mathrm{div}_x\, \psi(x,\xi)\Big|_{\xi = x/\varepsilon} dx \to -\int_{\Omega\times\square} (\chi_0(x,\xi), \psi(x,\xi))\, dx\, d\xi$$

as $\varepsilon \to 0$ because of (9.19) and (9.20).

On the other hand,

$$\int_\Omega u_\varepsilon(x)\, \mathrm{div}_x\, \psi(x,\xi)\, dx \to \int_{\Omega\times\square} u_0(x)\, \mathrm{div}_x\, \psi(x,\xi)\, dx\, d\xi$$

as $\varepsilon \to 0$. Thus,

$$\int_{\Omega\times\square} (\chi_0(x,\xi), \psi(x,\xi))\, dx\, d\xi = \int_{\Omega\times\square} u_0(x)\, \mathrm{div}_x\, \psi(x,\xi)\, dx\, d\xi$$

for any $\psi(x,\xi)$ such that $\mathrm{div}_\xi\, \psi(x,\xi) = 0$. The last relation implies $u_0(x) \in H^1(\Omega)$. Indeed, it can be regarded as a definition of the "action" of $u_0(x)$ on the derivatives of the test functions. It is not difficult to derive the estimate (check it!)

$$\left|\int_\Omega u_0(x)\frac{\partial\varphi}{\partial x_i}\, dx\right| \le C\|\varphi\|_{L_2(\Omega)}, \quad i = 1,\dots,d,$$

which implies that $u_0(x)$ belongs to the space $H^1(\Omega)$. Integrating by parts on the right-hand side of the last equality, we find

$$\int_{\Omega\times\square} ([\chi_0(x,\xi) - \nabla_x u_0(x)], \psi(x,\xi))\, dx\, d\xi = 0.$$

We see that the vector-valued function $\chi_0(x,\xi) - \nabla_x u_0(x)$ is orthogonal in $(L_2(\square))^d$ to the subspace of vector-valued functions $\Psi(\xi)$ such that $\mathrm{div}_\xi\, \Psi(\xi) = 0$. Therefore, there exists $u_j^{(1)}(x,\xi) \in L_2(\Omega, H^1_{\mathrm{per}}(\square))$ such that $\nabla_\xi u_j^{(1)}(x,\xi) = \chi_0(x,\xi) - \nabla_x u_0(x)$ for almost all $x \in \Omega$. The existence of $u_j^{(1)}(x,\xi)$ is established in the same way as in Subsection 4.2 in the proof of the De Rham theorem. However, if the functions under consideration are defined on a torus, the proof of the existence of a potential can be significantly simplified by using the Fourier expansion. Then the problem of finding the "potential" of $\overline{v}(\xi)$ orthogonal to all solenoidal vectors is reduced to the algebraic problem for the coefficients $A_n$ in the expansion $V(\xi) = \sum_{n\in\mathbb{Z}^d\setminus 0} A_n e^{i(\xi,n)}$ of the required potential $V(\xi)$ such that $\nabla V = \overline{v}(\xi)$ in $\square$ (check it!). Furthermore, it is easy to obtain the estimate $\|V\|_{H^1(\square)} \le C\|\overline{v}\|_{L_2(\square)}$ (the Poincaré inequality in a torus), where the constant $C > 0$ is independent of $\overline{v}(\xi)$.

By the Fubini theorem, the function $\|\overline{\chi}_0(x,\xi) - \nabla_x u_0(x)\|_{L_2(\square)}$ is square integrable with respect to $x \in \Omega$. Therefore, $u_j^{(1)}(x,\xi) \in L_2(\Omega, H^1(\square))$. Theorem 9.2 is proved. $\square$

PROOF OF THEOREM 9.3. By the uniform boundedness of $\varepsilon\nabla u_\varepsilon$ in $L_2(\Omega)$ and Theorem 9.1, we can extract a subsequence $\varepsilon'$ (the prime will be omitted) such that for any test function $\psi(x,\xi)$,

$$\int_\Omega \varepsilon\nabla u_\varepsilon \psi\Big(x, \frac{x}{\varepsilon}\Big)\, dx \to \int_{\Omega\times\square} \chi(x,\xi)\psi(x,\xi)\, dx\, d\xi \tag{9.21}$$

as $\varepsilon \to 0$, where $\chi(x,\xi) \in L_2(\Omega\times\square)$. On the other hand, integrating by parts on the left-hand side of (9.21), we find

$$\int_\Omega u_\varepsilon \Big(\operatorname{div}_\xi \psi\Big(x,\frac{x}{\varepsilon}\Big) + \varepsilon \operatorname{div}_x \psi\Big(x,\frac{x}{\varepsilon}\Big)\Big)\, dx \to \int_{\Omega\times\square} u_0(x,\xi) \operatorname{div}_\xi \psi(x,\xi)\, dx\, d\xi$$

as $\varepsilon \to 0$, where $u_0(x,\xi)$ is the two-scale limit of $u_\varepsilon(x)$. Therefore,

$$\int_{\Omega\times\square} (\chi(x,\xi) - \nabla_\xi u_0(x,\xi))\,\psi(x,\xi)\, dx\, d\xi = 0 \quad \forall\, \psi(x,\xi),$$

which implies the assertion of Theorem 9.3. Note that the last equality can be justified by methods of distribution theory in the same way as in the proof of Theorem 9.2. $\square$

PROOF OF THEOREM 9.4. By Theorem 9.1, from $u_\varepsilon$ we can extract a two-scale converging subsequence $u_{\varepsilon'}$ such that $u_{\varepsilon'}(x) \overset{2}{\to} u_0(x,\xi)$ (the prime will be omitted). Define $z_\varepsilon^i(x) \in L_2(\Omega)$ by the formula

$$z_\varepsilon^i(x) = \begin{cases} \frac{\partial u_\varepsilon}{\partial x_i} & \text{in } \square^h, \\ 0 & \text{in } \square^s. \end{cases}$$

Then for any $v \in L_2(\Omega)$,

$$\int_\Omega z_\varepsilon^i v\, dx = \int_{\Omega_\varepsilon^h} \frac{\partial u_\varepsilon}{\partial x_i} v\, dx. \tag{9.22}$$

By Theorem 9.1, from $z_\varepsilon^i$ we can extract a subsequence $z_{\varepsilon'}^i$ (the prime will be omitted) such that

$$\int_\Omega z_\varepsilon^i w^\varepsilon \varphi\, dx \to \int_{\Omega\times\square^h} z^i(x,\xi) w(\xi)\varphi(x)\, dx\, d\xi, \tag{9.23}$$

where $z^i(x,\xi) \in L_2(\Omega\times\square)$, $w^\varepsilon = w(\frac{x}{\varepsilon})$, and $w(\xi) \in C^\infty_{\rm per}(\square)$. Taking $w(\xi)$ with support in $\square^s$ as a test function, we see that $z^i(x,\cdot) = 0$ outside $\square^h$.

For a test function in (9.22) we take $v = \varepsilon w^\varepsilon \varphi$, where $\varphi(x) \in C_0^\infty(\Omega)$ and $\operatorname{supp} w(\xi) \in \square^h$, and integrate by parts:

$$\varepsilon \int_\Omega z_\varepsilon^i w^\varepsilon \varphi\, dx = -\int_\Omega u_\varepsilon \frac{\partial w}{\partial \xi_i}\Big(\frac{x}{\varepsilon}\Big)\varphi\, dx - \varepsilon \int_\Omega u_\varepsilon w^\varepsilon \frac{\partial \varphi}{\partial x_i}\, dx. \tag{9.24}$$

Passing to the limit as $\varepsilon \to 0$, we find

$$\int_{\Omega\times\square} u_0(x,\xi)\frac{\partial w}{\partial \xi_i}(\xi)\varphi(x)\, dx\, d\xi = 0 \quad \forall\, w(\xi) \in C_{\text{per}}^\infty(\square),\ \operatorname{supp} w \in \square^h.$$

Since $E$ is connected, we have $u_0(x,\xi) = u(x)$ for $\xi \in \square^h$.

EXERCISE 9.10. Prove that $u \in H^1(\Omega)$.

*Hint.* Let $\overline{\psi}(x,\xi) = (\varphi(x)\Phi_1(\xi), \dots, \varphi(x)\Phi_d(\xi))$, $\operatorname{supp}\overline{\psi}(x,\cdot) \in \square^h$ for all $x \in \Omega$, and $\operatorname{div}_\xi \overline{\psi}(x,\xi) = 0$. We set

$$\overline{\zeta} \equiv \int_{\square^h} \overline{\Phi}(\xi)\, d\xi, \quad \overline{\Phi}(\xi) = (\Phi_1(\xi), \dots, \Phi_d(\xi)).$$

Then the following equality holds (check it!):

$$\int_\Omega u(x)(\nabla\varphi(x), \overline{\zeta})\, dx = -\int_{\Omega\times\square^h} \big(\overline{z}(x,\xi), \overline{\psi}(x,\xi)\big)\, dx\, d\xi,$$

where $\overline{z}(x,\xi) = (z^1(x,\xi), \dots, z^d(x,\xi))$ is the two-scale limit of the vector-valued functions

$$\overline{z}_\varepsilon(x) = \begin{cases} \nabla u_\varepsilon(x), & x \in \Omega_\varepsilon^h, \\ 0, & x \in \Omega_\varepsilon^s, \end{cases}$$

$u_0(x,\xi) = u(x)$ for $\xi \in \square^h$, and $u_0(x,\xi)$ is the two-scale limit of $u_\varepsilon(x)$.

The last equality determines the "action" of the distribution $\nabla u(x)$ on smooth compactly supported functions; $u(x) \in H^1(\Omega)$ if $\overline{\zeta} \in \mathbb{R}^d$ can be chosen arbitrarily, which is possible because the set $E$ is connected (it has already been discussed in Section 8).

We prove that $z^i(x,\xi)$ is represented at $x \in \Omega$ in the form

$$z^i(x,\xi) = \frac{\partial u}{\partial x_i}(x) + \frac{\partial u^{(1)}}{\partial \xi_i}(x,\xi) \tag{9.25}$$

for $\xi \in \square^h$, where $u^{(1)}(x,\xi) \in L_2(\Omega, H_{\text{per}}^1(\square^h))$. As before, denote by $\overline{z}(x,\xi)$ a vector-valued function with components $z^1, \dots, z^d$. Suppose that $\overline{w}(\xi)$ is a vector-valued function in $H^1(\square^h)$ with support in $\square^h$ and $\operatorname{div}_\xi \overline{w}(\xi) = 0$. Then from (9.24) it follows that

$$\int_\Omega (\overline{z}_\varepsilon, \overline{w}^\varepsilon)\varphi\, dx = -\int_\Omega u_\varepsilon(\overline{w}^\varepsilon, \nabla_x\varphi)\, dx.$$

Passing to the limit as $\varepsilon \to 0$ and using the properties of two-scale convergence, we find

$$\int_\Omega \Big(\int_{\square^h} (\overline{z}(x,\xi), \overline{w}(\xi))\, d\xi\Big)\varphi(x)\, dx = -\int_\Omega u(x)\Big(\int_{\square^h} \overline{w}(\xi)\, d\xi\Big)\nabla_x\varphi(x)\, dx$$

or, taking into account that $u(x)$ belongs to $H^1(\Omega)$,

$$\int_{\Omega\times\square^h} [(\overline{z}(x,\xi) - \nabla_x u), \overline{w}(\xi)]\varphi(x)\, dx\, d\xi = 0.$$

Since $\varphi(x) \in C_0^\infty(\Omega)$ is arbitrary for almost all $x \in \Omega$, we have

$$\int_{\square^h} (\overline{z} - \nabla_x u)\overline{w}(\xi)\, d\xi = 0$$

for any $\overline{w}(\xi) \in L_2(\square^h)$ such that $\operatorname{div} \overline{w} = 0$. Therefore, for almost all $x \in \Omega$ there exists $u^{(1)}(x,\xi) \in H^1_{\text{per}}(\square^h)$ such that $\nabla_\xi u^{(1)}(x,\xi) = \overline{z} - \nabla_x u(x)$. Choosing, for every $x$, an additive constant such that $\int_{\square^h} u^{(1)}(x,\xi)\, d\xi = 0$ and using the Poincaré inequality, we have $\|u^{(1)}\|_{L_2(\square^h)} \le C\|\overline{z} - \nabla_x u\|_{L_2(\square^h)}$. Hence $u_j^{(1)}(x,\xi) \in L_2(\Omega, H^1_{\text{per}}(\square^h))$. Theorem 9.4 is proved. □

EXERCISE 9.11. Prove that $u \in \overset{\circ}{H}{}^1(\Omega)$ if $u_\varepsilon \in \overset{\circ}{H}{}^1(\Omega)$.
*Hint.* Passing to the limit as $\varepsilon \to 0$, prove the equality

$$\int_\Omega \frac{\partial u}{\partial x_i}\varphi\, dx = -\int_\Omega u\frac{\partial \varphi}{\partial x_i}\, dx \quad \forall \varphi \in C^\infty(\overline{\Omega}),$$

which implies $u \in \overset{\circ}{H}{}^1(\Omega)$.

## 10. Homogenization of thin structures and singular measures

The asymptotic behavior of thin structures, such that, e.g., shells, plates, thin films, rod structures, skeletons, and so on, is widely discussed in the mathematical, as well as physical and engineering literature. As a rule, the goal of an asymptotic analysis is to reduce dimension, i.e., to reduce the original problem to a problem on some structure of less dimension. For example, the equation on a thin three-dimensional plate is replaced with an equation on a two-dimensional domain, the system describing a rod construction is reduced to a family of ordinary differential equations, and so on.

The reduced system is usually simpler, especially from the point of view of numerical analysis. Indeed, with a smaller number of variables, we essentially decrease the amount of computations.

Of course, a procedure of dimension reduction must be justified. In the case of thin structures, it is also necessary to verify the commutativity of the corresponding diagram.

Asymptotic methods for such problems are well known in the literature, but, as a rule, they can be used under strict restrictions on the geometry and smoothness of the structure.

There are numerous works in mechanics where asymptotics for different special models was obtained at the physical level of rigor.

From the mathematical point of view, it is of great interest to construct a general homogenization theory of periodic measures, including singular measures, partially singular measures, as well as measures converging to singular ones. A successful attempt to create such a theory was made in [**132, 133, 25, 134**], and [**24**]. In this section, following these works, we define Sobolev spaces with an arbitrary measure, introduce the notion of convergence in variable spaces, and explain the meaning of equations in spaces with measure.

We present a natural sufficient condition for the existence of the homogenized operator and the validity of the homogenization theorem (the so-called connectedness or ergodicity of a measure). If this condition is satisfied, the two-scale convergence method can be adapted to the problems under consideration.

To clarify the idea of introducing Sobolev spaces with measure, we consider a simple example. Let $\mu$ be a positive finite Borel measure in a smooth bounded domain $G$. Consider the variational problem

$$\inf_{\varphi \in C_0^\infty(G)} \int_G (a(x)\nabla\varphi(x) \cdot \nabla\varphi(x) + \varphi^2(x) - 2f(x)\varphi(x))d\mu(x),$$

where $a(x)$ ia a continuous positive definite coefficient matrix in $\overline{G}$ and $f(x)$ is a continuous function in $\overline{G}$. Our goal is to introduce a Sobolev space with measure $\mu$ in such a way that the minimum is attained and a minimizer is found as a solution to the corresponding Euler equation.

## 10.1. Spaces with measure.

10.1.1. *Sobolev spaces.* Let $\mu(x)$ be a positive finite Borel measure on a standard $n$-dimensional torus $\mathbb{T}^n \equiv \mathbb{R}^n/\mathbb{Z}^n$. We identify $\mu$ with the corresponding periodic measure in $\mathbb{R}^n$. Without loss of generality, we can assume that

$$\int_{\mathbb{T}^n} d\mu(x) = 1.$$

Recall that $L_2(\mathbb{T}^n, \mu)$ is the space of Borel measurable functions $g(x)$ such that

$$\int_{\mathbb{T}^n} g^2(x)d\mu(x) < \infty.$$

The space $L_2(\mathbb{T}^n, \mu)$ equipped with the inner product

$$(g_1, g_2) = \int_{\mathbb{T}^n} g_1(x)g_2(x)d\mu(x), \quad g_1, g_2 \in L_2(\mathbb{T}^n, \mu),$$

is a Hilbert space.

Define the Sobolev space equipped with the measure $\mu$.

DEFINITION 10.1. We say that a function $u \in L_2(\mathbb{T}^n, \mu)$ belongs to the space $H^1(\mathbb{T}^n, \mu)$ if there exists a vector-valued function $z \in (L_2(\mathbb{T}^n, \mu))^n$ and a sequence $\varphi_k \in C^\infty(\mathbb{T}^n)$ such that

$$\varphi_k \longrightarrow u \quad \text{in } L_2(\mathbb{T}^n, \mu) \text{ as } k \to \infty,$$
$$\nabla\varphi_k \longrightarrow z \quad \text{in } (L_2(\mathbb{T}^n, \mu))^n \text{ as } k \to \infty.$$

The function $z(x)$ is called the *gradient* or *$\mu$-gradient* of $u(x)$ and is denoted by $\nabla^\mu u$.

Similarly, we can define the spaces $H^1(\mathbb{R}^n, \mu)$, $H^1_{\text{loc}}(\mathbb{R}^n, \mu)$, and also the space $H^1(G, \mu)$ for an arbitrary domain $G \subset \mathbb{R}^n$ and a (locally) finite Borel measure $\mu$ on $G$.

Generally speaking, the gradient of a function of class $H^1(\mathbb{T}^n, \mu)$ is not unique. In particular, the zero function may have a nontrivial gradient. To illustrate this fact, we consider the following example.

EXAMPLE 10.1. In the square $[-1/2, 1/2]^2$, consider the segment $\{-1/4 \le x_1 \le 1/4, x_2 = 0\}$ and introduce a measure $\mu(x)$ as Lebesgue measure on this segment:

$$d\mu = 2\chi(x_1)\, dx_1 \times \delta(x_2), \tag{10.1}$$

where $\chi(t)$ is the characteristic function of the segment $[-\frac{1}{4}, \frac{1}{4}]$ and $\delta(t)$ is the Dirac mass concentrated at zero. Let $\psi(x)$ be a smooth function with support in a neighborhood of the segment. Assume that $\psi(x)$ coincides with a function of the form $\theta(x_1)x_2$ in a sufficiently small neighborhood of the segment. Since $\psi = 0$ on the support of $\mu$, we have $\psi = 0$, where $\psi$ is regarded as an element of $L_2(\mathbb{T}^2, \mu)$. Choosing $\varphi_k(x) = \psi(x)$ for all $k$ in Definition 10.1, we find $z(x) = \nabla^\mu \psi(x) = (0, \theta(x_1))$. Thus, any vector-valued function of the form $(0, \theta(x_1))$ with smooth $\theta(s)$ serves as the $\mu$-gradient of the zero function. In fact, this assertion is valid for any $\theta(s)$ in $L_2$ (check it!).

EXERCISE 10.1. Prove that the gradients of zero form a closed subspace of $(L_2(\mathbb{T}^n, \mu))^n$. The set of the gradients of any $H^1(\mathbb{T}^n, \mu)$-function is the sum of one of its gradients and the set of the gradients of zero.

The subspace of the gradients of zero is denoted by $\Gamma_\mu(0)$.

EXERCISE 10.2. Prove that pairs of the form $\{(u, z)\}$, where $u \in H^1(\mathbb{R}^n, \mu)$ and $z = \nabla^\mu u$, form a closed subspace of $(L_2(\mathbb{R}^n, \mu))^{n+1}$. If $u_k \to u_0$ in $L_2(\mathbb{R}^n, \mu)$ and $z_k = \nabla^\mu u_k \to z_0$ in $(L_2(\mathbb{R}^n, \mu))^n$ as $k \to \infty$, then $u_0 \in H^1(\mathbb{R}^n, \mu)$ and $z_0 = \nabla^\mu u_0$.

This subspace, denoted by $H$, will be used in the further consideration. We set

$$H(\mathbb{R}^n, \mu) = \{(u, z) : u \in H^1(\mathbb{R}^n, \mu), z = \nabla^\mu u)\}. \tag{10.2}$$

The space $H(\mathbb{T}^n, \mu)$ is introduced in a similar way.

EXAMPLE 10.2 (Segment, Figure 10.1). Consider the space $H^1(\mathbb{T}^n, \mu)$ (or $H^1(\mathbb{R}^n, \mu)$) in the case where $\mu$ is the one-dimensional Lebesgue measure on a segment $I$. Without loss of generality, we assume that $I = \{x \in \mathbb{R}^n : 0 \leq x_1 \leq a, x_2 = x_3 = \cdots = x_n = 0\}$.

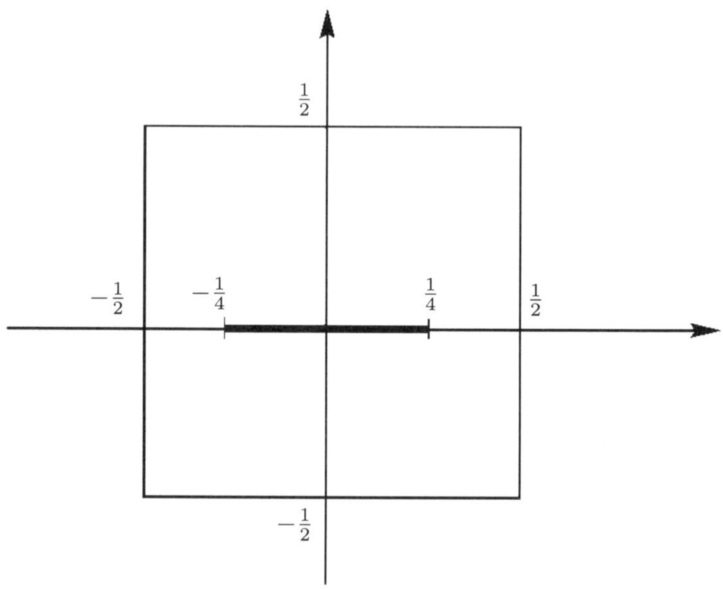


FIGURE 10.1

PROPOSITION 10.1. *The space $H^1(\mathbb{T}^n, \mu)$ consists of all Borel functions $u(x)$ such that $u(s, 0, 0, \ldots, 0) \in H^1(0, a)$. Moreover,*

$$\nabla^\mu u(x) = (u'_{x_1}(x_1, 0), \psi_2(x_1), \ldots, \psi_n(x_1)),$$

*where* $u'_{x_1} \equiv \frac{d}{ds}u(s,0,0,\dots,0)\Big|_{s=x_1}$, *and* $\psi_2, \psi_3, \dots, \psi_n$ *are arbitrary functions in* $L_2(0,a)$.

REMARK 10.1. Since the measure $\mu$ of the complement of the segment $I$ is equal to zero, it suffices to define $u$ as an element of $L_2(\mathbb{T}^n, \mu)$ only on the segment $I$.

EXERCISE 10.3. Prove Proposition 10.1.

Consider several examples where the structure of the space $H^1(\mathbb{T}^n, \mu)$ is easily determined. The proof of the corresponding assertions is left to the reader.

EXAMPLE 10.3 ("Hedgehog"; Figure 10.2). Consider segments $I_1, I_2, \dots, I_N$ starting at the origin and directed along vectors $v_1, v_2, \dots, v_N$. Suppose that $\frac{v_i}{|v_i|} \neq \frac{v_j}{|v_j|}$ for $i \neq j$. Let $\mu_1, \mu_2, \dots, \mu_N$ be the standard one-dimensional Lebesgue measures on the segments $I_1, \dots, I_N$ respectively, and let $\lambda_1, \dots, \lambda_N$ be arbitrary positive numbers. We set

$$\mu = \sum_{j=1}^{N} \lambda_j \mu_j.$$

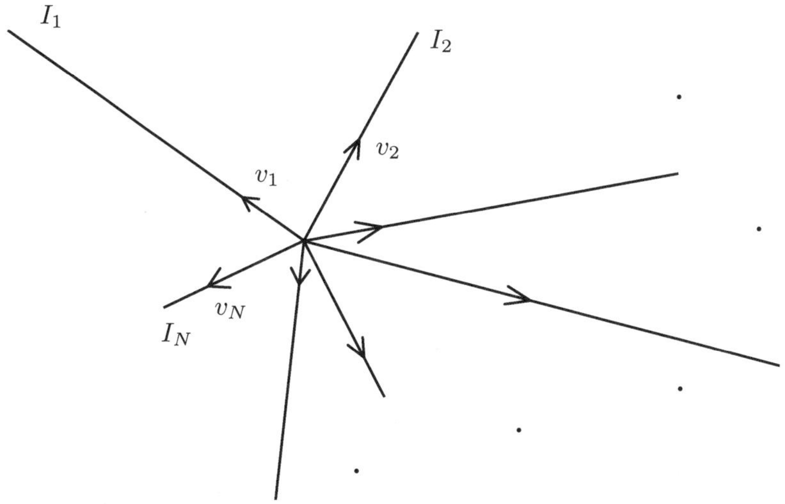


FIGURE 10.2

EXERCISE 10.4. Show that a Borel function $u(x)$ belongs to the space $H^1(\mathbb{T}^n, \mu)$ if and only if its restriction to each segment $I_j$ is an $H^1$-function of a single variable and the values of the restricted functions at the origin coincide for all segments (recall that an $H^1$-function of a single variable is continuous).

EXAMPLE 10.4 (Reinforced shells; Figure 10.3). Suppose a segment $I$ and a measure $\mu$ are the same as in Example 10.2. Let $dx$ be the standard Lebesgue measure on a two-dimensional torus $\mathbb{T}^2$. Introduce a measure $\tilde{\mu}$ on $\mathbb{T}^2$ by the equality $d\tilde{\mu}(x) = d\mu(x) + dx$.

EXERCISE 10.5. Prove that a Borel function $u(x)$ belongs to the space $H^1(\mathbb{T}^2, \tilde{\mu})$ if and only if $u \in H^1(\mathbb{T}^2)$ and the restriction (trace) of $u(x)$ to the segment $I$ is an $H^1$-function of a single variable.

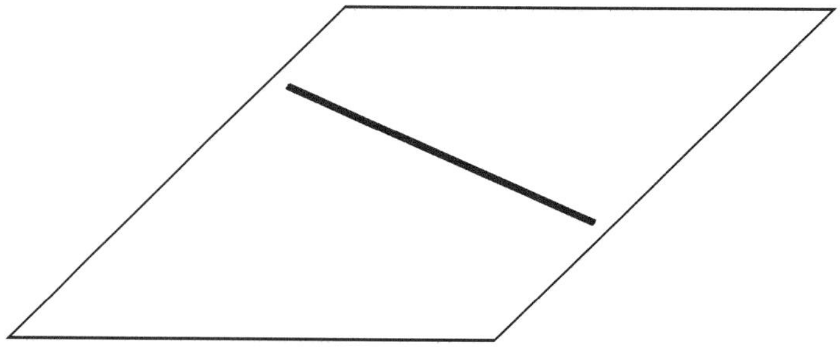

FIGURE 10.3

REMARK 10.2. Unlike the two-dimensional case, in the case of dimension greater than or equal to 3, the traces of an $H^1$-function on one-dimensional segments are not defined. Therefore, the space $H^1(\mathbb{T}^n, \tilde{\mu})$, $n \geq 3$, is isomorphic to the direct sum of the spaces $H^1(\mathbb{T}^n)$ and $H^1(\mathbb{T}^n, \mu)$ (check it!).

EXAMPLE 10.5 (Junction; Figure 10.4). Suppose that $G = [-\frac{1}{8}, \frac{1}{8}]^3$,

$$\Pi = \left\{x \in \left[-\frac{1}{2}, \frac{1}{2}\right]^3 : x_3 = 0, \ (x_1, x_2) \in \left[0, \frac{3}{8}\right] \times \left[-\frac{1}{8}, \frac{1}{8}\right]\right\},$$

$$I = \left\{x \in \left[-\frac{1}{2}, \frac{1}{2}\right]^3 : x_3 = x_2 = 0, \ \frac{1}{4} \leq x_1 \leq \frac{1}{2}\right\},$$

$\mu_1$ is the standard Lebesgue measure on the segment $I$, $\mu_2$ is the planar Lebesgue measure on $\Pi$, and $dx$ is the spatial Lebesgue measure in the cube $G$. Introduce the measure

$$d\mu(x) = d\mu_1(x) + d\mu_2(x) + dx.$$

Note that the trace of an $H^1(G)$-function is defined on $G \cap \Pi$ and is an element of the space $H^{1/2}(\Pi)$. The trace of an $H^1(\Pi)$-function on the segment $\Pi \cap I$ is also well defined.

EXERCISE 10.6. Describe the structure of the space $H^1(\mathbb{T}^3, \mu)$.

10.1.2. *Convergence in variable spaces.* When studying homogenization problems in spaces with measure, we deal with a family of measures obtained by scaling (sometimes, in addition, smoothing) a given periodic measure. We will need the definition of convergence of a sequence in variable function spaces.

Suppose that a sequence of measures $\mu_k$, $k = 1, 2 \ldots$, in $\mathbb{R}^n$ (or in a bounded domain $G$) weakly converges to a measure $\mu$ as $k \to \infty$.

DEFINITION 10.2. We say that a sequence of functions $\{g_k\}$, $g_k \in L_2(\mathbb{R}^n, \mu_k)$, *weakly converges* in $L_2(\mathbb{R}^n, \mu_k)$ to a function $g \in L_2(\mathbb{R}^n, \mu)$ as $k \to \infty$ if $\|g_k\|_{L_2(\mathbb{R}^n, \mu)} \leq C$ and for any $\varphi \in C_0^\infty(\mathbb{R}^n)$ the following limit relation holds:

$$\lim_{k \to \infty} \int_{\mathbb{R}^n} g_k(x)\varphi(x)d\mu_k(x) = \int_{\mathbb{R}^n} g(x)\varphi(x)d\mu(x).$$

A sequence $\{g_k\}$ is said to be *strongly converging* to $g(x)$ in $L_2(\mathbb{R}^n, \mu_k)$ as $k \to \infty$ if it weakly converges and

$$\lim_{k \to \infty} \int_{\mathbb{R}^n} g_k(x)h_k(x)d\mu_k(x) = \int_{\mathbb{R}^n} g(x)h(x)d\mu(x)$$

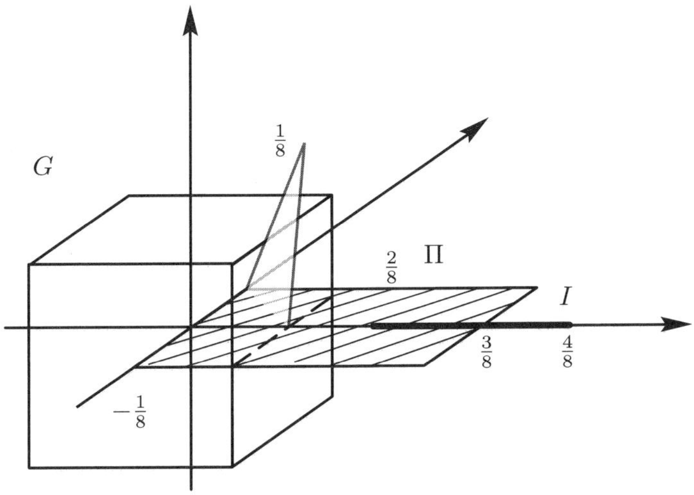


FIGURE 10.4

for any sequence $\{h_k(x)\}$ weakly converging to $h(x) \in L_2(\mathbb{R}^n, \mu)$ in $L_2(\mathbb{R}^n, \mu_k)$.

EXERCISE 10.7. Let $\{g_k\}$ weakly converge to $g(x)$ in $L_2(\mathbb{R}^n, \mu_k)$ as $k \to \infty$. Then the sequence $\{g_k\}$ strongly converges if and only if

$$\lim_{k\to\infty} \|g_k\|_{L_2(\mathbb{R}^n,\mu_k)} = \|g\|_{L_2(\mathbb{R}^n,\mu)}.$$

As an example we consider a sequence of measures obtained by scaling a fixed periodic measure. Let $\mu$ be an arbitrary periodic Borel measure such that $\mu(\mathbb{T}^n) = 1$. Denote $\mu_\varepsilon(x) = \varepsilon^n \mu\left(\frac{x}{\varepsilon}\right)$. By definition, $\mu_\varepsilon(B) = \varepsilon^n \mu(\varepsilon^{-1}B)$ for an arbitrary Borel set $B$.

EXERCISE 10.8. Prove that the measures $\mu_\varepsilon(x)$ weakly converge to the standard Lebesgue measure $dx$ in $\mathbb{R}^n$.

The important property of the weak compactness of a bounded set in a separable Hilbert space remains valid with respect to the convergence in variable spaces.

LEMMA 10.1. *Let $\{\mu_k\}$ be a sequence of Borel measures in $\mathbb{R}^n$ weakly converging to a measure $\mu$. Then from any sequence $g_k(x)$ that satisfies the estimate*

$$\|g_k\|_{L_2(\mathbb{R}^n,\mu_k)} \leq C$$

*we can extract a subsequence weakly converging to some function $g(x) \in L_2(\mathbb{R}^n, \mu)$.*

This assertion can be proved in the same way as in the classical case where the sequence is considered in a fixed space. The proof is left to the reader.

10.1.3. *Potential and solenoidal vector-valued functions.* In the study of homogenization problems, an important role is played by potential and solenoidal functions. Below we give definitions of these functions and discuss their main properties. Throughout this subsection, we assume that $\mu$ is a Radon measure in $\mathbb{R}^n$ or on a torus $\mathbb{T}^n$.

DEFINITION 10.3. The space $L_2^{\text{pot}}(\mathbb{R}^n, \mu)$ of potential vector-valued functions is the closure of the linear set $\{\nabla\varphi : \varphi \in C_0^\infty(\mathbb{R}^n)\}$ in the $(L_2(\mathbb{R}^n, \mu))^n$-norm. Similarly, $(L_2^{\text{pot}}(\mathbb{T}^n, \mu))$ is the closure of the set $\{\nabla\varphi : \varphi \in C^\infty(\mathbb{T}^n)\}$ in the $(L_2(\mathbb{T}^n, \mu))^n$-norm.

DEFINITION 10.4. The space $L_2^{\text{sol}}(\mathbb{R}^n, \mu)$ of solenoidal vector-valued functions is the orthogonal complement of the space $L_2^{\text{pot}}(\mathbb{R}^n, \mu)$ in $(L_2(\mathbb{T}^n, \mu))^n$. Similarly, the space $L_2^{\text{pot}}(\mathbb{T}^n, \mu)$ is the orthogonal complement of $L_2^{\text{pot}}(\mathbb{T}^n, \mu)$ in $(L_2(\mathbb{T}^n, \mu))^n$.

10.1.4. *Smoothing a periodic measure. Approximations.* Let $K(x)$ be a nonnegative $C_0^\infty$-function such that $\int_{\mathbb{R}^n} K(x)dx = 1$ and $K(-x) = K(x)$. For an arbitrary Radon measure $\mu(x)$ in $\mathbb{R}^n$ or on $\mathbb{T}^n$ (the measure on $\mathbb{T}^n$ can be represented as a periodic measure in $\mathbb{R}^n$) we set

$$d\mu^\delta(x) \equiv d\mathcal{K}^\delta(\mu)(x) = \rho^\delta(x)dx, \tag{10.3}$$

where

$$\rho^\delta(x) = \delta^{-n} \int_{\mathbb{R}^n} K\Big(\frac{x-y}{\delta}\Big) d\mu(y). \tag{10.4}$$

The measures $\mu^\delta$ locally weakly converge in $\mathbb{R}^n$ to the measure $\mu$. Indeed, for an arbitrary continuous function $\varphi(x)$ with compact support,

$$\begin{aligned}
&\Big|\int_{\mathbb{R}^n} \varphi(x)d\mu^\delta(x) - \int_{\mathbb{R}^n} \varphi(x)d\mu(x)\Big| \\
&\quad = \Big|\int_{\mathbb{R}^n} \delta^{-n}\varphi(x)K\Big(\frac{x-y}{\delta}\Big)d\mu(y)dx - \int_{\mathbb{R}^n} \delta^{-n}\varphi(y)K\Big(\frac{x-y}{\delta}\Big)d\mu(y)dx\Big| \\
&\quad \le \int_{\mathbb{R}^n} \delta^{-n}|\varphi(x) - \varphi(y)|\, K\Big(\frac{x-y}{\delta}\Big)d\mu(y)dx \\
&\quad \le \max_{|x-y|\le C_K\delta} |\varphi(x) - \varphi(y)|
\end{aligned}$$

as $\delta \to 0$, where the constant $C_K$ is chosen in such a way that $K(y) = 0$ for $|y| > C_K$. The right-hand side of the last formula converges to zero as $\delta \to 0$ since $\varphi(x)$ is uniformly continuous.

We also introduce an ordinary smoothing operator by the equality

$$\varphi^\delta(x) \equiv (\tilde{\mathcal{K}}^\delta\varphi)(x) = \delta^{-n} \int_{\mathbb{R}^n} K\Big(\frac{y}{\delta}\Big)\varphi(x-y)dy. \tag{10.5}$$

Then

$$\int_{\mathbb{R}^n} \varphi^\delta(x)d\mu(x) = \int_{\mathbb{R}^n} \varphi(x)d\mu^\delta(x) \tag{10.6}$$

(check it!). The following assertion allows us to define a special smoothing operator agreed with the measure $\mu$.

LEMMA 10.2. *For every $v \in L_2(\mathbb{R}^n, \mu)$ there is $v_\delta \in L_2(\mathbb{R}^n, \mu^\delta)$ such that*

$$\int_{\mathbb{R}^n} v_\delta(x)\varphi(x)d\mu^\delta(x) = \int_{\mathbb{R}^n} v(x)\varphi^\delta(x)d\mu(x) \tag{10.7}$$

*for all $\varphi \in C_0(\mathbb{R}^n)$. The family $v_\delta(x)$ strongly converges to $v(x)$ in $L_2(\mathbb{R}^n, \mu^\delta)$ as $\delta \to 0$.*

PROOF. Denote

$$F_\delta(\varphi) = \int_{\mathbb{R}^n} v(x)\varphi^\delta(x)d\mu(x).$$

Then for all $\varphi \in C_0(\mathbb{R}^n)$,

$$F_\delta^2(\varphi) \le \left(\int_{\mathbb{R}^n} v^2(x)d\mu(x)\right)\int_{\mathbb{R}^n} (\varphi^\delta)^2 d\mu(x).$$

Denote

$$C(v) = \int_{\mathbb{R}^n} v^2(x)d\mu(x).$$

By the Jensen integral inequality,

$$\int_{\mathbb{R}^n} (\varphi^\delta(x))^2 d\mu(x) \le \int_{\mathbb{R}^n} (\varphi^2(x))^\delta d\mu(x).$$

Hence

$$F_\delta^2(\varphi) \le C(v)\int_{\mathbb{R}^n} (\varphi^2(x))^\delta d\mu(x) = C(v)\int_{\mathbb{R}^n} \varphi^2(x)d\mu^\delta(x).$$

We used the equality (10.6). Therefore, the linear functional $F_\delta(\varphi)$ is bounded in the space $L_2(\mathbb{R}^n, \mu^\delta)$ and, by the Riesz representation theorem, $F_\delta(\varphi)$ is representable in the form

$$F_\delta(\varphi) = \int_{\mathbb{R}^n} v_\delta(x)\varphi(x)d\mu^\delta(x), \quad v_\delta \in L_2(\mathbb{R}^n, \mu^\delta);$$

moreover,

$$\int_{\mathbb{R}^n} v_\delta^2(x)d\mu^\delta(x) \le \int_{\mathbb{R}^n} v^2(x)d\mu(x). \tag{10.8}$$

Since for every $\varphi \in C_0(\mathbb{R}^n)$ the functions $\varphi^\delta(x)$ uniformly converge to $\varphi(x)$, we have

$$\int_{\mathbb{R}^n} v_\delta(x)\varphi(x)d\mu^\delta(x) = \int_{\mathbb{R}^n} v(x)\varphi^\delta(x)d\mu(x) \longrightarrow \int_{\mathbb{R}^n} v(x)\varphi(x)d\mu(x),$$

i.e., $v_\delta \rightharpoonup v$ weakly in $L_2(\mathbb{R}^n, \mu^\delta)$. The strong convergence immediately follows from the inequality (10.8). □

We define the divergence with respect to the measure $\mu$.

DEFINITION 10.5. Suppose that $g \in L_2(\mathbb{R}^n, \mu)$ and $v \in (L_2(\mathbb{R}^n, \mu))^n$. We say that $g(x) = \operatorname{div}^\mu v(x)$ if

$$\int_{\mathbb{R}^n} g(x)\varphi(x)d\mu(x) = -\int_{\mathbb{R}^n} v(x)\cdot\nabla\varphi(x)d\mu(x)$$

for any $\varphi \in C_0^\infty(\mathbb{R}^n)$.

It is easy to see that in Definition 10.5 instead of smooth functions $\varphi$ one can take functions $\varphi \in H^1(\mathbb{R}^n, \mu)$.

10.1.5. *Elliptic equations connected with the measure* $\mu$. Let $a(x) = \{a_{ij}(x)\}$ be a continuous function with values in the space of symmetric $n \times n$-matrices satisfying the uniform ellipticity condition

$$\Lambda|\xi|^2 \leq a_{ij}(x)\xi_i\xi_j \leq \Lambda^{-1}|\xi|^2, \quad \Lambda > 0,\ \xi \in \mathbb{R}^n,$$

for all $x \in \mathbb{R}^n$. Suppose that $f \in L_2(\mathbb{R}^n, \mu)$ and $\lambda > 0$.

DEFINITION 10.6. We say that a pair $(u, \nabla^\mu u)$, where $u \in H^1(\mathbb{R}^n, \mu)$ and $\nabla^\mu u$ is a $\mu$-gradient of $u$, satisfies the equation

$$-\operatorname{div}^\mu(a(x)\nabla^\mu u(x)) + \lambda u(x) = f(x) \tag{10.9}$$

in $L_2(\mathbb{R}^n, \mu)$ if for any function $v \in H^1(\mathbb{R}^n, \mu)$ and any gradient $\nabla^\mu v$ of $v$ we have

$$\int_{\mathbb{R}^n} a(x)\nabla^\mu u(x) \cdot \nabla^\mu v(x)d\mu(x) + \lambda \int_{\mathbb{R}^n} u(x)v(x)d\mu(x) = \int_{\mathbb{R}^n} f(x)v(x)d\mu(x). \tag{10.10}$$

A function $u \in H^1(\mathbb{R}^n, \mu)$ is called a *solution* to equation (10.9) if (10.10) holds for some of the gradients of $u$.

Similarly, we can define a solution to the elliptic equation on a torus $\mathbb{T}^n$. Boundary value problems in bounded domains will be discussed later.

Note that in Definition 10.6, instead of functions in $H^1(\mathbb{R}^n, \mu)$, one can take test functions in $C_0^\infty(\mathbb{R}^n)$.

LEMMA 10.3. *Equation* (10.9) *has a unique solution* $(u, \nabla^\mu u)$, $u \in H^1(\mathbb{R}^n, \mu)$. *Moreover, the choice of a* $\mu$*-gradient of* $u$ *is uniquely determined by the condition of orthogonality of the vector* $a(x)\nabla^\mu u(x)$ *and the subspace* $\Gamma_\mu(0)$ *of the gradients of zero.*

PROOF. Since the matrix $a(x)$ is positive definite, the left-hand side of (10.10) is the inner product in $H(\mathbb{R}^n, \mu)$, whereas the right-hand side of (10.10) is a continuous linear functional in $H(\mathbb{R}^n, \mu)$. By the Riesz representation theorem, there exists $(u, \nabla^\mu u) \in H(\mathbb{R}^n, \mu)$ satisfying (10.10). Taking for a test function in (10.10) the pair $(0, z)$, $z \in \Gamma_\mu(0)$, we see that $a(x)\nabla^\mu u$ is orthogonal to $\Gamma_\mu(0)$. Using the Riesz theorem again, we conclude that there is a unique element $z$ of the set of $\mu$-gradients of $u$ satisfying the orthogonality condition. This means the uniqueness of a solution $u \in H^1(\mathbb{R}^n, \mu)$. □

In a special case where $a(x)$ is the identity matrix, (10.10) takes the form

$$\int_{\mathbb{R}^n} \nabla^\mu u(x) \cdot \nabla^\mu v(x)d\mu(x) + \lambda \int_{\mathbb{R}^n} u(x)v(x)d\mu(x) = \int_{\mathbb{R}^n} f(x)v(x)d\mu(x). \tag{10.11}$$

The expression $\operatorname{div}^\mu \nabla^\mu u$ is called the $\mu$*-Laplacian* of $u$.

By the *tangent gradient* of a function $u \in H^1(\mathbb{R}^n, \mu)$ we mean a $\mu$-gradient $z$ of $u$ that is orthogonal to $\Gamma_\mu(0)$, i.e., $z$ is the projection of an arbitrary $\mu$-gradient of $u$ on $(\Gamma_\mu(0))^\perp$ (the orthogonal complement to $\Gamma_\mu(0)$ in $(L_2(\mathbb{R}^n, \mu)^n)$). By Lemma 10.3, the gradient chosen by a solution to the problem (10.11), is tangent.

10.1.6. *Ergodicity of measures and consequences.* The property of 2-conectedness or ergodicity of a measure is crucial for the homogenization of periodic measures and the study of properties of variational problems and equations connected with such measures.

DEFINITION 10.7. A periodic measure $\mu$ is said to be *ergodic* or 2-*connected* if any function $u \in H^1(\mathbb{T}^n, \mu)$ such that some of its gradients vanishes, satisfies the condition $u = \text{const}$ $\mu$-almost surely.

An important consequence of the 2-connectedness of measures is the approximation property.

LEMMA 10.4. *Let a measure* $\mu$ *be* 2-*connected. Then the set* $\{g(x) \in L_2(\mathbb{T}^n, \mu) : g(x) = \operatorname{div}^\mu v(x)\}$ *is dense in the space* $\{u \in L_2(\mathbb{T}^n) : \int_{\mathbb{T}^n} u(x)d\mu(x) = 0)\}$.

PROOF. Assume the contrary. Then there exists a nonzero function $\zeta \in L_2(\mathbb{T}^n, \mu)$, $\int_{\mathbb{R}^n} \zeta(x)d\mu(x) = 0$, which is orthogonal to any function of the form $\operatorname{div}^\mu v(x)$. Taking $\zeta$ for the right-hand side of (10.11), setting $\lambda = 1$, and substituting $u(x)$ for a test function, we obtain the equality

$$\int_{\mathbb{T}^n} |\nabla^\mu u(x)|^2 d\mu(x) + \int_{\mathbb{T}^n} (u(x))^2 d\mu(x) = \int_{\mathbb{T}^n} u(x)\zeta(x)d\mu(x).$$

This equality, together with the Cauchy–Schwarz–Bunyakovskii inequality, yields the estimate

$$\int_{\mathbb{T}^n} (u(x))^2 d\mu(x) \leq \int_{\mathbb{T}^n} (\zeta(x))^2 d\mu(x).$$

On the other hand, equation (10.11) can be written as

$$\int_{\mathbb{T}^n} (\zeta(x) - u(x))v(x)d\mu(x) = \int_{\mathbb{T}^n} \nabla^\mu u(x) \cdot \nabla^\mu v(x)d\mu(x)$$

for all $v \in H^1(\mathbb{T}^n, \mu)$. Hence

$$\zeta - u = \operatorname{div}^\mu u.$$

By the choice of $\zeta$, we have

$$\int_{\mathbb{T}^n} (\zeta - u)\zeta d\mu(x) = 0.$$

Therefore,

$$0 \leq \int_{\mathbb{T}^n} (\zeta - u)^2 d\mu(x) = -\int_{\mathbb{T}^n} (\zeta - u)u\, d\mu(x).$$

Finally, in view of the equation, we find

$$\int_{\mathbb{T}^n} |\nabla^\mu u(x)|^2 d\mu(x) = \int_{\mathbb{T}^n} (\zeta - u)u\, d\mu(x) \leq 0,$$

which yields the equality $\nabla^\mu u = 0$ and, since the measure $\mu$ is ergodic, the equality $u = \text{const}$. Consequently, $\zeta = \text{const}$. Since the mean value of $\zeta$ vanishes, we have $\zeta = 0$, which contradicts our assumptions that $\zeta$ is nontrivial. □

We give several examples of 2-connected measures and measures that are not 2-connected. We begin with a simple example of a perforated domain.

EXERCISE 10.9. Let $Q$ be an open connected set on a torus $\mathbb{T}^n$, and let $\mu$ be a measure given by the formula $d\mu(x) = \chi_Q dx$. Then the measure $\mu$ is 2-connected.

EXERCISE 10.10. Suppose that $Q = Q_1 \cup Q_2$ and $Q_1$, $Q_2$ are open sets in $\mathbb{T}^n$ such that $\operatorname{dist}(\overline{Q}_1, \overline{Q}_2) > 0$. Then the measure $\mu$ defined by the equality $d\mu(x) = \chi_Q dx$ is not 2-connected.

10.1.7. *Construction of a self-adjoint operator.* Suppose that the coefficient matrix $a_{ij}(x)$ of equation (10.9) is symmetric and positive definite.

LEMMA 10.5. *The set of solutions to the equation*

$$-\operatorname{div}^\mu(a(x)\nabla^\mu u(x)) + u(x) = f, \quad f \in L_2(\mathbb{R}^n, \mu), \tag{10.12}$$

*is dense in the space* $L_2(\mathbb{R}^n, \mu)$.

PROOF. Denote by $\mathcal{D}$ the set of solutions to equation (10.12). Assume that there is a nontrivial element $g \in L_2(\mathbb{R}^n, \mu)$ that is orthogonal to $\mathcal{D}$. Denote by $u_0(x)$ a solution to the equation

$$-\operatorname{div}^\mu(a(x)\nabla^\mu u_0(x)) + u_0(x) = g.$$

Taking $u_0(x)$ for a test functions in (10.12) and $u(x)$ for a test function in the last equation, and taking the difference of the obtained integral identities, we see that

$$\int_{\mathbb{R}^n} f(x)u_0(x)d\mu(x) = 0$$

for all $f \in L_2(\mathbb{R}^n, \mu)$. Therefore, $u_0 = 0$ and $g = 0$, which contradicts the assumption that $g$ is nontrivial. □

For an element $u \in \mathcal{D}$, i.e., a solution to equation (10.12), we set

$$Au = f - u.$$

Then the operator $(A+I)^{-1}$ sends a function $f \in L_2(\mathbb{R}^n, \mu)$ to the corresponding solution to equation (10.12). This operator is nonnegative, bounded, and symmetric. Therefore, the operator $A$ is self-adjoint. The domain of this operator is denoted by $\mathcal{D}(A)$. Equation (10.9) can be written in the operator form $Au + \lambda u = f$.

10.1.8. *Variational problem.* Equation (10.9) or the equivalent equation $Au + \lambda u = f$ is an Euler equation for the corresponding variational problem

$$\inf_{u\in H^1(\mathbb{R}^n,\mu)} \left\{ \int_{\mathbb{R}^n} \Big(a(x)\nabla^\mu u(x) \cdot \nabla^\mu u(x) + \lambda u^2(x)\Big) d\mu(x) - \int_{\mathbb{R}^n} 2f(x)u(x)d\mu(x) \right\}. \tag{10.13}$$

The following assertion can be proved in the same way as in the case of classical variational problems in the space $H^1(\mathbb{R}^n)$.

PROPOSITION 10.2. *Let* $f \in L_2(\mathbb{R}^n, \mu)$. *Then for every* $\lambda > 0$ *the problem* (10.13) *has a unique minimum point in the space* $H^1(\mathbb{R}^n, \mu)$ *and it is a solution to equation* (10.9).

EXERCISE 10.11. Prove Proposition 10.2.

In the same way we can analyze the variational problem

$$\inf_{u\in H^1(\mathbb{R}^n,\mu)} \left\{ \int_{\mathbb{R}^n} \Big(a(x)\nabla^\mu u(x) \cdot \nabla^\mu u(x) + c(x)u^2(x)\Big) d\mu(x) - \int_{\mathbb{R}^n} 2f(x)u(x)d\mu(x) \right\}, \tag{10.14}$$

where $c(x)$ satisfies the estimate $\Lambda \le c(x) \le \Lambda^{-1}$ for all $x \in \mathbb{R}^n$, $\Lambda > 0$, and the corresponding equation

$$-\operatorname{div}^\mu(a(x)\nabla^\mu u(x)) + c(x)u(x) = f(x). \tag{10.15}$$

The same assertions can be proved for similar variational problems and equations on a torus.

10.1.9. *Boundary value problems in spaces with measures.* The technique developed allows us to study not only problems in the entire space $\mathbb{R}^n$ but also various boundary value problems. We illustrate this by an example of the Dirichlet problem.

Let $G$ be a bounded Lipschitz domain in $\mathbb{R}^n$, and let $\mu(dx)$ be a positive finite Borel measure on $G$.

DEFINITION 10.8. We say that $u \in L_2(G, \mu)$ belongs to the space $\overset{\circ}{H}{}^1 (G, \mu)$, and $z \in (L_2(G, \mu))^n$ is a gradient of $u$ if there is a sequence $\varphi_k \in C_0^\infty(G)$ such that

$$\begin{aligned} \varphi_k &\longrightarrow u \quad \text{in } L_2(G, \mu) \text{ as } k \to \infty, \\ \nabla \varphi_k &\longrightarrow z \quad \text{in } (L_2(G, \mu))^n \text{ as } k \to \infty. \end{aligned}$$

Consider the Dirichlet problem

$$\begin{aligned} -\operatorname{div} a(x) \nabla^\mu u(x) + c(x) u(x) &= f(x) \quad \text{in } L_2(G, \mu), \\ u|_{\partial G} &= 0. \end{aligned} \tag{10.16}$$

As before, we assume that the matrix $a(x)$ is symmetric and satisfies the uniform ellipticity condition in $G$, the function $c(x)$ is bounded and is bounded away from zero by a positive constant for all $x \in G$, and $f(x) \in L_2(G, \mu)$.

DEFINITION 10.9. We say that $u \in \overset{\circ}{H}{}^1(G, \mu)$ is a solution to the problem (10.16) if for any $v \in \overset{\circ}{H}{}^1 (G, \mu)$,

$$\int_G a(x) \big( \nabla^\mu u(x) \cdot \nabla^\mu v(x) + c(x) u(x) v(x) \big) d\mu(x) = \int_G f(x) v(x) d\mu(x).$$

The existence and uniqueness of a solution to the problem (10.16) can be established in the same way as in the proof of Lemma 10.3. We recall that a gradient of the solution that satisfies the above integral identity is chosen in a unique way.

10.1.10. *Dual definition of Sobolev spaces.* To analyze Sobolev spaces with measure, it is convenient to use the dual definition, which is equivalent to Definition 10.1.

DEFINITION 10.10. We say that $u(x)$ belongs to the space $H^1(\mathbb{R}^n, \mu)$, and that $z(x) \in (L_2(\mathbb{R}^n, \mu))^n$ is a $\mu$-gradient of $u(x)$ if

$$\int_{\mathbb{R}^n} u(x) g(x) d\mu(x) = -\int_{\mathbb{R}^n} z(x) \cdot v(x) d\mu(x), \tag{10.17}$$

where $g(x) = \operatorname{div}^\mu v(x)$.

PROPOSITION 10.3. *Definitions* 10.10 *and* 10.1 *are equivalent.*

PROOF. Let $u \in H^1(\mathbb{R}^n, \mu)$ in the sense of Definition 10.1. Since for test functions in Definition 10.5 we can take arbitrary $u \in H^1(\mathbb{R}^n, \mu)$, the pair $(u, \nabla^\mu u)$ satisfies (10.17) and thereby $u \in H^1(\mathbb{R}^n, \mu)$ in the sense of Definition 10.10.

To prove the inverse implication, we note that functions $(u, z)$ satisfying (10.17) form a closed subspace of $(L_2(\mathbb{R}^n, \mu))^{n+1}$. Assume that the set of such functions is larger than $\{(u, \nabla^\mu u) : u \in H^1(\mathbb{R}^n, \mu)\}$. Then some nonzero element $(u_0, z_0) \in (L_2(\mathbb{R}^n, \mu))^{n+1}$ satisfies (10.17) and simultaneously

$$\int_{\mathbb{R}^n} u_0(x) u(x) d\mu(x) = -\int_{\mathbb{R}^n} z_0(x) \cdot \nabla^\mu u(x) d\mu(x), \quad \forall\, u \in H^1(\mathbb{R}^n, \mu).$$

(Prove the existence of such an element!) By the definition of the $\mu$-divergence operator, this means that $u_0 = \operatorname{div}^\mu z_0$. Taking $(u_0, z_0)$ for a test function in (10.17), we arrive at the relation

$$\int_{\mathbb{R}^n} u_0^2 d\mu(x) = -\int_{\mathbb{R}^n} |z_0|^2 d\mu(x),$$

which contradicts the assumption that $(u_0, z_0)$ is nontrivial. □

10.1.11. *Approximation by smoothing.* Due to the dual definition of Sobolev spaces, we can approximate singular measures by smoothing. For an arbitrary Radon measure $\mu$ in $\mathbb{R}^n$ (on a torus) we introduce the measure $\mu^\delta$ by the equalities (10.3)–(10.4). Recall that such measures $\mu^\delta$ locally weakly converge to $\mu$ as $\delta \to 0$.

LEMMA 10.6. *Let $g(x) = \operatorname{div}^\mu v(x)$. Then there are $g_\delta \in L_2(\mathbb{R}^n, \mu^\delta)$ and $v_\delta \in (L_2(\mathbb{R}^n, \mu^\delta))^n$ such that*

$$\operatorname{div}^{\mu^\delta} v_\delta = g_\delta \tag{10.18}$$

*and*

$$\begin{aligned} g_\delta \to g \quad & \text{strongly in } L_2(\mathbb{R}^n, \mu^\delta) \text{ as } \delta \to 0, \\ v_\delta \to v \quad & \text{strongly in } (L_2(\mathbb{R}^n, \mu^\delta))^n \text{ as } \delta \to 0. \end{aligned} \tag{10.19}$$

PROOF. We show that it suffices to take the same $g_\delta$ and $v_\delta$ as in Lemma 10.2. The strong convergence is guaranteed by this lemma. Let us check the relation (10.18). For $\varphi \in C_0^\infty(\mathbb{R}^n, \mu)$ we have

$$\begin{aligned} \int_{\mathbb{R}^n} v_\delta(x) \cdot \nabla\varphi(x) d\mu^\delta(x) &= \int_{\mathbb{R}^n} v(x) \cdot \nabla(\varphi^\delta)(x) d\mu(x) \\ &= \int_{\mathbb{R}^n} g(x)\varphi^\delta(x) d\mu(x) = \int_{\mathbb{R}^n} g_\delta(x)\varphi(x) d\mu^\delta(x). \end{aligned}$$

□

In the following assertion, the $\mu^\delta$ are also defined by the equalities (10.3)–(10.4), but the family $u_\delta \in H^1(\mathbb{R}^n, \mu^\delta)$ is arbitrary and is not related to the smoothing of any fixed function.

THEOREM 10.1. *Let $u_\delta \in H^1(\mathbb{R}^n, \mu^\delta)$, and let*

$$\begin{aligned} u_\delta \rightharpoonup u \quad & \text{weakly in } L_2(\mathbb{R}^n, \mu^\delta) \text{ as } \delta \to 0, \\ \nabla^{\mu^\delta} u_\delta \rightharpoonup z \quad & \text{weakly in } (L_2(\mathbb{R}^n, \mu^\delta))^n \text{ as } \delta \to 0. \end{aligned}$$

*Then $u \in H^1(\mathbb{R}^n, \mu)$ and $z = \nabla^\mu u$.*

PROOF. For $g(x)$ and $v(x)$ such that $g(x) = \operatorname{div}^\mu v(x)$ we use the approximations $g_\delta$ and $v_\delta$ constructed in Lemma 10.6, so that $g_\delta(x) = \operatorname{div}^{\mu^\delta} v_\delta(x)$. Then

$$\int_{\mathbb{R}^n} u_\delta(x) g_\delta(x) d\mu^\delta(x) = \int_{\mathbb{R}^n} \nabla^{\mu^\delta} u_\delta(x) v_\delta(x) d\mu^\delta(x).$$

Passing to the limit as $\delta \to 0$ and taking into account the properties of strong convergence of $g_\delta$ and $v_\delta$, we obtain the equality

$$\int_{\mathbb{R}^n} u(x) g(x) d\mu(x) = \int_{\mathbb{R}^n} z(x) \cdot v(x) d\mu(x).$$

Using the dual definition of Sobolev spaces, we find that $u \in H^1(\mathbb{R}^n, \mu)$ and $z = \nabla^\mu u$. □

The previous assertions allow us to establish an important result on the approximation and convergence of solutions of elliptic equations.

Let $\mu$ be a Radon measure in $\mathbb{R}^n$, and let $\mu^\delta$ be smoothed measures. Consider the elliptic equation

$$-\mathrm{div}^\mu a(x)\nabla^\mu u + \lambda u = f \quad \text{in } L_2(\mathbb{R}^n, \mu), \tag{10.20}$$

and the family of approximating equations of the form

$$-\mathrm{div}^{\mu^\delta} a_\delta(x)\nabla^{\mu^\delta} u + \lambda u = f_\delta \quad \text{in } L_2(\mathbb{R}^n, \mu^\delta). \tag{10.21}$$

THEOREM 10.2. *Suppose that*

$$\Lambda|\xi|^2 \le a(x)\xi\cdot\xi \le \Lambda^{-1}|\xi|^2, \quad \Lambda|\xi|^2 \le a_\delta(x)\xi\cdot\xi \le \Lambda^{-1}|\xi|^2 \quad \forall\, \xi \in \mathbb{R}^n,\ x \in \mathbb{R}^n,$$

$a_\delta(x) \to a(x)$ *strongly in* $L_2(\mathbb{R}^n, \mu^\delta)$, *and* $f_\delta(x) \to f(x)$ *strongly in* $L_2(\mathbb{R}^n, \mu^\delta)$. *Then the solutions* $u_\delta(x)$ *of equation* (10.21) *strongly converge to the solution* $u(x)$ *of equation* (10.20) *in* $L_2(\mathbb{R}^n, \mu)$ *as* $\delta \to 0$.

PROOF. Since the $(u_\delta, \nabla^{\mu^\delta} u_\delta)$ satisfy the estimate

$$\|u_\delta\|_{L_2(\mathbb{R}^n,\mu^\delta)} + \|\nabla^{\mu^\delta} u_\delta\|_{(L_2(\mathbb{R}^n,\mu^\delta))^n} \le C,$$

for some subsequence $\delta_k \to 0$ we have the weak convergence $(u_\delta, \nabla^{\mu^\delta} u_\delta) \to (u_0, z_0)$ in $(L_2(\mathbb{R}^n, \mu^\delta))^{n+1}$. By Theorem 10.1, $u_0 \in H^1(\mathbb{R}^n, \mu^\delta)$ and $v_0 = \nabla^\mu u_0$. It is easy to verify that

$$a_\delta(x)\nabla^{\mu^\delta} u^\delta(x) \rightharpoonup a(x)\nabla^\mu u_0(x) \quad \text{weakly in } (L_2(\mathbb{R}^n, \mu^\delta))^n.$$

Passing to the limit in the integral identity

$$\int_{\mathbb{R}^n} \big(a_\delta(x)\nabla^{\mu^\delta} u_\delta(x)\cdot\nabla\varphi(x) + \lambda u_\delta(x)\varphi(x)\big)d\mu^\delta(x) = \int_{\mathbb{R}^n} f_\delta(x)\varphi(x)d\mu^\delta(x),$$

we conclude that $(u_0(x), \nabla^\mu u_0(x))$ is a solution to equation (10.20). The strong convergence can be proved by using a trick based on the energy convergence. Taking $(u_\delta, \nabla^{\mu^\delta} u_\delta)$ for a test function in the integral identity for equation (10.21) and taking into account the strong convergence of $f_\delta$, we obtain the equalities

$$\begin{aligned}\lim_{\delta\to 0}\int_{\mathbb{R}^n} \Big(a_\delta(x)\nabla^{\mu^\delta} u_\delta(x)\cdot\nabla^{\mu^\delta} u_\delta(x) + \lambda u_\delta^2\Big)d\mu^\delta(x) &= \lim_{\delta\to 0}\int_{\mathbb{R}^n} u_\delta(x) f_\delta(x)d\mu^\delta(x)\\ = \int_{\mathbb{R}^n} u(x)f(x)d\mu(x) &= \int_{\mathbb{R}^n}\Big(a(x)\nabla^\mu u(x)\cdot\nabla^\mu u(x) + \lambda u^2\Big)d\mu(x).\end{aligned}$$

Both quadratic terms on the left-hand side of this equality are lower semicontinuous under the weak convergence:

$$\liminf_{\delta\to 0}\int_{\mathbb{R}^n} a_\delta(x)\nabla^{\mu^\delta} u_\delta(x)\cdot\nabla^{\mu^\delta} u_\delta(x)d\mu^\delta(x) \ge \int_{\mathbb{R}^n} a(x)\nabla^\mu u(x)\cdot\nabla^\mu u(x)d\mu(x),$$

$$\liminf_{\delta\to 0}\int_{\mathbb{R}^n} \lambda u_\delta^2(x)d\mu^\delta(x) \ge \int_{\mathbb{R}^n} \lambda u^2(x)d\mu(x).$$

Therefore, the above relation can be valid only if both last inequalities are, in fact, equalities. This implies the strong convergence. □

10.1.12. *Nonsingularity of periodic measures.* Let $\mu$ be a periodic ergodic Radon measure in $\mathbb{R}^n$. As usual, we identify it with a measure on a torus $\mathbb{T}^n$. For every $\xi \in \mathbb{R}^n$ we consider the variational problem

$$\widehat{A}\xi \cdot \xi = \min_{v \in L_2^{\mathrm{pot}}(\mathbb{T}^n)} \int_{\mathbb{T}^n} (\xi + v(x)) \cdot (\xi + v(x)) d\mu(x).$$

It is easy to verify that the minimum of this variational problem is a nonnegative quadratic form in $\mathbb{R}^n$. The matrix of this quadratic form, denoted by $\widehat{A}$, is said to be *effective.*

DEFINITION 10.11. A periodic Radon measure $\mu$ is said to be *nonsingular* if the matrix $\widehat{A}$ is positive definite, i.e., $\widehat{A}\xi \cdot \xi \geq c|\xi|^2$, $c > 0$.

The matrix $\widehat{A}$ can be singular along certain directions or everywhere. The subspace of vectors forming the kernel of $\widehat{A}$ is denoted by $\mathcal{K}^\mu$.

EXERCISE 10.12. Let $a(x)$ be a periodic matrix such that

$$\Lambda|\xi|^2 \leq a(x)\xi \cdot \xi \leq \Lambda^{-1}|\xi|^2 \quad \mu\text{-almost everywhere.}$$

Let

$$\widehat{A}_1\xi \cdot \xi = \min_{v \in L_2^{\mathrm{pot}}(\mathbb{T}^n)} \int_{\mathbb{T}^n} a(x)(\xi + v(x)) \cdot (\xi + v(x)) d\mu(x). \tag{10.22}$$

Prove that the kernel of $\widehat{A}_1$ coincides with the kernel of $\widehat{A}$.

The matrix $\widehat{A}_1$ is called the *effective matrix* of the operator $-\mathrm{div}^\mu(a(x)\nabla^\mu \cdot)$.

The Euler equation of the variational problem (10.22) takes the form

$$\text{Find } v_\xi(x) \in L_2^{\mathrm{pot}}(\mathbb{T}^n, \mu) \text{ such that } a(x)(\xi + v_\xi(x)) \in L_2^{\mathrm{sol}}(\mathbb{T}^n, \mu). \tag{10.23}$$

Denote by $\Pi_{\mathrm{pot}}$ the operator of orthogonal projection from $(L_2(\mathbb{T}^n, \mu))^n$ onto the subspace $L_2^{\mathrm{pot}}(\mathbb{T}^n, \mu)$. Then (10.23) can be written as follows:

$$\text{Find } v_\xi(x) \in L_2^{\mathrm{pot}}(\mathbb{T}^n, \mu) \text{ such that } \Pi_{\mathrm{pot}}\big(a(x)v_\xi(x)\big) = -\Pi_{\mathrm{pot}}\big(a(x)\xi\big).$$

EXERCISE 10.13. Prove that the operator that sends $v \in L_2^{\mathrm{pot}}(\mathbb{T}^n, \mu)$ to $\Pi_{\mathrm{pot}}\big(a(x)v_\xi(x)\big)$ is coercive in $L_2^{\mathrm{pot}}(\mathbb{T}^n, \mu)$.

This implies the unique solvability of the problem (10.23). Then the matrix $\widehat{A}_1$ can be written in the form

$$\widehat{A}_1\xi = \int_{\mathbb{T}^n} a(x)(v_\xi(x) + \xi)) d\mu(x), \quad \xi \in \mathbb{R}^n. \tag{10.24}$$

Denote by $V(x)$ the matrix whose columns are formed by vector-valued functions $v_{e_1}(x), \ldots, v_{e_n}(x)$ (the $e_j$ are the unit vectors in $\mathbb{R}^n$) and write the above equality in the matrix form

$$\widehat{A}_1 = \int_{\mathbb{T}^n} a(x)(\mathrm{Id} + V(x)) d\mu(x). \tag{10.25}$$

PROPOSITION 10.4. *The kernel $\mathcal{K}^\mu$ of the effective matrix $\widehat{A}$ (or $\widehat{A}_1$) coincides with the set of constant potential vectors, i.e., $\xi \in \mathcal{K}^\mu$ if and only if a vector-valued function equal to $\xi$ for $\mu$-almost all $x$ is potential.*

*A vector $\eta \in \mathbb{R}^n$ belongs to the orthogonal complement of $\mathcal{K}^\mu$ if and only if there is $v \in L_2^{\rm sol}(\mathbb{T}^n, \mu)$ such that*

$$\int_{\mathbb{T}^n} v(x)d\mu(x) = \eta.$$

PROOF. The first assertion immediately follows from the variational formula (10.22).

To prove the second assertion, we assume that $\eta = \int_{\mathbb{T}^n} v(x)d\mu(x)$ for some $v \in L_2^{\rm sol}(\mathbb{T}^n, \mu)$. If $\xi \in \mathcal{K}^\mu$, then the vector $\xi$ is potential. Therefore,

$$0 = \int_{\mathbb{T}^n} \xi \cdot v(x)d\mu(x) = \xi \cdot \eta.$$

To obtain the inverse implication, we note that the set $\widehat{A}_1\xi$, $\xi \in \mathbb{R}^n$, coincides with the orthogonal complement of $\mathcal{K}^\mu$. By (10.24), any vector of the form $\widehat{A}_1\xi$ is the $\mu$-mean value of some function in $L_2^{\rm sol}(\mathbb{T}^n, \mu)$. □

Consider the modified cell problem,

(10.26) Find $v_\xi^+(x) \in L_2^{\rm pot}(\mathbb{T}^n, \mu)$ such that $a(x)(\Pi^{\rm eff}\xi + v_\xi^+(x)) \in L_2^{\rm sol}(\mathbb{T}^n, \mu)$,

where $\Pi^{\rm eff}$ is the operator of orthogonal projection onto the orthogonal complement of $\mathcal{K}^\mu$. The existence and uniqueness of a solution to this problem can be established in the same way as for the problem (10.23). From Proposition 10.4 we obtain the following assertion.

COROLLARY 10.1. *The following equality holds*:

$$a(x)(\xi + v_\xi(x)) = a(x)(\Pi^{\rm eff}\xi + v^+).$$

*The effective matrix $\widehat{A}_1$ introduced in* (10.24) *can be determined by the formula*

$$\widehat{A}_1\xi = \int_{\mathbb{T}^n} a(x)(v_\xi^+(x) + \Pi^{\rm eff}\xi))d\mu(x), \quad \xi \in \mathbb{R}^n. \tag{10.27}$$

**10.2. Two-scale convergence in variable spaces.** The technique of two-scale convergence (cf. Section 9) introduced in [**100, 3**] turned out to be very effective for studying many homogenization problems. Now, we extend the classical notion of two-scale convergence to the case of a variable space with a periodic rapidly oscillating measure. From the mathematical point of view, these topics were discussed in [**24**] and [**134, 136**].

Let $\mu$ be a periodic Radon measure in $\mathbb{R}^n$ with period 1 along every coordinate direction, and let $\varepsilon$ be an arbitrary positive number. Hereinafter, we set $\mu_\varepsilon(dx) = \varepsilon^n\mu(\frac{dx}{\varepsilon})$, i.e.,

$$\mu_\varepsilon(B) = \varepsilon^n\mu(\varepsilon^{-1}B)$$

for any Borel set $B \subset \mathbb{R}^n$.

We recall (cf. Exercise 10.8) that the measures $\mu_\varepsilon$ weakly converge to a measure $\mu(\square)dx$, $\square = [0,1)^n$, as $\varepsilon \to 0$. In particular, $\mu(\square) = 1$ implies that the $\mu_\varepsilon$ converge to the standard Lebesgue measure.

Let $G$ be a Jordan domain in $\mathbb{R}^n$ (we consider only the case of $\mathbb{R}^n$).

DEFINITION 10.12. We say that $u_\varepsilon \in L_2(G, \mu_\varepsilon)$ two-scale converge in $L_2(G, \mu_\varepsilon)$ to $u(x,y) \in L_2(G \times \square, dx \times \mu(y))$ as $\varepsilon \to 0$ if

$$\|u_\varepsilon\|_{L_2(G,\mu_\varepsilon)} \le C, \quad \varepsilon > 0,$$

and

$$\int_G u_\varepsilon(x)\phi(x)\psi\left(\frac{x}{\varepsilon}\right) d\mu_\varepsilon(x) \longrightarrow \int_G \int_\square u(x,y)\varphi(x)\psi(y)dxd\mu(y) \quad \text{as } \varepsilon \to 0$$

for any $\varphi \in C_0^\infty(G)$ and $\psi \in C_{\text{per}}^\infty(\square)$.

REMARK 10.3. As was mentioned in Section 9, we cannot take an arbitrary function of class $L_2(G\times\square, dx\times\mu(y))$ for a test function in this definition. However, the class of test functions in the definition of two-scale convergence can be extended. In particular, functions in $C(\overline{G}, C_{\text{per}}(\square))$ can be used (check it!).

Another admissible class is the set of functions

$$\psi(x,y) = \sum_1^N \varphi_i(x)g_i(y), \tag{10.28}$$

where $\varphi_i \in C^\infty(G)$, $g_i \in L_2(\square, \mu)$, and $N$ is an arbitrary number (check it!).

Under the additional condition of uniform integrability, Carathéodory functions can also be taken for test functions (cf. [**136**, Lemma 3.1 and Theorem 3.3]. However, we will not use them, restricting ourselves to test functions in one of the above-mentioned classes.

The following simple technical assertion will be repeatedly used in the further consideration. It is often referred to as the *mean-value property*.

LEMMA 10.7. *Suppose that $g(x,y) \in C(\overline{G}; C_{\text{per}}(\square))$, i.e., $g(x,y)$ is continuous with respect to both variables in $\overline{G}\times\square$ and is $\square$-periodic with respect to $y$. Then*

$$\lim_{\varepsilon\to 0} \int_G g\Big(x, \frac{x}{\varepsilon}\Big) d\mu_\varepsilon(x) = \int_{G\times\square} g(x,y)dxd\mu(y). \tag{10.29}$$

EXERCISE 10.14. Prove Lemma 10.7.

A similar assertion for functions of the form (10.28) is trivial.

We begin our description of the properties of two-scale convergence with a result on the weak compactness of a family of bounded functions in $L_2(G, \mu_\varepsilon)$.

PROPOSITION 10.5. *Suppose that $u^\varepsilon \in L_2(G, \mu_\varepsilon)$ satisfy the estimate*

$$\|u^\varepsilon\|_{L_2(G,\mu_\varepsilon)} \le C.$$

*Then, on a subsequence $\varepsilon_k \to 0$, the functions $u^\varepsilon$ two-scale converge in $L_2(G, \mu_\varepsilon)$ to some function $u(x,y) \in L_2(G\times\square, dx\times\mu(y))$.*

PROOF. Consider an arbitrary countable family of functions $\{g_j(x,y)\}$, $j = 1, 2, \dots$, everywhere dense in $C(\overline{G}; C_{\text{per}}(\square))$. This family is also dense in $L_2(G\times\square, dx\times\mu(y))$. We set

$$\mathcal{H}_\varepsilon(\psi) \equiv \int_G u^\varepsilon(x)\psi\left(x, \frac{x}{\varepsilon}\right) \mu_\varepsilon(x).$$

By Lemma 10.7, for any continuous functions $\psi(x,y)$,

$$\begin{aligned} \limsup_{\varepsilon\to 0} \mathcal{H}_\varepsilon^2(\psi) &\le \limsup_{\varepsilon\to 0} \|u^\varepsilon\|^2_{L_2(G,\mu_\varepsilon)} \int_G \psi^2\Big(x, \frac{x}{\varepsilon}\Big) d\mu_\varepsilon(x) \\ &\le C\|\psi(x,y)\|^2_{L_2(G\times\square, dx\times\mu(y))}. \end{aligned} \tag{10.30}$$

We choose a subsequence $\varepsilon_k$ such that $\mathcal{H}_{\varepsilon_k}(\psi_1)$ converge, and denote the limit by $\mathcal{H}_0(\psi_1)$. Further, from the sequence $\varepsilon_k$ we extract a subsequence, denoted again by $\varepsilon_k$, such that the $\mathcal{H}_{\varepsilon_k}(\psi_2)$ converge; denote the limit by $\mathcal{H}_0(\psi_2)$. Continuing the process and using the diagonal procedure, we construct a sequence $\varepsilon_k \to 0$ such that the $\mathcal{H}_{\varepsilon_k}(\psi_j)$ converge to $\mathcal{H}_0(\psi_j)$ for all $j = 1, 2, \dots$. By construction, in view of (10.30), the mapping $\mathcal{H}_0(\psi)$, $\psi \in \{\psi_j\}$, is extended to a continuous linear operator in $L_2(G \times \square, dx \times \mu(y))$; moreover, for any $\psi \in C(\overline{G}; C_{\text{per}}(\square))$,

$$\lim_{k\to\infty} \mathcal{H}_{\varepsilon_k}(\psi) = \mathcal{H}_0(\psi).$$

Using the Riesz theorem, we find $u(x,y) \in L_2(G \times \square, dx \times \mu(y))$ such that

$$\mathcal{H}_0(\psi) = \int_{G\times\square} \psi(x,y)u(x,y)\,dx\,d\mu(y)$$

for any $\psi(x,y) \in L_2(G \times \square, dx \times \mu(y))$, which completes the proof. □

Another important property of two-scale convergence is the lower semicontinuity of the $L_2$-norm in the corresponding topology.

PROPOSITION 10.6. *Suppose that $u^\varepsilon(x)$ two-scale converge in $L_2(G, \mu_\varepsilon)$ to a function $u(x,y)$. Then*

$$\liminf_{\varepsilon\to 0} \|u^\varepsilon\|_{L_2(G,\mu_\varepsilon)} \geq \|u(x,y)\|_{L_2(G\times\square, dx\times\mu(y))}.$$

PROOF. Generally speaking, we cannot take $u(x,y)$ for a test function since it is not necessarily continuous. Therefore, we consider a continuous $y$-periodic function $u_\delta(x,y)$ approximating $u(x,y)$ in the $L_2$-norm, so that

$$\|u(x,y) - u_\delta(x,y)\|_{L_2(G\times\square, dx\times\mu(y))} \leq \delta.$$

Using the definition of two-scale convergence and Lemma 10.7, we find that

$$\begin{aligned}
0 &\leq \liminf_{\varepsilon\to 0} \int_G \Big(u^\varepsilon(x) - u_\delta\Big(x, \frac{x}{\varepsilon}\Big)\Big)^2 d\mu_\varepsilon(x) \\
&= \liminf_{\varepsilon\to 0} \int_G (u^\varepsilon(x))^2 d\mu_\varepsilon(x) - 2\lim_{\varepsilon\to 0} \int_G u^\varepsilon(x) u_\delta\Big(x, \frac{x}{\varepsilon}\Big) d\mu_\varepsilon(x) \\
&\quad + \lim_{\varepsilon\to 0} \int_G u_\delta^2\Big(x, \frac{x}{\varepsilon}\Big) d\mu_\varepsilon(x) \\
&= \liminf_{\varepsilon\to 0} \int_G (u^\varepsilon(x))^2 d\mu_\varepsilon(x) - 2\int_{G\times\square} u(x,y)u_\delta(x,y)\,dx\,d\mu(y) \\
&\quad + \int_{G\times\square} u_\delta^2(x,y)\,dx\,d\mu(y).
\end{aligned}$$

Hence

$$\liminf_{\varepsilon\to 0} \int_G (u^\varepsilon(x))^2 d\mu_\varepsilon(x) \geq 2\int_{G\times\square} u(x,y)u_\delta(x,y)\,dx\,d\mu(y) - \int_{G\times\square} u_\delta^2(x,y)\,dx\,d\mu(y).$$

In view of the choice of $u_\delta$, the expression on the right-hand side converges to $\|u(x,y)\|^2_{L_2(G\times\square, dx\times\mu(y))}$ as $\delta \to 0$. Since the left-hand side is independent of $\delta$, we can pass to the limit with respect to $\delta$ and arrive at the required inequality. □

We introduce the notion of strong two-scale convergence.

DEFINITION 10.13. We say that $u^\varepsilon(x) \in L_2(G, \mu_\varepsilon)$ strongly two-scale converge to $u(x,y) \in L_2(G \times \square, dx \times \mu(y))$ in $L_2(G, \mu_\varepsilon)$ if the $u^\varepsilon(x)$ two-scale converge in

$L_2(G,\mu_\varepsilon)$ to $u(x,y)$ and for any $v^\varepsilon(x)$ two-scale converging in $L_2(G,\mu_\varepsilon)$ to $v(x,y)$ the following limit relation holds:

$$\int_G u^\varepsilon(x)v^\varepsilon(x)\,d\mu_\varepsilon(x) \longrightarrow \int_{G\times\square} u(x,y)v(x,y)dxd\mu(y) \quad \text{as } \varepsilon\to 0.$$

It is possible to introduce an equivalent definition that does not use an auxiliary family.

DEFINITION 10.14. We say that $u^\varepsilon(x)\in L_2(G,\mu_\varepsilon)$ strongly two-scale converge to $u(x,y)\in L_2(G\times\square, dx\times\mu(y))$ in $L_2(G,\mu_\varepsilon)$ if the $u^\varepsilon(x)$ two-scale converge to $u(x,y)$ in $L_2(G,\mu_\varepsilon)$ and

$$\lim_{\varepsilon\to 0}\int_G |u^\varepsilon(x)|^2 d\mu_\varepsilon(x) = \int_{G\times\square}|u(x,y)|^2 dxd\mu(y).$$

EXERCISE 10.15. Prove that Definitions 10.14 and 10.13 are equivalent.

The following assertions concerning the structure of the two-scale limit generalize the results of Section 9 to the case of variable spaces.

PROPOSITION 10.7. *Suppose that* $u^\varepsilon(x)\in H^1(G,\mu_\varepsilon)$ *and*

$$\|u^\varepsilon\|_{L_2(G,\mu_\varepsilon)}\le C, \quad \lim_{\varepsilon\to 0}\varepsilon\|\nabla^\mu u^\varepsilon(x)\|_{(L_2(G,\mu_\varepsilon))^n} = 0.$$

*Then, on a subsequence, the* $u^\varepsilon$ *two-scale converge in* $L_2(G,\mu_\varepsilon)$ *to some function* $u^0(x)$ *which is independent of* $y$.

PROOF. By Proposition 10.5, the functions $u^\varepsilon(x)$ two-scale converge on a subsequence to some $u(x,y)\in L_2(G\times\square, dx\times\mu(y))$ in $L_2(G,\mu_\varepsilon)$. Assume that $g(y) = \operatorname{div}^\mu v(y)$ for $v(y)\in(L_2(G,\mu_\varepsilon))^n$ and $g(y)\in L_2(G,\mu_\varepsilon)$. For arbitrary $\varphi(x)\in C_0^\infty(G)$,

$$\int_G u^\varepsilon(x)\varphi(x)g\Big(\frac{x}{\varepsilon}\Big)d\mu_\varepsilon(x) = -\varepsilon\int_G \nabla^\mu\big(u^\varepsilon(x)\varphi(x)\big)\cdot v\Big(\frac{x}{\varepsilon}\Big)d\mu_\varepsilon(x).$$

Passing to the two-scale limit, we find that

$$\int_{G\times\square} u(x,y)\varphi(x)g(y)dxd\mu(y) = 0.$$

Since $\varphi(x)$ is an arbitrary $C_0^\infty(G)$-function, we have

$$\int_{G\times\square} u(x,y)g(y)d\mu(y) = 0$$

for almost all $x$. Based on Lemma 10.4, we conclude that

$$\int_{G\times\square} u(x,y)g(y)d\mu(y) = 0$$

for any functions $g(y)\in L_2(\square,\mu)$ with zero mean relative to the measure $\mu$. Therefore, $u(x,y)$ is independent of $y$, i.e., $u(x,y) = u(x)$. □

Recall that $\mathcal{K}^\mu$ denotes the kernel of the matrix $\widehat{A}$. Denote by $\Pi^{\text{eff}}$ the operator of orthogonal projection from $\mathbb{R}^n$ onto the orthogonal complement of $\mathcal{K}^\mu$. We also set $\nabla^{\text{eff}} = \Pi^{\text{eff}}\nabla$ and

$$H^{\text{eff}}(G) = \{u\in L_2(G) : \Pi^{\text{eff}}\nabla u\in(L_2(G))^n\}.$$

THEOREM 10.3. *Suppose that functions* $u^\varepsilon$ *satisfy the estimates*

$$\|u^\varepsilon\|_{L_2(G,\mu_\varepsilon)} \le C, \quad \|\nabla^{\mu_\varepsilon} u^\varepsilon\|_{(L_2(G,\mu_\varepsilon))^n} \le C.$$

*Then, on a subsequence,*

$$u^\varepsilon(x) \xrightarrow{2} u^0(x) \quad \textit{two-scale in } L_2(G,\mu_\varepsilon) \textit{ as } \varepsilon \to 0, \tag{10.31}$$

$$\nabla^\mu u^\varepsilon(x) \xrightarrow{2} \nabla^{\text{eff}} u^0(x) + u_1(x,y) \quad \textit{two-scale in } (L_2(G,\mu_\varepsilon)) \textit{ as } \varepsilon \to 0; \tag{10.32}$$

*moreover,* $u^0 \in H^{\text{eff}}(G)$ *and* $u_1 \in L_2(G; L_2^{\text{pot}}(\square,\mu))$.

REMARK 10.4. For nonsingular measures, $H^{\text{eff}}(G)$ coincides with $H^1(G)$ and $\nabla^{\text{eff}} = \nabla$. In particular, if $\mu(y)$ is a nonsingular measure under the assumptions of Theorem 10.32, then $u^0 \in H^1(G)$ and $\nabla^{\text{eff}} u^0 = \nabla u^0$.

PROOF. The independence of $u^0$ of $y$ has been established before. Denote by $q(x,y)$ the two-scale limit of $\nabla^{\mu_\varepsilon} u^\varepsilon$ in $(L_2(G,\mu_\varepsilon))^n$. The existence of this limit is guaranteed by Proposition 10.5. We choose a test function of the form $\varphi(x)v(\frac{x}{\varepsilon})$, where $\varphi(x) \in C_0^\infty(G)$ and $v(y) \in L_2^{\text{sol}}(\square,\mu)$, and set

$$\xi = \int_\square v(y)\,d\mu(y).$$

Then

$$\begin{aligned}
&\int_G \nabla^{\mu_\varepsilon} u^\varepsilon(x) \cdot v\Big(\frac{x}{\varepsilon}\Big)\varphi(x)\,d\mu_\varepsilon(x) \\
&\quad = \int_G \nabla^{\mu_\varepsilon}\big(u^\varepsilon(x)\varphi(x)\big) \cdot v\Big(\frac{x}{\varepsilon}\Big)d\mu_\varepsilon(x) - \int_G u^\varepsilon(x)\nabla\varphi(x) \cdot v\Big(\frac{x}{\varepsilon}\Big)d\mu_\varepsilon(x) \\
&\quad = -\int_G u^\varepsilon(x)\nabla\varphi(x) \cdot v\Big(\frac{x}{\varepsilon}\Big)d\mu_\varepsilon(x).
\end{aligned}$$

Passing to the two-scale limit, we obtain the equality

$$\begin{aligned}
\int_{G\times\square} q(x,y) \cdot v(y)\varphi(x)\,dx\,d\mu(y) &= -\int_{G\times\square} u^0(x)v(y) \cdot \nabla\varphi(x)\,dx\,d\mu(y) \\
&= -\xi \cdot \int_G u^0(x)\nabla\varphi(x)\,dx.
\end{aligned}$$

Using the definition of a weak derivative, we conclude that

$$\xi \cdot \nabla u^0(x) = -\int_\square q(x,y) \cdot v(y)\,d\mu(y) \tag{10.33}$$

for almost all $x \in G$. Since $q \in L_2(G \times \square, dx \times \mu(y))$, we have $\xi \cdot \nabla u^0(x) \in L_2(G)$. The inclusion $\nabla^{\text{eff}} u^0 \in (L_2(G))^n$ follows from Proposition 10.4.

It remains to prove that $u_1(x,y) \in L_2(G; L_2^{\text{pot}}(\square))$. For this purpose, we transform (10.33) as follows:

$$\int_\square \big(q(x,y) - \nabla^{\text{eff}} u^0(x)\big) \cdot v(y)\,d\mu(y) = 0$$

for almost all $x$. Consequently, $(q(x,y) - \nabla^{\text{eff}} u^0(x)) \in L_2(G; L_2^{\text{pot}}(\square))$. ☐

THEOREM 10.4. *If*

$$\|u^\varepsilon\|_{L_2(G,\mu_\varepsilon)} \le C, \quad \varepsilon\|\nabla^{\mu_\varepsilon} u^\varepsilon\|_{(L_2(G,\mu_\varepsilon))^n} \le C,$$

*then there are subsequences* $\varepsilon_k \to 0$ *and* $u^0(x,y) \in L_2(G; H^1_{\mathrm{per}}(\square,\mu))$ *such that*

$$u^\varepsilon(x) \xrightarrow{2} u^0(x,y) \quad \textit{two-scale in } L_2(G,\mu_\varepsilon), \tag{10.34}$$

$$\varepsilon\nabla^\mu u^\varepsilon(x) \xrightarrow{2} \nabla^\mu_y u^0(x,y) \quad \textit{two-scale in } (L_2(G,\mu_\varepsilon))^n. \tag{10.35}$$

PROOF. Let $g(y) = \operatorname{div}^\mu v(y)$. Then $g(\frac{x}{\varepsilon}) = \varepsilon \operatorname{div}^{\mu_\varepsilon} v(\frac{x}{\varepsilon})$. Therefore,

$$\int_G u^\varepsilon(x)\varphi(x) g\Big(\frac{x}{\varepsilon}\Big) d\mu_\varepsilon(x) = -\int_G \varepsilon\nabla^{\mu_\varepsilon}\big(u^\varepsilon(x)\varphi(x)\big)\cdot v\Big(\frac{x}{\varepsilon}\Big) d\mu_\varepsilon(x)$$

for any $\varphi(x) \in C_0^\infty(G)$. Denote by $q(x,y)$ the two-scale limit of the sequence $\varepsilon\nabla^{\mu_\varepsilon} u^\varepsilon(x)$ in $(L_2(G,\mu_\varepsilon))^n$. Passing to the two-scale limit in the previous equality, we find that

$$\int_{G\times\square} u^0(x,y)\varphi(x)g(y)dx\,\mu(dy) = -\int_{G\times\square} q(x,y)\varphi(x)v(y)dx\,\mu(dy).$$

Therefore, for almost all $x$,

$$\int_\square u^0(x,y)g(y)\mu(dy) = -\int_\square q(x,y)v(y)\mu(dy).$$

Consequently, $q(x,y) = \nabla^\mu u^0(x,y)$, where we used the dual definition of the space $H^1(\mathbb{T}^n,\mu)$. $\square$

In the study of boundary value problems, a question arises whether the boundary conditions are satisfied by the two-scale limit of the sequence of solutions.

Let $\mathcal{K}^\mu$ be the kernel of the effective matrix $\widehat{A}$ of a periodic Radon measure $\mu$ in $\mathbb{R}^n$, and let $\Pi^{\mathrm{eff}}$ be the orthogonal projection from $\mathbb{R}^n$ onto the orthogonal complement of $\mathcal{K}^\mu$. We introduce the norm

$$\|u\|^2_{H^{\mathrm{eff}}(G)} = \int_G \big(u^2(x) + |\Pi^{\mathrm{eff}}\nabla u(x)|^2\big)dx$$

and define the space $H_0^{\mathrm{eff}}(G)$ as the closure of the set $C_0^\infty(G)$ with respect to this norm. It is known (cf. [**58**, Chapter X, § 2, Lemma 2.6]) that this space coincides with the set of functions $u(x)$ with support in $G$ such that

$$\int_{\mathbb{R}^n} \big(u^2(x) + |\Pi^{\mathrm{eff}}\nabla u(x)|^2\big)dx \le \infty.$$

By Theorem 10.3, the following assertion holds.

COROLLARY 10.2. *Let all the assumptions of Theorem* 10.3 *be satisfied, and let* $u^\varepsilon \in \overset{\circ}{H}{}^1(G;\mu_\varepsilon)$. *Then the limit function* $u^0(x)$ *belongs to* $H_0^{\mathrm{eff}}(G)$. *In particular, if* $\mu$ *is nonsingular, then* $u^0 \in \overset{\circ}{H}{}^1(G)$.

**10.3. Homogenization.** Now we apply the results on the two-scale convergence and other results obtained above to proving the homogenization theorem for periodic elliptic equations in variable spaces.

Let $\mu$ be a periodic Radon measure in $\mathbb{R}^n$, and let $\mu_\varepsilon = \varepsilon^n\mu(\frac{dx}{\varepsilon})$. We begin by considering an elliptic equation of the form

$$-\operatorname{div}^{\mu_\varepsilon}\Big(a\Big(\frac{x}{\varepsilon}\Big)\nabla^{\mu_\varepsilon}u\Big) + c\Big(\frac{x}{\varepsilon}\Big)u = f^\varepsilon(x) \quad \text{in } L_2(\mathbb{R}^n,\mu_\varepsilon) \tag{10.36}$$

in the entire space. As before, we assume that $a(y)$ is uniformly elliptic, i.e.,

$$\Lambda|\xi|^2 \le a(y)\xi\cdot\xi \le \Lambda^{-1}|\xi|^2$$

for $\mu$-almost all $y \in \square$ and all $\xi \in \mathbb{R}^n$. We also assume that $0 < c_0 \le c(y) \le c_1$.

Let $v_\xi(y)$ be a solution to problem (10.23), and let $\widehat{A}_1$ be the matrix defined by formula (10.25) or (10.22). We set

$$\widehat{c} = \int_\square c(y)d\mu(y).$$

The equation

$$-\operatorname{div}\big(\widehat{A}_1\nabla u\big) + \widehat{c}u = f(x), \quad x \in \mathbb{R}^n, \tag{10.37}$$

is said to be *homogenized*. The solution to equation (10.37) is denoted by $u^0(x)$. Under our assumptions, this equation is coercive and, consequently, has a unique solution in $L_2(\mathbb{R}^n)$.

THEOREM 10.5. *Let $f^\varepsilon(x)$ converge strongly (weakly) in $L_2(\mathbb{R}^n,\mu_\varepsilon)$ to a function $f(x) \in L_2(\mathbb{R}^n)$. Under the above assumptions, the following limit relation holds*:

$$u^\varepsilon(x) \longrightarrow u^0(x) \quad \textit{strongly (weakly) in } L_2(\mathbb{R}^n,\mu_\varepsilon) \textit{ as } \varepsilon\to 0,$$

*where $u^\varepsilon(x)$ is a solution to equation* (10.36). *Moreover, the flux convergence holds*:

$$a\Big(\frac{x}{\varepsilon}\Big)\nabla^{\mu_\varepsilon}u^\varepsilon \rightharpoonup \widehat{A}_1\nabla^{\text{eff}}u^0(x) \quad \textit{weakly in } (L_2(\mathbb{R}^n,\mu_\varepsilon))^n \textit{ as } \varepsilon\to 0,$$

*and in the case of strong convergence of $f^\varepsilon$ to $f$, the energy convergence holds*:

$$\lim_{\varepsilon\to 0}\int_{\mathbb{R}^n} a\Big(\frac{x}{\varepsilon}\Big)\nabla^{\mu_\varepsilon}u^\varepsilon(x)\cdot\nabla^{\mu_\varepsilon}u^\varepsilon(x)d\mu_\varepsilon(x) = \int_{\mathbb{R}^n}\widehat{A}_1\nabla^{\text{eff}}u^0\cdot\nabla^{\text{eff}}u^0dx.$$

PROOF. Taking $u^\varepsilon(x)$ for a test function in the integral identity for equation (10.36), we obtain the following a priori estimate:

$$\|u^\varepsilon\|_{L_2(\mathbb{R}^n,\mu_\varepsilon)} + \|\nabla^{\mu_\varepsilon}u^\varepsilon\|_{(L_2(\mathbb{R}^n,\mu_\varepsilon))^n} \le C\|f^\varepsilon\|_{L_2(\mathbb{R}^n,\mu_\varepsilon)}.$$

By Theorem 10.3, we can choose a subsequence $\varepsilon_k \to 0$ such that

$$u^\varepsilon \rightharpoonup u^0(x) \quad \text{weakly in } L_2(\mathbb{R}^n,\mu_\varepsilon) \text{ as } \varepsilon_k\to 0,$$

$$\nabla^{\mu_\varepsilon}u^\varepsilon \xrightarrow{\ 2\ } \nabla^{\text{eff}}u^0(x) + u_1(x,y) \quad \text{two-scale in } L_2(\mathbb{R}^n,\mu_\varepsilon) \text{ as } \varepsilon_k\to 0;$$

moreover, $u^0 \in H^{\text{eff}}(\mathbb{R}^n)$ and $u_1 \in L_2(G; L_2^{\text{pot}}(\square,\mu))$. Substituting a test function of the form $\varepsilon\varphi(x)g(\frac{x}{\varepsilon})$, $\varphi\in C_0^\infty(\mathbb{R}^n)$, $g(y)\in H^1(\square,\mu)$, in the integral identity for equation (10.36), we find that

$$\begin{aligned}\varepsilon\int_{\mathbb{R}^n} a\Big(\frac{x}{\varepsilon}\Big)\nabla^{\mu_\varepsilon}u^\varepsilon(x)\cdot\Big\{\varphi(x)\nabla^{\mu_\varepsilon}g\Big(\frac{x}{\varepsilon}\Big) + g\Big(\frac{x}{\varepsilon}\Big)\nabla\varphi(x)\Big\}d\mu_\varepsilon(x)\\ +\varepsilon\int_{\mathbb{R}^n} c\Big(\frac{x}{\varepsilon}\Big)u^\varepsilon(x)\varphi(x)g\Big(\frac{x}{\varepsilon}\Big)d\mu_\varepsilon(x) = \varepsilon\int_{\mathbb{R}^n} f^\varepsilon(x)\varphi(x)g\Big(\frac{x}{\varepsilon}\Big)d\mu_\varepsilon(x).\end{aligned}$$

Passing to the two-scale limit in this equality, we obtain the relation

$$\int_{\mathbb{R}^n\times\square} a(y)\big(\nabla^{\text{eff}}u^0(x) + u_1(x,y)\big)\cdot\varphi(x)\nabla^\mu g(y)dxd\mu(y) = 0,$$

which implies

$$\int_\square a(y)\big(\nabla^{\text{eff}}u^0(x) + u_1(x,y)\big)\nabla^\mu g(y)\mu(y) = 0$$

for almost all $x \in \mathbb{R}^n$. Since a solution to (10.27) is a linear function of $\xi$ and $\nabla^{\text{eff}} u^0(x) = \Pi^{\text{eff}} \nabla u^0(x)$, we have $u_1(x,y) = v^+_{(\nabla^{\text{eff}} u^0(x))}(y)$.

Taking $\varphi \in C_0^\infty(x)$ for a test function, we see that

$$\int_{\mathbb{R}^n} a\Big(\frac{x}{\varepsilon}\Big) \nabla^{\mu_\varepsilon} u^\varepsilon(x) \cdot \nabla\varphi(x) d\mu_\varepsilon(x) + \int_{\mathbb{R}^n} c\Big(\frac{x}{\varepsilon}\Big) u^\varepsilon(x)\varphi(x) d\mu_\varepsilon(x)$$
$$= \int_{\mathbb{R}^n} f^\varepsilon(x)\varphi(x) d\mu_\varepsilon(x).$$

Passing to the two-scale limit, we find that

$$\begin{aligned}(10.38)\qquad &\int_{\mathbb{R}^n\times\square} a(y)\big(\nabla^{\text{eff}} u^0(x) + v^+_{(\nabla^{\text{eff}} u^0(x))}(y)\big) \cdot \nabla\varphi(x) dx d\mu(y)\\ &+ \int_{\mathbb{R}^n\times\square} c(y) u^0(x)\varphi(x) dx d\mu(y) = \int_{\mathbb{R}^n\times\square} f(x)\varphi(x) dx d\mu(y).\end{aligned}$$

By Corollary 10.1,

$$\int_{\mathbb{R}^n\times\square} a(y)\big(\nabla^{\text{eff}} u^0(x) + v^+_{(\nabla^{\text{eff}} u^0(x))}(y)\big) d\mu(y) = \widehat{A}_1 \nabla^{\text{eff}} u^0(x).$$

Therefore, equation (10.38) takes the form

$$\int_{\mathbb{R}^n} \widehat{A}_1 \nabla^{\text{eff}} u^0(x) \cdot \nabla\varphi(x) dx + \int_{\mathbb{R}^n} \widehat{c} u^0(x)\varphi(x) dx = \int_{\mathbb{R}^n} f(x)\varphi(x) dx.$$

It remains to note that the $u^\varepsilon(x)$ converge to $u^0(x)$ due to the uniqueness of a solution to the limit problem.

We prove that the $u^\varepsilon(x)$ strongly converge to $u^0(x)$ in $L_2(\mathbb{R}^n, \mu_\varepsilon)$ if the $f^\varepsilon$ strongly converge. For this purpose, we consider the auxiliary equation

$$(10.39)\qquad -\text{div}^{\mu_\varepsilon}\Big(a\Big(\frac{x}{\varepsilon}\Big) \nabla^{\mu_\varepsilon} z^\varepsilon\Big) + c\Big(\frac{x}{\varepsilon}\Big) z^\varepsilon = u^\varepsilon(x) \quad \text{in } L_2(\mathbb{R}^n, \mu_\varepsilon).$$

Since $u^\varepsilon$ weakly converge to $u^0(x)$ in $L_2(\mathbb{R}^n, \mu_\varepsilon)$, the above arguments show that $z^\varepsilon(x)$ converge in the same space to the solution $z^0(x)$ of the equation

$$-\text{div}\,(\widehat{A}_1 \nabla^{\text{eff}} z^0(x)) + \widehat{c} z^0 = u^0(x).$$

Substituting $u^\varepsilon$ for a test function in the integral identity for equation (10.39) and $z^\varepsilon(x)$ for a test function in the integral identity for equation (10.36), and subtracting the obtained equalities, we find that

$$\int_{\mathbb{R}^n} (u^\varepsilon(x))^2 d\mu_\varepsilon(x) = \int_{\mathbb{R}^n} f^\varepsilon(x) z^\varepsilon(x) d\mu_\varepsilon(x).$$

Handling the limit equations in the same way, we obtain the equality

$$\int_{\mathbb{R}^n} (u^0(x))^2 dx = \int_{\mathbb{R}^n} f(x) z^0(x) dx.$$

Passing to the limit in the last two relations, we see that

$$\lim_{\varepsilon\to 0} \int_{\mathbb{R}^n} (u^\varepsilon(x))^2 d\mu_\varepsilon(x) = \int_{\mathbb{R}^n} f(x) z^0(x) dx = \int_{\mathbb{R}^n} (u^0(x))^2 dx.$$

To prove the flux convergence, we recall that

$$a\Big(\frac{x}{\varepsilon}\Big) \nabla^{\mu_\varepsilon} u^\varepsilon \xrightarrow{2} a(y)(\nabla^{\text{eff}} u^0(x) + v^+_{\nabla^{\text{eff}} u^0(x)})$$

two-scale in $L_2(\mathbb{R}^n, \mu_\varepsilon)$. Hence we obtain the required flux convergence since

$$\int_\square \big(a(y)(\nabla^{\mathrm{eff}} u^0(x) + v^+_{\nabla^{\mathrm{eff}} u^0(x)})\big)\,d\mu(y) = \widehat{A}_1 \nabla^{\mathrm{eff}} u^0(x).$$

The proof of the energy convergence is left to the reader. □

The homogenization theorem is also valid for boundary value problems. We illustrate this by an example of a Dirichlet problem. Let $G$ be a Lipschitz bounded domain. Consider the Dirichlet problem

$$-\mathrm{div}^{\mu_\varepsilon}\Big(a\Big(\frac{x}{\varepsilon}\Big)\nabla^{\mu_\varepsilon} u^\varepsilon\Big) + c\Big(\frac{x}{\varepsilon}\Big)u^\varepsilon = f^\varepsilon(x) \quad \text{in } L_2(G, \mu_\varepsilon), \tag{10.40}$$

$$u^\varepsilon \in \overset{\circ}{H}{}^1 (G, \mu_\varepsilon).$$

The assertion below can be proved in the same way as the previous theorem.

THEOREM 10.6. *Suppose that the $f^\varepsilon(x)$ strongly (weakly) converge in $L_2(G, \mu_\varepsilon)$ to some function $f(x) \in L_2(G)$. Then, under the assumptions of Theorem 10.5, the following limit relation holds:*

$$u^\varepsilon(x) \longrightarrow u^0(x) \quad \textit{strongly (weakly) in } L_2(G, \mu_\varepsilon) \textit{ as } \varepsilon \to 0,$$

*where $u^\varepsilon(x)$ is a solution to equation (10.40) and $u^0(x)$ is a solution to the homogenized problem*

$$-\mathrm{div}\,(\widehat{A}_1 \nabla^{\mathrm{eff}} u^0) + \widehat{c} u^0 = f \quad \textit{in } G,$$
$$u^0 \in H_0^{\mathrm{eff}}(G).$$

*Moreover, the flux convergence holds:*

$$a\Big(\frac{x}{\varepsilon}\Big)\nabla^{\mu_\varepsilon} u^\varepsilon \rightharpoonup \widehat{A}_1 \nabla^{\mathrm{eff}} u^0(x) \quad \textit{weakly in } (L_2(G, \mu_\varepsilon))^n \textit{ as } \varepsilon \to 0$$

*and, in the case of strong convergence of $f^\varepsilon$, the energy convergence holds:*

$$\lim_{\varepsilon \to 0} \int_G a\Big(\frac{x}{\varepsilon}\Big)\nabla^{\mu_\varepsilon} u^\varepsilon(x) \cdot \nabla^{\mu_\varepsilon} u^\varepsilon(x) d\mu_\varepsilon(x) = \int_G \widehat{A}_1 \nabla^{\mathrm{eff}} u^0 \cdot \nabla^{\mathrm{eff}} u^0 dx.$$

**10.4. Comments.** The developed technique can be successfully used in the study of homogenization problems for high contrast singular and thin structures, for example, singular double porosity problems (cf. [**134, 26**]); the homogenization of parabolic problems in variable spaces, elasticity problems for thin frames (cf. [**135, 139, 112**]); nonlinear operators in variable spaces (cf. [**136, 83**]); and many other problems. It is of interest not only for homogenization theory, but also in relation to close topics such as the central limit theorem, spectral problems, the commutativity of diagrams under the limit passage with respect to the period size and the thickness of the structure (cf. [**42, 135**]), and many other aspects.

## 11. Method of matching asymptotic expansions

**11.1. Simple example.** We begin with a simple example from [**71**]. This example is not related to boundary value problems but well illustrates the method itself. The idea is that a function (usually, a solution to a singularly perturbed boundary value problem) depending on a small parameter is approximated by one asymptotic series (*outer expansion*) everywhere in the domain except for a small neighborhood of some set, and by another series (*inner expansion*) in a small neighborhood of this set, called a *boundary layer*. From these two expansions we construct a *complete expansion* as the sum of intervals of outer and inner expansions minus their common part.

We study the behavior of the function

$$f_\varepsilon(x) = \frac{e^x}{x+\varepsilon} \tag{11.1}$$

on $[0,1]$ as $\varepsilon \to 0$. For fixed $\varepsilon$ the function $f_\varepsilon(x)$ has no singularities on $[0,1]$. However, the singular point of $f_\varepsilon(x)$ approaches $[0,1]$ as $\varepsilon \to 0$. Therefore, as $\varepsilon \to 0$, one can expect some difficulties with convergence of asymptotic power series with respect to $\varepsilon$ in a neighborhood of $x = 0$.

For any $x > 0$ the function (11.1) is expanded into a power series with respect to $\varepsilon$:

$$f_\varepsilon(x) = \frac{e^x}{x}\left(1 - \frac{\varepsilon}{x} + \frac{\varepsilon^2}{x^2} - \frac{\varepsilon^3}{x^3} + \cdots\right), \tag{11.2}$$

and this asymptotics is uniform with respect to $x$ for $x \geq \delta > 0$. Indeed, for $x \geq \delta$ we have

$$\left| f_\varepsilon(x) - \frac{e^x}{x}\sum_{i=0}^{n}(-1)^i\frac{\varepsilon^i}{x^i}\right| = \frac{e^x}{x+\varepsilon}\,\frac{\varepsilon^{n+1}}{x^{n+1}} \leq \frac{3\varepsilon^{n+1}}{\delta^{n+2}}. \tag{11.3}$$

On the other hand, for small $x$ the series (11.2) does not approximate the function $f_\varepsilon(x)$. It is easy to estimate the difference (11.3), for example, for $x \leq \varepsilon$. We have

$$\left| f_\varepsilon(x) - \frac{e^x}{x}\sum_{i=0}^{n}(-1)^i\frac{\varepsilon^i}{x^i}\right| = \frac{e^x}{x+\varepsilon}\,\frac{\varepsilon^{n+1}}{x^{n+1}} > \frac{1}{2\varepsilon}.$$

To represent $f_\varepsilon(x)$ as a series in a neighborhood of $x = 0$, we introduce "fast" variables $\xi = \frac{x}{\varepsilon}$ and expand $f_\varepsilon(x)$ as the asymptotic series

$$f_\varepsilon(\varepsilon\xi) = \frac{e^{\varepsilon\xi}}{\varepsilon(1+\xi)} = \frac{1}{\varepsilon(1+\xi)}\left(1 + \varepsilon\xi + \varepsilon^2\frac{\xi^2}{2} + \cdots\right). \tag{11.4}$$

In this case, it is easy to estimate the difference between $f_\varepsilon$ and a partial sum of the series:

$$\left| f_\varepsilon(\varepsilon\xi) - \frac{1}{\varepsilon(1+\xi)}\sum_{i=0}^{k}\frac{(\varepsilon\xi)^i}{i!}\right| = \frac{3}{\varepsilon(1+\xi)}\,\frac{(\varepsilon\xi)^{k+1}}{(k+1)!}. \tag{11.5}$$

Therefore, for $\xi \ll 1/\varepsilon$ the series (11.4) well approximates $f_\varepsilon(\varepsilon\xi)$.

We describe the intervals where the series (11.2) and (11.4) well approximate the function $f_\varepsilon(x)$ (cf. Figure 11.1).

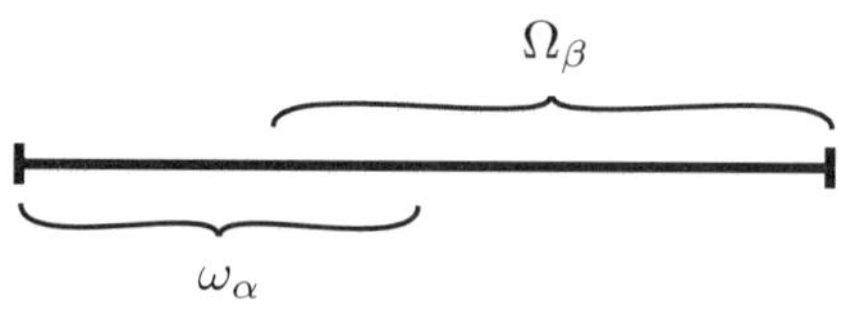


FIGURE 11.1. Sets where the series well approximate the function.

Let $0 < \alpha < \beta < 1$. Consider the intervals $\Omega = \Omega_\beta = \{x : \varepsilon^\beta \leq x \leq 1\}$ and $\omega = \omega_\alpha = \{x : 0 \leq x \leq \varepsilon^\alpha\}$. The series (11.2) well approximates $f_\varepsilon(x)$ in $\Omega$ by

(11.3), and the series (11.4) well approximates $f_\varepsilon(x)$ in $\omega$ by (11.5). Moreover, on the intersection $\{x : \varepsilon^\beta \le x \le \varepsilon^\alpha\}$ of $\Omega$ and $\omega$, both series well approximate $f_\varepsilon(x)$ and, in particular, they are close to each other. Therefore, we can construct a new (complete) asymptotic series approximating $f_\varepsilon(x)$ on the entire segment $[0, 1]$.

Introduce the notation

$$S_n^x := \sum_{i=0}^{n} (-1)^i \frac{e^x}{x^{i+1}} \varepsilon^i, \quad \sigma_m^\xi := \sum_{i=0}^{m+1} \frac{\xi^i}{(1+\xi)\, i!} \varepsilon^{i-1}.$$

As usual, setting $x = \varepsilon\xi$, we note that $S_n^x$ and $\sigma_m^\xi$ are close in $\Omega \cap \omega$ for small $x$ and large $\xi$.

We compare these series as follows. Expanding all the terms of the partial sum $S_n^x$ into power series with respect to $x$ in a neighborhood of $x = 0$, replacing $x$ with $\varepsilon\xi$, and denoting by $S_{m,n}^{\xi,x}$ the sum of all terms with $\varepsilon^s$, $s \le m$, we obtain

$$S_{m,n}^{\xi,x} = \sum_{j=0}^{n} \sum_{i=0}^{m+1} (-1)^j \frac{\xi^{-j-1+i}}{i!} \varepsilon^{i-1}.$$

Expanding all the terms of the partial sum $\sigma_m^\xi$ into power series with respect to $\xi$ in a neighborhood of $+\infty$, replacing $\xi$ with $x/\varepsilon$, and denoting by $\sigma_{n,m}^{x,\xi}$ the sum of all terms with $\varepsilon^s$, $s \le n$, we find

$$\sigma_{n,m}^{x,\xi} = \sum_{j=0}^{m+1} \sum_{i=0}^{n} (-1)^i \frac{x^{j-i-1}}{j!} \varepsilon^i.$$

Substituting $x/\varepsilon$ for $\xi$ in $S_{m,n}^{\xi,x}$, we see that

$$S_{m,n}^{\xi,x} \equiv \sigma_{n,m}^{x,\xi}. \tag{11.6}$$

This is the common part of the asymptotic series (11.2) and (11.4), and the equality (11.6) is an algebraic expression of the matching condition for the outer and inner expansions.

Expanding all the terms of the partial sum $\sigma_m^\xi$ into power series with respect to $\xi$ in a neighborhood of $+\infty$, replacing $\xi$ with $x/\varepsilon$, and denoting by $\sigma_{n,m}^{x,\xi}$ the sum of all terms with $\varepsilon^s$, $s \le n$, we find

$$\sigma_{n,m}^{x,\xi} = \sum_{j=0}^{m+1} \sum_{i=0}^{n} (-1)^i \frac{x^{j-i-1}}{j!} \varepsilon^i.$$

We construct a complete asymptotic expansion in the form

$$\begin{aligned} f_\varepsilon^n(x) &= (S_n^x + \sigma_n^\xi - S_{n,n}^{\xi,x})\big|_{\xi = x/\varepsilon} \\ &= \sum_{i=0}^{n} (-1)^i \frac{e^x}{x^{i+1}} \varepsilon^i + \sum_{i=0}^{n+1} \frac{x^i}{(\varepsilon + x)\, i!} - \sum_{j=0}^{n} \sum_{i=0}^{n+1} (-1)^j \frac{x^{i-j-1}}{i!} \varepsilon^j. \end{aligned} \tag{11.7}$$

The function $f_\varepsilon^n(x)$ uniformly approximates $f_\varepsilon(x)$ everywhere on $[0, 1]$ (cf. Figure 11.2).

Indeed, on $\omega$, the function $f_\varepsilon(x)$ is close to $\sigma_n^\xi$ and the difference $S_n^x - S_{n,n}^{\xi,x}$ is small. On $\Omega$, the function $f_\varepsilon(x)$ is close to $S_n^x$ and the difference $\sigma_n^\xi - S_{n,n}^{\xi,x} \equiv \sigma_n^\xi - \sigma_{n,n}^{x,\xi}$ is small. Thus, the series (11.7) well approximates $f_\varepsilon(x)$ on the entire interval $[0, 1]$.

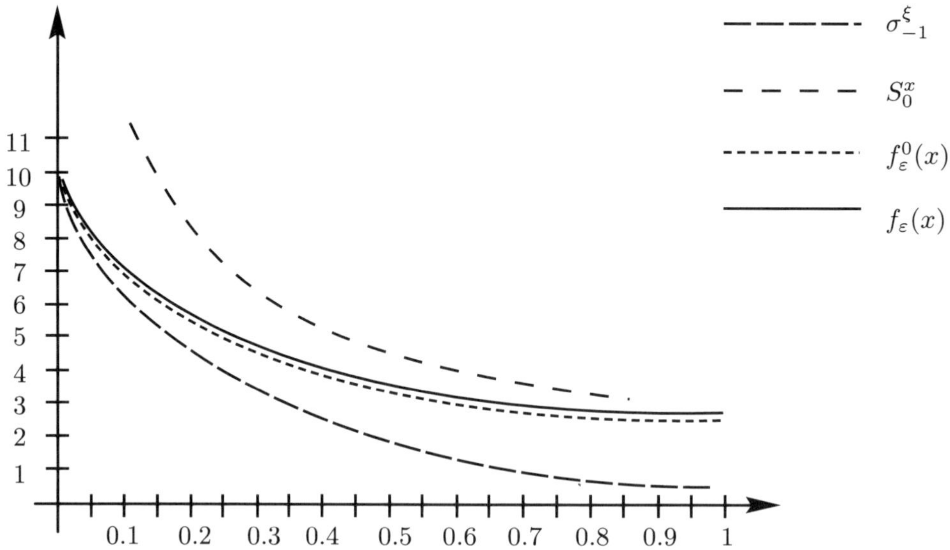


FIGURE 11.2. The graphs of functions and partial sums of series with $\varepsilon = 1/10$; here, $|f_{1/10}(x) - f^0_{1/10}(x)| < 0.07$.

**11.2. The Dirichlet problem in a domain with small cavity.** Let $\Omega$ and $\omega$ be bounded domains in $\mathbb{R}^3$ that have infinitely differentiable boundary and contain the origin. Suppose $0 < \varepsilon \ll 1$ is a small parameter, $\omega_\varepsilon = \{x : \varepsilon^{-1}x \in \omega\}$, $\Omega_\varepsilon = \Omega \backslash \overline{\omega_\varepsilon}$, $\Gamma = \partial\Omega$, $\Gamma_\varepsilon = \partial\Omega_\varepsilon$, $\gamma = \partial\omega$, $\gamma_\varepsilon = \partial\omega_\varepsilon$, and $f \in C^\infty(\Gamma)$. An example of such a domain is represented in Figure 11.3.

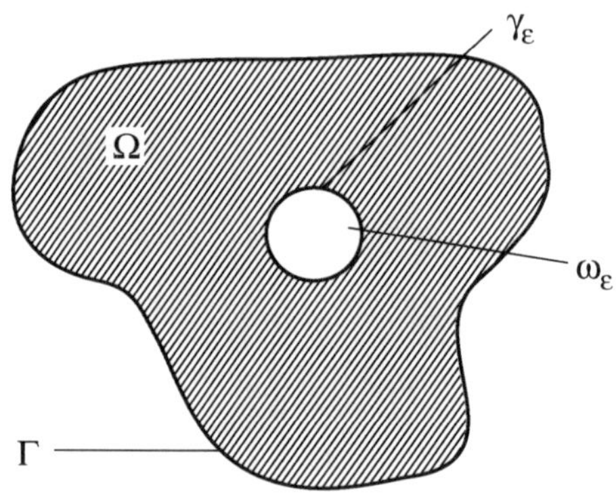


FIGURE 11.3

Our goal is to construct asymptotic solutions $u_\varepsilon$ as $\varepsilon \to 0$ to the boundary value problem

$$\Delta u_\varepsilon = 0 \quad \text{for } x \in \Omega_\varepsilon, \qquad u_\varepsilon = f \quad \text{for } x \in \Gamma, \tag{11.8}$$

$$u_\varepsilon = 0 \quad \text{for } x \in \gamma_\varepsilon. \tag{11.9}$$

We show that the outer expansion has the form

$$u_\varepsilon(x) = u_0(x) + \sum_{i=1}^{\infty} \varepsilon^i u_i(x), \tag{11.10}$$

where $u_0$ is a solution to the problem

$$\Delta u_0 = 0 \quad \text{for } x \in \Omega, \qquad u_0 = f \quad \text{for } x \in \Gamma. \tag{11.11}$$

A polynomial $Y(x)$, $x \in \mathbb{R}^d$, is said to be *harmonic* if $\Delta Y(x) = 0$ and *homogeneous of degree* $i$ if it is the sum of monomials of the same degree $i$.

EXERCISE 11.1. Prove that a harmonic function $u_0(x)$ admits the expansion

$$u_0(x) = \sum_{i=0}^{\infty} Y_i^{(0)}(x)$$

in a neighborhood of $x = 0$, where $Y_i^{(0)}(x)$ are homogeneous harmonic polynomials of degree $i$.

EXERCISE 11.2. Let $u_0(x) = \sum_{i=1}^{\infty} \sum_{|\alpha|=i} \frac{D^\alpha u_0(x)}{\alpha!}\Big|_{x=0} x^\alpha$ be the Taylor series of $u_0(x)$ at the origin. Are the polynomials $Y_i(x) \equiv \sum_{|\alpha|=i} \frac{D^\alpha u_0(x)}{\alpha!}\Big|_{x=0} x^\alpha$ harmonic? Here, $\alpha = (\alpha_1, \dots, \alpha_d)$, $\alpha! = \alpha_1! \cdots \alpha_d!$, $D^\alpha = \frac{\partial^{\alpha_1}}{\partial x_1^{\alpha_1}} \cdots \frac{\partial^{\alpha_d}}{\partial x_d^{\alpha_d}}$, $x^\alpha = x_1^{\alpha_1} \cdots x_d^{\alpha_d}$, $|\alpha| = \alpha_1 + \cdots + \alpha_d$.

Let $u_0$ be a solution to the problem (11.11). As is known, $u_0$ belongs to $C^\infty(\overline{\Omega})$ (recall that $f \in C^\infty(\Gamma)$) and, in a neighborhood of the origin, $u_0$ is represented as the series (cf. Exercise 11.1)

$$u_0(x) = \sum_{i=0}^{\infty} Y_i^{(0)}(x). \tag{11.12}$$

Hereinafter, $Y_i^{(m)}$ and $X_i^{(q,p)}$ denote homogeneous harmonic polynomials of degree $i$. It is obvious that $Y_0^{(0)} = u_0(0)$. Since $u_0$ satisfies (11.8) and the hole $\omega_\varepsilon$ is small, it is natural to assume that $u_\varepsilon$ is close to $u_0$. The obvious drawback of $u_0$ is that it does not satisfy the boundary condition (11.9). Therefore, in a neighborhood of the small hole $\omega_\varepsilon$, we look for an asymptotic formula for the solution to the problem (11.8), (11.9) in some other form. Furthermore, equation (11.8) and the boundary condition (11.9) must be satisfied in a neighborhood of $\omega_\varepsilon$. Since $\omega_\varepsilon$ is obtained by contracting a fixed domain $\omega$ by a factor of $\varepsilon^{-1}$, it is natural to seek a solution in a neighborhood of $\omega_\varepsilon$ in the form of a series $v_\varepsilon$ with coefficients depending on the "inner" variables $\xi = x\varepsilon^{-1}$.

We begin with an informal explanation of the idea of the matching method. Let $r = |x|$. If $0 < q < 1$, then for $\varepsilon^q < r < 2\varepsilon^q$ and $\rho = |\xi|$ we have $\varepsilon^{-1+q} < \rho < 2\varepsilon^{-1+q}$, i.e., in the indicated domain, $x$ is small and $\xi$ is large. The asymptotic series $u_\varepsilon(x)$ (we have only the zeroth term $u_0(x)$ yet) and the asymptotic series $v_\varepsilon(\xi)$ (the structure of which is not yet known) must coincide in the intermediate zone, where $x$ is small and $\xi$ is large. Then it is possible to "jump" from one expansion to the other in this zone. For definiteness, we set $q = 1/2$.

Let us clarify the structure of the series $v_\varepsilon$ and $u_\varepsilon$. Passing to the variables $\xi$ in (11.12), we find that for $\varepsilon^{1/2} < r < 2\varepsilon^{1/2}$ (or, which is the same, for $\varepsilon^{-1/2} < \rho <$

$2\varepsilon^{-1/2}$),

$$u_0(x) = \sum_{i=0}^{\infty} \varepsilon^i Y_i^{(0)}(\xi). \tag{11.13}$$

From (11.13) and the matching condition it follows that the terms $v_i(\xi)$ of the expansion $v_\varepsilon(\xi) = \sum_{i=0}^{\infty} \varepsilon^i v_i(\xi)$ admit the following asymptotics:

$$v_i(\xi) \sim Y_i^{(0)}(\xi) \quad \text{as } \rho \to \infty. \tag{11.14}$$

Substituting (11.14) in (11.8) and (11.9), passing to the variables $\xi$, and collecting terms with equal power of $\varepsilon$, we obtain the following boundary value problem:

$$\Delta_\xi v_i = 0 \quad \text{in } \mathbb{R}^3 \backslash \overline{\omega}, \quad v_i = 0 \quad \text{on } \gamma. \tag{11.15}$$

A solution to the problem (11.15) is not unique since there are no conditions at infinity. It is obvious that for any homogeneous harmonic polynomial $Y_i^{(0)}(\xi)$ there exists a solution $v_i = V_i^{(0)}$ to the problem (11.15) with the leading term $Y_i^{(0)}(\xi)$ in the asymptotic expansion at infinity. Indeed, representing $V_i^{(0)}$ as the sum of $Y_i^{(0)}$ and a harmonic function in $\mathbb{R}^3 \backslash \overline{\omega}$ (decreasing at infinity) such that its boundary value on $\gamma$ is equal to $-Y_i^{(0)}(\xi)\big|_\gamma$, we obtain the solution to the problem (11.15) with the following asymptotic expansion at infinity:

$$V_i^{(0)}(\xi) = Y_i^{(0)}(\xi) + \sum_{j=0}^{\infty} X_j^{(i,0)}(\xi)\rho^{-2j-1}. \tag{11.16}$$

Therefore, this solution satisfies (11.14).

EXERCISE 11.3. Prove the uniqueness of a solution to the Dirichlet problem

$$\Delta V = 0 \quad \text{in } \mathbb{R}^3 \backslash \omega, \quad V|_{\partial\omega} = \varphi, \quad V \to 0 \text{ as } |\xi| \to 0.$$

EXERCISE 11.4. Prove that a harmonic function in $\mathbb{R}^3 \backslash \omega$ decreasing at infinity can be represented as a convergent series $V(\xi) = \sum_{j=0}^{\infty} X_j(\xi)\rho^{-2j-1}$ for $|\xi| > R_0$, where $X_j(\xi)$ is a homogeneous harmonic polynomial of degree $j$.

EXERCISE 11.5. Prove that this expansion is unique.

It is obvious that for $i \geq 1$ the function $V_i^{(0)} + \sum_{j=0}^{i-1} a_j V_j^{(0)}$ also satisfies (11.14) and (11.15) with any $a_i$. We set $v_0 = V_0^{(0)}$. The remaining terms $v_i$ will be specified below.

We write the asymptotics (11.16) for $v_0$ in terms of the "outer" variables $x$. For $\varepsilon^{1/2} < r < 2\varepsilon^{1/2}$ (or, which is the same, for $\varepsilon^{-1/2} < \rho < 2\varepsilon^{-1/2}$) we have

$$v_0\left(\frac{x}{\varepsilon}\right) = Y_0^{(0)} + \sum_{j=0}^{\infty} \varepsilon^{j+1} X_j^{(0,0)}(x) r^{-2j-1}. \tag{11.17}$$

From (11.17) and the matching condition it follows that the coefficients of the series $u_\varepsilon(x) = \sum_{i=0}^{\infty} \varepsilon^i u_i(x)$ (cf. (11.10)) admit the following asymptotic expansions:

$$u_0(x) \sim Y_0^{(0)}, \quad u_{j+1}(x) \sim X_j^{(0,0)}(x) r^{-2j-1} \quad \text{as} \quad r \to 0. \tag{11.18}$$

Substituting (11.10) in (11.8) and collecting terms with equal power of $\varepsilon$, we obtain the boundary value problem

$$\Delta u_i = 0 \quad \text{in } \Omega \backslash \{0\}, \qquad u_i = \delta_i^0 f \quad \text{for } x \in \Gamma, \tag{11.19}$$

where $\delta_i^j$ is the Kronecker symbol. Note that the $u_j$, $j \geq 1$, have singularities at the origin, and consequently the origin is not a removable singular point. It is obvious that $u_0$ satisfies (11.18), (11.19) and for any homogeneous harmonic polynomial there exists a solution to the boundary value problem (11.19) with the leading asymptotic term (11.18) in a neighborhood of the origin. Indeed, representing $u_{j+1}$ as the sum of $X_j^{(0,0)} r^{-2j-1}$ and a harmonic function in $\Omega$, we obtain the solution to the problem (11.19) with the asymptotics (11.18) at the origin. However, for $i \geq 1$ the function $u_{i+1} + \sum_{j=0}^{i-1} a_j u_{j+1}$ also satisfies (11.18) and (11.19) for any constants $a_i$. Therefore, the form of lower order terms in the asymptotic formula at the origin for $u_{i+1}$, $i \geq 1$, will be specified below (unlike $u_0$, which is completely determined at this step).

Now, we rigorously describe the method of matching asymptotic expansions.

Suppose that $U_i(x)$ is represented as the Laurent series

$$U_i(x) = \sum_{j=-\infty}^{+\infty} \Phi_{ij}(x), \tag{11.20}$$

where $\Phi_{ij}$ are homogeneous polynomials of degree $j \geq 0$ of the form

$$\Phi_{ij}(x) = \frac{\widetilde{\Phi}_{ij}(x)}{r^{2j+1}}, \tag{11.21}$$

where $\widetilde{\Phi}_{ij}(x)$ are homogeneous polynomials of degree $-j-1$ for $j < 0$. On the set of functions $U(x, \varepsilon) = \sum_{i=0}^{\infty} \varepsilon^i U_i(x)$, we define the operator $K_i(U(x, \varepsilon))$ by the formula

$$K_i(U(x, \varepsilon)) \equiv \sum_{l=-i}^{\infty} \Phi_{l,i-l}(\xi).$$

This operator can be obtained by replacing the coefficients of the series $U(x, \varepsilon)$ with their expansions (11.20) and (11.21) at the origin, making the change of variables $\xi = \frac{x}{\varepsilon}$, and taking the sum of terms with $\varepsilon^i$. Let

$$\mathcal{K}_N(U(x, \varepsilon)) := \sum_{i \leq N} K_i(U(x, \varepsilon)).$$

Suppose that $V_i(\xi)$ is represented as the Laurent series

$$V_i(\xi) = \sum_{j=-\infty}^{+\infty} W_{ij}(\xi), \tag{11.22}$$

where $W_{ij}$ are homogeneous polynomials of degree $j \geq 0$ of the form

$$W_{ij}(\xi) = \frac{\widetilde{W}_{ij}(\xi)}{\rho^{2j+1}}, \tag{11.23}$$

where $\widetilde{W}_{ij}(\xi)$ are homogeneous polynomials of degree $-j-1$ for $j < 0$. On the set of functions $V(\xi, \varepsilon) = \sum_{i=0}^{\infty} \varepsilon^i V_i(\xi)$, we introduce the operator $M_i(V(\xi, \varepsilon))$ by the formula

$$M_i(V(\xi, \varepsilon)) \equiv \sum_{l=-i}^{\infty} W_{l,-i+l}(\xi).$$

This operator is obtained by replacing the coefficients of the series $V(\xi, \varepsilon)$ with their expansions (11.22) and (11.23) at infinity, making the change of variables $x = \varepsilon\xi$, and taking the sum of terms with $\varepsilon^i$ in the double series obtained. We set

$$\mathcal{M}_N(V(\xi,\varepsilon)) = \sum_{i \le N} M_i(V(\xi,\varepsilon)).$$

We say that a function $u(x)$ is *homogeneous of degree* $i$ if $u(\kappa x) = \kappa^i x$ for any $x \in \mathbb{R}^d$ and $\kappa > 0$ ($i$ may be a real number).

EXERCISE 11.6. Let $u(x)$ be homogeneous of degree $i$ and differentiable for $x \neq 0$. Is the function $\frac{\partial u}{\partial x_k}(x)$ homogenous? If the answer is "yes," what is its homogeneity degree?

Denote by $\widehat{u}_{\varepsilon,N}$ and $\widehat{v}_{\varepsilon,N}$ partial sums of the series $u_\varepsilon$ and $v_\varepsilon$, respectively, and by $H_i^{(m)}(x)$ and $h_i^{(q)}(\xi)$, homogeneous functions of degree $i$.

DEFINITION 11.1. We say that a function $u(x)$ admits a *differential asymptotics* as $x \to 0$ if there exist homogeneous functions $U_j(x)$ of degree $j$ ($j$ is an integer) such that

$$D^\alpha\left(u(x) - \sum_{j=-i}^{N} U_j(x)\right) = O(|x|^{1+N-|\alpha|}) \quad \forall N \ge i$$

for $|\alpha| \ge 0$.

Asymptotics at infinity and (asymptotic) equalities can be introduced in a similar way.

LEMMA 11.1. *Suppose that the coefficients of the series $u_\varepsilon$ belong to $C^\infty(\overline{\Omega}\backslash\{0\})$ and have the following differential asymptotics as $r \to 0$:*

$$u_i(x) = \sum_{j=-i}^{\infty} H_j(x),$$

*whereas the coefficients of the series $v_\varepsilon$ (cf. (11.14)) belong to $C^\infty(\mathbb{R}^3\backslash\omega)$ and have the following differential asymptotics as $\rho \to \infty$:*

$$v_i(\xi) = \sum_{j=-\infty}^{i} h_j(\xi).$$

*Then for any $N \ge 0$ we have*

$$\widehat{u}_{\varepsilon,N}(x) - \mathcal{K}_N(\widehat{u}_{\varepsilon,N}(x)) = O(r^{N+1} + \varepsilon^N r) \quad \text{as } r \to 0,$$
$$\widehat{v}_{\varepsilon,N}(\xi) - \mathcal{M}_N(\widehat{v}_{\varepsilon,N}(\xi)) = O(\rho^{-N-1} + \varepsilon^N \rho^{-1}) \quad \text{as } \rho \to \infty.$$

Lemma 11.1 immediately follows from the definition of the operators $\mathcal{K}_N$, $\mathcal{M}_N$ and the asymptotics of $u_i$ and $v_i$. However, it is not connected with the boundary value problems for these coefficients or with matching of these series.

Let $\chi(t)$ be an infinitely differentiable cut-off function vanishing for $t < 1$ and equal to 1 for $t > 2$ (cf. Figure 11.4). We set

$$\widetilde{u}_{\varepsilon,N}(x) = \chi(r\varepsilon^{-1/2})\widehat{u}_{\varepsilon,N}(x) + (1 - \chi(r\varepsilon^{-1/2}))\widehat{v}_{\varepsilon,N}(\xi).$$

Figure 11.5 presents the zones of action of the terms on the right-hand side.

From the definition of $\widetilde{u}_{\varepsilon,N}(x)$ we immediately obtain the following assertion.

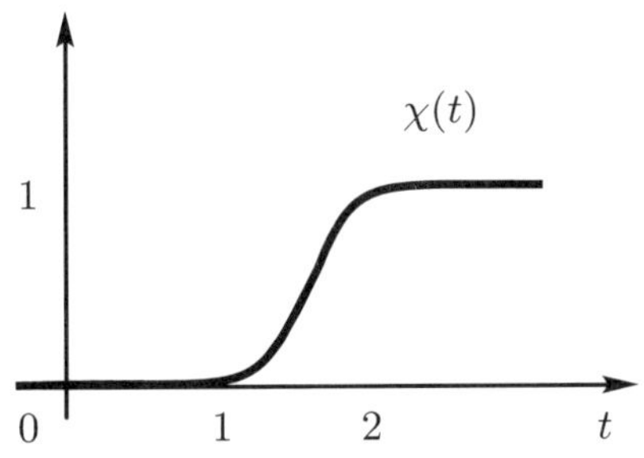


FIGURE 11.4. Cut-off function.

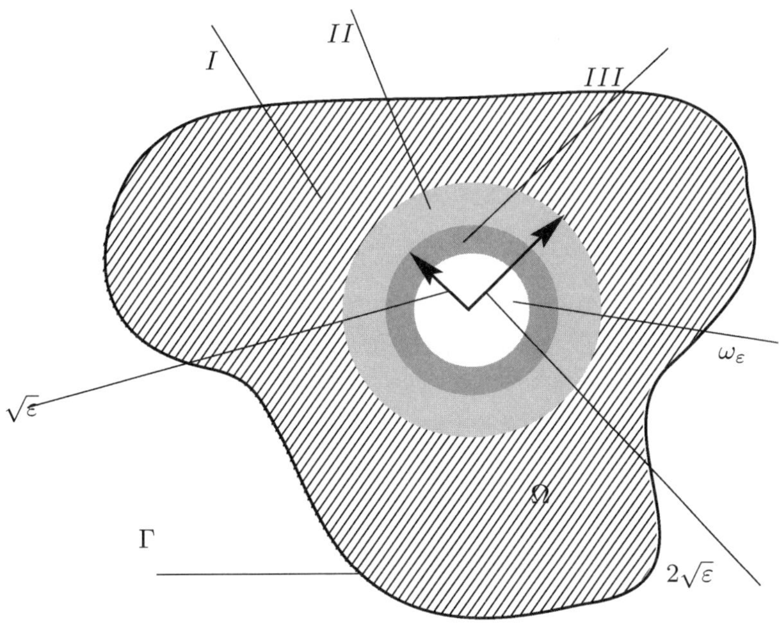


FIGURE 11.5. $I$ is the zone of the action of the series $\widehat{u}_{\varepsilon,N}(x)$, $II$ is the matching zone, and $III$ is the zone of the action of the series $\widehat{v}_{\varepsilon,N}(\xi)$.

LEMMA 11.2. *Suppose that the coefficients of the series $u_\varepsilon$ belong to $C^\infty(\overline{\Omega}\backslash\{0\})$ and solve the boundary value problem* (11.19), *whereas the coefficients of the series $v_\varepsilon$ belong to $C^\infty(\mathbb{R}^3\backslash\omega)$ and solve the boundary value problem* (11.15). *Then $\widetilde{u}_{\varepsilon,N}(x)$ belongs to $C^\infty(\overline{\Omega_\varepsilon})$ and is a solution to the boundary value problem*

$$\Delta\widetilde{u}_{\varepsilon,N} = F_N \textit{ for } x \in \Omega_\varepsilon, \quad \widetilde{u}_{\varepsilon,N} = f \textit{ for } x \in \Gamma, \quad \widetilde{u}_{\varepsilon,N} = 0 \textit{ for } x \in \gamma_\varepsilon,$$

*where*

$$F_N(x,\varepsilon) = 2\sum_{j=1}^{3} \chi_{x_j}(r\varepsilon^{-1/2})(\widehat{u}_{\varepsilon,N}(x) - \widehat{v}_{\varepsilon,N}(\xi))_{x_j} + (\widehat{u}_{\varepsilon,N}(x) - \widehat{v}_{\varepsilon,N}(\xi))\Delta\chi(r\varepsilon^{-1/2}). \tag{11.24}$$

We emphasize that the last assertion is connected with neither the asymptotic expansions of the coefficients of the series $u_\varepsilon$ and $v_\varepsilon$ (at zero and at infinity respectively) nor the compatibility condition which consists in proving the existence of series such that the "smallness order" of the right-hand side of (11.24) unboundedly increases as $N$ increases.

The following assertion is the key point of the matching of series.

THEOREM 11.1. *Suppose that the assumptions of Lemma* 11.2 *are satisfied. Let*

$$\mathcal{K}_N(\widehat{u}_{\varepsilon,N}(x)) = \mathcal{M}_N(\widehat{v}_{\varepsilon,N}(\xi)). \tag{11.25}$$

*Then*

$$|F_N| \le C_N \varepsilon^{N-1/2}. \tag{11.26}$$

PROOF. From (11.25) and Lemma 11.1 it follows that for $\varepsilon^{1/2} < r < 2\varepsilon^{1/2}$,

$$\widehat{u}_{\varepsilon,N}(x) - \widehat{v}_{\varepsilon,N}(\xi) = O(\varepsilon^{N+1/2}), \quad (\widehat{u}_{\varepsilon,N}(x) - \widehat{v}_{\varepsilon,N}(\xi))_{x_j} = O(\varepsilon^{N/2}). \tag{11.27}$$

Since the supports of $\chi_{x_i}(r\varepsilon^{-1/2})$ and $\Delta\chi(r\varepsilon^{-1/2})$ belong to the annuli $\varepsilon^{1/2} < r < 2\varepsilon^{1/2}$ and $\varepsilon^{1/2}|\chi_{x_j}(r\varepsilon^{-1/2})| + \varepsilon|\Delta\chi(r\varepsilon^{-1/2})| \le C$, from (11.27) and (11.24) we obtain (11.26). □

If $u_0$ and $v_0$ are already chosen, we thereby have (11.25) for $N = 0$. We proceed by describing a formal procedure for constructing the remaining coefficients. We look for $v_i$ in the form

$$v_i(\xi) = \sum_{j=0}^{i} V_i^{(j)}(\xi), \tag{11.28}$$

where $V_i^{(0)}$ was defined above, and the remaining harmonic terms will be found at the following steps of the procedure. Let

$$v_i^{(n)} = \sum_{j=0}^{\min\{n,i\}} V_i^{(j)}(\xi), \quad v_\varepsilon^{(n)}(\xi) = \sum_{i=0}^{\infty} \varepsilon^i v_i^{(n)}(\xi).$$

Denote by $\widehat{v}_{\varepsilon,N}^{(n)}(\xi)$ the partial sums of the series $v_\varepsilon^{(n)}(\xi)$. By definition, $\widehat{v}_{\varepsilon,N}^{(N)} = \widehat{v}_{\varepsilon,N}$. By (11.13) and (11.16), we have

$$\mathcal{K}_N(\widehat{u}_{\varepsilon,0}(x)) = \mathcal{M}_0(\widehat{v}_{\varepsilon,N}^{(0)}(\xi)). \tag{11.29}$$

By the definition of $\mathcal{M}_q$,

$$\mathcal{M}_1(\widehat{v}_{\varepsilon,N}^{(0)}(\xi)) = \mathcal{M}_0(\widehat{v}_{\varepsilon,N}^{(0)}(\xi)) + M_1(\widehat{v}_{\varepsilon,N}^{(0)}(\xi)), \tag{11.30}$$

where

$$M_1(\widehat{v}_{\varepsilon,N}^{(0)}(\xi)) = \varepsilon X_0^{(0,0)} r^{-1} \tag{11.31}$$

in view of (11.28) and (11.16). To "eliminate" $\varepsilon X_0^{(0,0)} r^{-1}$, we already introduced $u_1(x)$ as the series in a neighborhood of the origin:

$$u_1(x) = X_0^{(0,0)} r^{-1} + \sum_{i=0}^{\infty} Y_i^{(1)}(x). \tag{11.32}$$

By the definition of $\mathcal{K}_q$ and (11.32), we have

$$\mathcal{K}_N(\widehat{u}_{\varepsilon,1}(x)) = \mathcal{K}_N(\widehat{u}_{\varepsilon,0}(x)) + \mathcal{K}_N(\varepsilon u_1(x)), \tag{11.33}$$

$$\mathcal{K}_N(\varepsilon u_1(x)) = X_0^{(0,0)} \rho^{-1} + \varepsilon \sum_{i=0}^{N-1} \varepsilon^i Y_i^{(1)}(\xi). \tag{11.34}$$

Note that the first term on the right-hand side of (11.34) already agrees with $v_0 = V_0^{(0)}$. To eliminate the remaining residuals on the right-hand side of (11.34) we "correct" $v_i$ for $i \geq 1$ as follows: add $\varepsilon^i V_i^{(1)}$ to $v_\varepsilon^{(0)}$ and require that, first, $V_i^{(1)}$ satisfy (11.15) (in this case, $v_i^{(1)}$ is still a solution of (11.15)) and, second, $V_i^{(1)}$ must have the following asymptotics at infinity:

$$V_i^{(1)}(\xi) \sim Y_{i-1}^{(1)}(\xi), \tag{11.35}$$

which is required for eliminating the residuals on the right-hand side of (11.34). It is obvious that such a solution to problem (11.15) exists. Furthermore, it can be represented at infinity in the form $V_i^{(1)}(\xi) = Y_{i-1}^{(1)}(\xi) + \sum_{j=0}^{\infty} X_j^{(i,1)}(\xi)\rho^{-2j-1}$. By (11.29)–(11.35), we have

$$\mathcal{K}_N(\widehat{u}_{\varepsilon,m}(x)) = \mathcal{M}_1(\widehat{v}_{\varepsilon,N}^{(m)}(\xi)) \tag{11.36}$$

for $m = 1$, which implies (11.25) for $N = 1$ since $\widehat{v}_{\varepsilon,m}^{(m)}(\xi) = \widehat{v}_{\varepsilon,m}(\xi)$. Thus, after two steps (referred to as the zeroth and the first steps), we construct $u_0(x)$, $u_1(x)$, and $v_j^{(1)}(\xi)$, $j \leq N$ (hence $v_0$ and $v_{H^1(\Omega)}$ can be found from (11.28)), such that (11.36) holds for $m = 1$ and (11.25) for $N = 1$. The further consideration is similar. Eliminating the residual from the outer expansion at the $n$th step, we find $u_n$. To eliminate the residuals caused by $u_n$ in the inner expansion, we introduce $V_i^{(n)}$ and obtain (11.36) for $m = n$, and consequently (11.25) for $N = n$.

Now, we proceed with rigorous formulations.

Consider harmonic polynomials $Y_j^{(0)}(x)$ of degree $j$ in the expansion $u_0(x) = \sum_{j=1}^{\infty} Y_j^{(0)}(\xi)$ of the solution $u_0$ to the boundary value problem

$$\Delta u_0 = 0 \text{ in } \Omega, \quad u_0 = f \text{ on } \partial\Omega.$$

For every $Y_j^{(0)}$, we introduce a harmonic function $V_j^{(0)}(\xi)$ in $\mathbb{R}^3 \backslash \overline{\omega}$ as the sum of $Y_j^{(0)}(\xi)$ and a harmonic function $\widetilde{V}_j^{(0)}(\xi)$ decreasing at infinity so that $Y_j^{(0)}(\xi) + \widetilde{V}_j^{(0)}(\xi) = 0$ on $\omega$. Note that $\widetilde{V}_j^{(0)}(\xi)$ is uniquely determined by these conditions. Expand $\widetilde{V}_j^{(0)}(\xi)$ into a series for $|\xi| > R_0$ with sufficiently large $R_0$:

$$\widetilde{V}_j^{(0)}(\xi) = \sum_{i=0}^{\infty} X_i^{(j,0)}(\xi)\rho^{-2i-1},$$

where $X_i^{(j,0)}(\xi)$ are harmonic polynomials of degree $i$. Since this expansion is unique, the harmonic polynomials $X_i^{(j,0)}(\xi)$ are uniquely determined. We define the function $\widetilde{u}_1(x)$ in $\Omega\backslash\{0\}$ with singularity at $\{0\}$ by the formula $\widetilde{u}_1(x) = X_0^{(0,0)} r^{-1}$ and construct a harmonic function $\overline{u}_1(x)$ in $\Omega$ such that $\overline{u}_1(x) + \widetilde{u}_1(x) = 0$ on $\partial\Omega$. We expand $\overline{u}_1(x)$ into a series with respect to homogeneous harmonic polynomials in the $\delta_0$-neighborhood of the origin:

$$\overline{u}_1(x) = \sum_{j=0}^{\infty} Y_j^{(1)}(x).$$

This expansion determines the harmonic polynomials $Y_j^{(1)}(x)$. To construct the harmonic polynomials $X_i^{(j,q)}(\xi)$ and $Y_j^{(q+1)}(x)$ for $q = 1$, we repeat the induction step. Namely, as above, we introduce the harmonic functions $\widetilde{V}_j^{(q)}(\xi)$ for $Y_j^{(j,q)}(\xi)$ such that $\widetilde{V}_j^{(q)}(\xi) + Y_j^{(j,q)}(\xi) = 0$ on $\omega$. Then $X_i^{(j,q)}(\xi)$ can be determined by $\widetilde{V}_j^{(q)}(\xi) = \sum_{i=0}^{\infty} X_i^{(j,q)}(\xi)\rho^{-2i-1}$. We set

$$\widetilde{u}_{q+1}(x) = \sum_{j=0}^{q} \sum_{t=0}^{q-j} X_j^{(q-j,t)}(x) r^{-2j-1},$$

where $X_j^{(q,t)}(x)$ have been introduced before, and we define a harmonic function $\overline{u}_{q+1}(x)$ in $\Omega$ such that $\overline{u}_{q+1}(x) + \widetilde{u}_{q+1}(x) = 0$ on $\partial\Omega$. We expand $\overline{u}_{q+1}(x)$ into a series with respect to homogeneous polynomials in the $\delta_0$-neighborhood of the origin:

$$\overline{u}_{q+1}(x) = \sum_{j=0}^{\infty} Y_j^{(q+1)}(x).$$

Note that the harmonic polynomials $Y_j^{(q+1)}(x)$ are uniquely determined. This step can be repeated for $q = 0, 1, 2, \dots$. Hence the harmonic polynomials $Y_j^{(q)}(x)$ and $X_j^{(q,t)}(\xi)$ are uniquely determined for all $q$.

THEOREM 11.2. *Let $N$ be an arbitrary natural number, and let $n \le N$. There exists a solution $u_q(x) \in C^\infty(\overline{\Omega}\backslash\{0\})$, $0 \le q \le n$, to the boundary value problem* (11.19) *admitting the asymptotic expansion* (11.12) *for $q = 0$ and*

$$u_q(x) = \sum_{j=0}^{q-1} \Bigg( \sum_{t=0}^{q-1-j} X_j^{(q-1-j,t)}(x) \Bigg) r^{-2j-1} + \sum_{j=0}^{\infty} Y_j^{(q)}(x)$$

*for $q > 0$. There exists a solution $V_i^{(t)}(\xi) \in C^\infty(\mathbb{R}^3\backslash\omega)$, $0 \le t \le i \le n$, to the boundary value problem* (11.15) *admitting the following asymptotic expansion at infinity:*

$$V_i^{(t)}(\xi) = Y_{i-t}^{(t)}(\xi) + \sum_{j=0}^{\infty} X_j^{(i,t)}(\xi)\rho^{-2j-1}.$$

*The solutions $u_q$ and $V_i^{(t)}$ are independent of $n$ and $N$.*

The following assertion is directly verified.

COROLLARY 11.1. *Let $u_q$ and $V_i^{(t)}$ be the same as in Theorem* 11.2, *and let $v_i$ be defined by* (11.28). *Then they are solutions to the problems* (11.15) *and* (11.19); *moreover,* (11.25) *holds.*

By Theorems 11.1 and 11.2, Corollary 11.1, Lemma 11.2, and the maximum principle, we conclude that

$$|u_\varepsilon(x) - \widetilde{u}_{\varepsilon,N}(x)| \le C_N \varepsilon^{N_1}, \tag{11.37}$$

where $N_1 = (N-1)/2$. Since $N$ is arbitrary, we have a complete asymptotic expansion of the solution to the problem (11.8), (11.9). Since $|\widetilde{u}_{\varepsilon,L}(x) - \widetilde{u}_{\varepsilon,L+1}(x)| = O(\varepsilon^{L/2})$ by construction, the estimate (11.37) remains valid for $N_1 = N/2$.

EXERCISE 11.7. Estimate the difference between $u_\varepsilon(x)$ and $\widetilde{u}_{\varepsilon,N}(x)$ in the $H^1(\Omega_\varepsilon)$-norm.

The method of matching asymptotic expansions is presented in [**71**] in detail. For other versions of this method we refer the reader to [**87, 61**]. In recent years, the method of matching asymptotic expansions is actively used for constructing asymptotic formulas in homogenization (cf., for example, [**64**]).

**11.3. Perturbation of the Dirichlet problem by changing the type of boundary condition on a small part of the boundary.** Let $\Omega$ be a bounded domain in $\mathbb{R}^2$ with infinitely differentiable boundary. Assume that $\Omega$ coincides with the half-plane $x_2 > 0$ in a neighborhood of the origin. Introduce the notation: $\nu$ is the outward normal, $\gamma$ and $\gamma_\varepsilon$ are the intervals $(-1,1)$ and $(-\varepsilon,\varepsilon)$, respectively, on the $x_{H^1(\Omega)}$-axis, $0 < \varepsilon \ll 1$ is a small parameter, $\Gamma_\varepsilon = \partial\Omega\backslash\overline{\gamma_\varepsilon}$, and $f \in C^\infty(\overline{\Omega})$. We construct an asymptotic representation of a solution to the boundary value problem

$$\Delta u_\varepsilon = f \ \text{ for } x \in \Omega, \quad u_\varepsilon = 0 \ \text{ for } x \in \Gamma_\varepsilon, \quad \frac{\partial u_\varepsilon}{\partial \nu} = 0 \ \text{ for } x \in \gamma_\varepsilon, \tag{11.38}$$

as $\varepsilon \to 0$. We look for the outer expansion of $u_\varepsilon$ in the form

$$u_\varepsilon(x) = u_0(x) + \sum_{i=1}^{\infty} \varepsilon^i u_i(x), \tag{11.39}$$

where $u_0$ is the solution to the problem

$$\Delta u_0 = f \text{ in } \Omega, \quad u_0 = 0 \text{ on } \partial\Omega. \tag{11.40}$$

For simplicity, we assume that $f$ vanishes in a neighborhood of the origin. Then the solution $u_0$ to the Dirichlet problem (11.40) can be extended through the boundary $\{x_2 = 0\}$ to the lower half-plane in such a way that the extended function is harmonic in a neighborhood of the origin and, consequently, is represented there as the series

$$u_0(x) = \sum_{j=1}^{\infty} a_j^{(0)} r^j \sin(j\theta), \tag{11.41}$$

where $(r,\theta)$ are polar coordinates and the $a_i^{(0)}$ are constant coefficients of the Taylor series.

EXERCISE 11.8. Prove (11.41).

Furthermore, $u_0 \in C^\infty(\overline{\Omega})$ in view of the properties of solutions to elliptic equations. Since $u_0$ satisfies the equation in (11.38) and the length of $\gamma_\varepsilon$ is small, it is natural to assume that $u_\varepsilon$ is close to $u_0$. The obvious drawback of $u_0$ is that it does not satisfy the boundary condition on $\gamma_\varepsilon$. Therefore, we look for another asymptotic representation of the solution to problem (11.38) in a neighborhood of $\gamma_\varepsilon$. Since $\gamma_\varepsilon$ is obtained by contracting a fixed domain $\gamma$ by a factor of $\varepsilon^{-1}$, it is natural to represent the solution in a neighborhood of $\gamma_\varepsilon$ as a series $v_\varepsilon$ with coefficients depending on the "inner" variables $\xi = x\varepsilon^{-1}$.

If $0 < q < 1$, then for $\varepsilon^q < r < 2\varepsilon^q$, $\rho = |\xi|$ we have $\varepsilon^{-1+q} < \rho < 2\varepsilon^{-1+q}$, i.e., $x$ is small and $\xi$ is large there. The asymptotic series for $u_\varepsilon(x)$ (we know only the zeroth term $u_0(x)$) and $v_\varepsilon(\xi)$ (we do not know its structure yet) must coincide in the intermediate zone where $x$ is small and $\xi$ is large. Then, roughly speaking, we can pass from one expansion to the other in this zone. For definiteness, we set $q = 1/2$.

Based on these arguments, we begin with the structure of the series $v_\varepsilon$ and $u_\varepsilon$. Passing to the variables $\xi$ in (11.41), for $\varepsilon^{1/2} < r < 2\varepsilon^{1/2}$ (or, which is the same, for $\varepsilon^{-1/2} < \rho < 2\varepsilon^{-1/2}$) we find that

$$u_0(x) = \sum_{j=1}^{\infty} \varepsilon^j a_j^{(0)} \rho^j \sin(j\theta). \tag{11.42}$$

From (11.42) and the matching condition it follows that the coefficients of the "inner" expansion

$$v_\varepsilon(\xi) = \sum_{j=1}^{\infty} \varepsilon^j v_j(\xi) \tag{11.43}$$

satisfy the conditions

$$v_j(\xi) \sim a_j^{(0)} \rho^j \sin(j\theta) \quad \text{as } \rho \to \infty. \tag{11.44}$$

Substituting (11.43) in (11.38), passing to the variables $\xi$, and collecting terms with equal power of $\varepsilon$, we obtain the boundary value problem

$$\Delta_\xi v_j = 0 \ \text{ for } \xi_2 > 0, \quad v_j = 0 \ \text{ for } \xi \in \Gamma, \quad \frac{\partial v_j}{\partial \xi_2} = 0 \ \text{ for } \xi \in \gamma, \tag{11.45}$$

where $\Gamma = \{\xi \in \mathbb{R}^2 : \xi_2 = 0\} \backslash \gamma$ (cf. Figure 11.6).

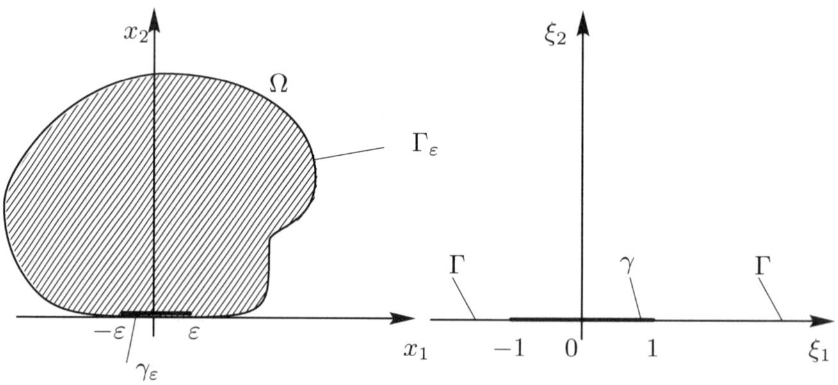


FIGURE 11.6

For $j \geq 1$ we set $X_j(\xi) = \operatorname{Im}(z^{j-1}\sqrt{z^2-1})$, where $z = \xi_{H^1(\Omega)} + \mathrm{i}\,\xi_2$, Im is the imaginary part of a complex number, and i is the imaginary unit.

EXERCISE 11.9. Prove that $X_j(\xi)$ is a solution to problem (11.45) and admits the following asymptotic representation as $\rho \to \infty$:

$$X_j(\xi) = \rho^j \sin(j\theta) + \sum_{k=1}^{j-1} \widetilde{\beta}^{(j)}_{-k} \rho^k \sin(k\theta) + \sum_{k=1}^{\infty} \widetilde{\beta}^{(j)}_k \rho^{-k} \sin(k\theta).$$

By Exercise 11.9, there exist linear combinations $V_j(\xi)$ of $X_j(\xi)$ that are solutions to the problem (11.45) and admit the asymptotics $Y_j(\xi) = \rho^j \sin(j\theta) + \sum_{k=1}^{\infty} \beta^{(j)}_k \rho^{-k} \sin(k\theta)$ as $\rho \to \infty$. Thus, the functions

$$\begin{aligned} v_1(\xi) &= a_1^{(0)} Y_1(\xi), \\ v_2(\xi) &= a_2^{(0)} Y_2(\xi), \\ v_n(\xi) &= a_n^{(0)} Y_n(\xi) + \sum_{j=1}^{n-2} A_j^{(n-j)} Y_j(\xi) \quad \text{for } n \geq 3 \end{aligned} \tag{11.46}$$

satisfy (11.44) and (11.45), where $A_j^{(k)}$, $k \geq 2$, are arbitrary constants to be specified below. We emphasize that $a_j^{(0)}$ is found from (11.41). Consequently, by (11.46), $v_1$ and $v_2$ are uniquely determined.

From (11.46) it follows that for $\varepsilon^{1/2} < r < 2\varepsilon^{1/2}$ (or, which is the same, for $\varepsilon^{-1/2} < \rho < 2\varepsilon^{-1/2}$) we have

$$v_n(\xi) = a_n^{(0)} \rho^n \sin(n\theta) + \sum_{j=1}^{n-2} A_j^{(n-j)} \rho^j \sin(j\theta) + \sum_{k=1}^{\infty} b_k^{(n)} \rho^{-k} \sin(k\theta); \tag{11.47}$$

moreover, since $v_1$ and $v_2$ have already been found, we also have $b_k^{(1)}$ and $b_k^{(2)}$.

Note that if we know $A_j^{(n-j)}$ in (11.46) for fixed $n$, then we can actually determine $v_i$ and consequently all $b_k^{(n)}$ in (11.47).

Passing to the outer variables $x$ in (11.47) and substituting (11.47) in (11.43), from the matching condition for $u_\varepsilon(x)$ and $v_\varepsilon(\xi)$ we find the following asymptotic representations for the terms of the series (11.39) as $r \to 0$:

$$u_0(x) \sim \sum_{j=1}^{\infty} a_j^{(0)} r^j \sin(j\theta), \quad u_{k+1}(x) \sim \sum_{j=1}^{k} b_j^{(k+1-j)} r^{-j} \sin(j\theta). \tag{11.48}$$

Substituting (11.39) in (11.38) and collecting terms with equal power of $\varepsilon$, we obtain the boundary value problem

$$\Delta u_j = \delta_j^0 f \ \text{ for } x \in \Omega, \quad u_j = 0 \ \text{ for } x \in \partial\Omega \backslash \{0\}, \tag{11.49}$$

where $\delta_i^j$ is the Kronecker symbol. By construction, $u_0$ satisfies (11.48) and (11.49). Denote by $\widetilde{Z}_k(x)$ a harmonic function in $\Omega$ that is equal to $-r^{-k}\sin(k\theta)$ on $\partial\Omega\backslash\{0\}$ and is extended by zero to $\{0\}$. Note that $r^{-k}\sin(k\theta)$ vanishes everywhere on the axis $x_2 = 0$, and thereby it provides a smooth boundary condition on $\partial\Omega$. It is clear that $Z_k(x) = r^{-k}\sin(k\theta) + \widetilde{Z}_k(x)$ is a solution to the problem (11.49) and it admits the asymptotic expansion $Z_k(x) = r^{-k}\sin(k\theta) + \sum_{j=1}^{\infty} \alpha_j^{(k)} r^j \sin(j\theta)$ as $r \to 0$.

It is obvious that the function

$$u_{k+1}(x) = \sum_{j=1}^{k} b_j^{(k+1-j)} Z_j(x), \quad k \geq 1, \tag{11.50}$$

satisfies (11.48) and (11.49). Moreover, since $b_j^{(1)}$ and $b_j^{(2)}$ have already been found (from formula (11.47) for $v_1$ and $v_2$), $u_2$ and $u_3$ are also determined in view of (11.50). The remaining $b_j^{(i)}$ (and thereby $u_m$) will be specified below.

We return to the basic ideas of the method of matching asymptotic expansions. As before, we introduce an operator $K_j(U(x,\varepsilon))$ on the set of functions $U(x,\varepsilon)$ of the form (11.39) as follows. We replace the coefficients of the series $U(x,\varepsilon)$ with their asymptotic expressions as $r \to 0$, pass to the variables $\xi$, and leave only terms of the form $\varepsilon^j \Phi(\xi)$ in the double series obtained. The resulting sum is denoted by $K_j(U(x,\varepsilon))$. We set

$$\mathcal{K}_N(U(x,\varepsilon)) \overset{\text{def}}{=} \sum_{j=1}^{N} K_j(U(x,\varepsilon)).$$

On the set of functions $V(\xi,\varepsilon)$ of the form (11.43), we introduce an operator $M_j(V(\xi,\varepsilon))$ as follows. We replace the coefficients of the series $V(\xi,\varepsilon)$ with their asymptotic expressions at infinity, pass to the variables $x$, and leave only terms of the form $\varepsilon^j \Psi(x)$ in the double series obtained. The resulting sum is denoted by $M_j(V(\xi,\varepsilon))$. We set

$$\mathcal{M}_N(V(\xi,\varepsilon)) = \sum_{j=1}^{N} M_j(V(\xi,\varepsilon)).$$

Denote by $H_j^{(m)}(x)$ and $h_j^{(q)}(\xi)$ homogeneous functions of degree $j$, and by $\widehat{u}_{\varepsilon,N}$ and $\widehat{v}_{\varepsilon,N}$ the partial sums of the series $u_\varepsilon$ and $v_\varepsilon$ respectively. We set $\mathbb{R}^2_+ = \{\xi : \xi_2 > 0\}$.

LEMMA 11.3. *Suppose the coefficients of the series $u_\varepsilon$ belong to $C^\infty(\overline{\Omega}\backslash\{0\})$ and have the asymptotics*

$$u_k(x) = \sum_{j=-k+1+\kappa_k}^{\infty} H_j^{(k)}(x)$$

*as $r \to 0$ ($\kappa_k$ are integers). Suppose the coefficients of the series $v_\varepsilon$ belong to $C^\infty(\mathbb{R}^2_+)$ and have the differentiable asymptotics*

$$v_k(\xi) = \sum_{j=-\infty}^{k} h_j^{(k)}(\xi)$$

*as $\rho \to \infty$. Then for any $N \geq 0$ we have*

$$\widehat{u}_{\varepsilon,N}(x) - \mathcal{K}_N(\widehat{u}_{\varepsilon,N}(x)) = O(r^{N+1} + \varepsilon^N r) \quad \text{as } r \to 0,$$
$$\widehat{v}_{\varepsilon,N}(\xi) - \mathcal{M}_N(\widehat{v}_{\varepsilon,N}(\xi)) = O(\rho^{-N-1} + \varepsilon^N \rho^{-1}) \quad \text{as } \rho \to \infty.$$

Lemma 11.3 immediately follows from the definition of $\mathcal{K}_N$, $\mathcal{M}_N$ and the asymptotic formulas for $u_i$ and $v_i$. However, this lemma is related to neither boundary value problems nor the matching condition.

Let $\chi(t)$ be an infinitely differentiable cut-off function vanishing for $t < 1$ and equal to 1 for $t > 2$. Let

$$\widetilde{u}_{\varepsilon,N}(x) = \chi(r\varepsilon^{-1/2}) u_{\varepsilon,N}(x) + (1 - \chi(r\varepsilon^{-1/2})) v_{\varepsilon,N}(\xi).$$

By the definition of this function, we immediately obtain the following assertion.

LEMMA 11.4. *Suppose that the coefficients of the series $u_\varepsilon$ belong to $C^\infty(\overline{\Omega}\backslash\{0\})$ and solve the boundary value problem (11.49). Suppose that the coefficients of the series $v_\varepsilon$ belong to $C^\infty(\overline{\mathbb{R}^2_+}\backslash\gamma) \cap H^1_{\rm loc}(\mathbb{R}^2_+)$ and solve the boundary value problem (11.45). Then $\widetilde{u}_{\varepsilon,N}(x)$ belongs to $C^\infty(\overline{\Omega}\backslash\gamma_\varepsilon) \cap H^1(\Omega)$ and is a solution to the boundary value problem*

$$\Delta\widetilde{u}_{\varepsilon,N} = F_N + f \quad \textit{for } x \in \Omega_\varepsilon, \quad u_{\varepsilon,N} = 0 \quad \textit{for } x \in \Gamma_\varepsilon, \quad \frac{\partial u_{\varepsilon,N}}{\partial \nu} = 0 \quad \textit{for } x \in \gamma_\varepsilon,$$

*where*

$$\begin{aligned} F_N(x,\varepsilon) = 2\sum_{j=1}^{2} \chi_{x_j}(r\varepsilon^{-1/2})(\widehat{u}_{\varepsilon,N}(x) - \widehat{v}_{\varepsilon,N}(\xi))_{x_j} \\ + (\widehat{u}_{\varepsilon,N}(x) - \widehat{v}_{\varepsilon,N}(\xi))\Delta\chi(r\varepsilon^{-1/2}). \end{aligned} \tag{11.51}$$

Note that the last assertion is connected with neither asymptotics of the coefficients of the series $u_\varepsilon$ and $v_\varepsilon$ (at zero and at infinity, respectively) nor the matching condition which requires the existence of series such that the "smallness order" of the right-hand side of (11.51) unboundedly increases as $N$ increases.

The following assertion is the key point of the matching of series.

THEOREM 11.3. *Under the assumptions of Lemmas 11.3 and 11.4, let*

$$\mathcal{K}_N(\widehat{u}_{\varepsilon,N}(x)) = \mathcal{M}_N(\widehat{v}_{\varepsilon,N}(\xi)). \tag{11.52}$$

*Then*

$$|F_N| \le C_N \varepsilon^{N-1/2}. \tag{11.53}$$

PROOF. By Lemma 11.3 and (11.52), for $\varepsilon^{1/2} < r < 2\varepsilon^{1/2}$ we have

$$\widehat{u}_{\varepsilon,N}(x) - \widehat{v}_{\varepsilon,N}(\xi) = O(\varepsilon^{N+1/2}), \quad (\widehat{u}_{\varepsilon,N}(x) - \widehat{v}_{\varepsilon,N}(\xi))_{x_j} = O(\varepsilon^{N/2}). \tag{11.54}$$

Since the supports of $\chi_{x_i}(r\varepsilon^{-1/2})$ and $\Delta\chi(r\varepsilon^{-1/2})$ belong to the annuli $\varepsilon^{1/2} < r < 2\varepsilon^{1/2}$ and $\varepsilon^{1/2}|\chi_{x_j}(r\varepsilon^{-1/2})| + \varepsilon|\Delta\chi(r\varepsilon^{-1/2})| \le C$, from (11.51) and (11.52) we obtain (11.53). □

Thus, for a formal construction of the asymptotics of the solution to the problem (11.38), it suffices to show that the coefficients $A_j^{(k)}$ in (11.46) (consequently, $b_k^{(n)}$ in (11.48) for $n \ge 2$) can be chosen so that (11.52) holds for any $n$. From (11.50) it follows that

$$u_{k+1}(x) = \sum_{j=1}^{k} b_j^{(k+1-j)} r^{-j}\sin(j\theta) + \sum_{j=1}^{\infty} a_j^{(k+1)} r^j \sin(j\theta) \quad \text{as } r \to 0; \tag{11.55}$$

moreover, since $u_2$ and $u_3$ have already been found, the coefficients $a_j^{(2)}$ and $a_j^{(3)}$ are also determined.

Note that if for fixed $k$ we already know all the coefficients $b_j^{(k+1-j)}$ in (11.55), we can determine $u_{k+1}$ and all $a_j^{(k+1)}$.

Passing to the inner variables in (11.55) and substituting (11.55) and the asymptotic expression (11.48) of $u_0$ in (11.43), we find that for $\varepsilon^{-1/2} < \rho < 2\varepsilon^{-1/2}$,

$$u_\varepsilon(x) = \sum_{n=1}^{\infty} \varepsilon^n V_n(\xi), \tag{11.56}$$

$$V_n(\xi) = a_n^{(0)} \rho^n \sin(n\theta) + \sum_{j=1}^{n-2} a_j^{(n-j)} \rho^j \sin(j\theta) + \sum_{k=1}^{\infty} b_k^{(n)} \rho^{-k} \sin(k\theta).$$

From the definition of $\mathcal{M}_N$ and $\mathcal{K}_N$ it follows that the equality (11.52) holds if the series (11.47) and (11.56) coincide for any $N$.

Since $a_j^{(3)}$ are already determined, we set $A_j^{(3)} = a_j^{(3)}$ and find $v_3$. Then we can find the terms $v_k$, $k \geq 4$, in (11.46) that increase at infinity like $O(\rho^{k-2})$. If we find $v_3$, we can determine all $b_j^{(3)}$. In turn, if we have $b_j^{(3)}$, we can completely determine $u_4$ and find the terms for $u_k$, $k \geq 5$, in (11.55) that increase like $O(r^{-k+3})$ at the origin. Finally, if we have $u_4$, we thereby can determine all $a_j^{(4)}$. Further, the chain is repeated:

$$A_j^{(4)} = a_j^{(4)} \Rightarrow v_4 \Rightarrow b_j^{(4)} \Rightarrow u_5 \Rightarrow a_j^{(5)} \Rightarrow A_j^{(5)} = a_j^{(5)} \Rightarrow \cdots .$$

Based on these arguments, we can prove the following assertion.

THEOREM 11.4. *There exist series* (11.43) *and* (11.39) *that satisfy the assumptions of Theorem* 11.3 *and their coefficients satisfy* (11.46) *and* (11.50).

Thus, we formally constructed a complete asymptotic expansion for the solution to the boundary value problem (11.38). Now, we justify this asymptotics. Denote by $H^1(\Omega; \Gamma_\varepsilon)$ the closure of the set of functions in $C^\infty(\overline{\Omega})$ vanishing in a neighborhood of $\Gamma_\varepsilon$ in the $H^1(\Omega)$-norm. By a solution to the problem (11.38) we mean a function $u_\varepsilon \in H^1(\Omega; \Gamma_\varepsilon)$ satisfying the integral identity

$$-(\nabla u_\varepsilon, \nabla v)_{L_2(\Omega)} = (f, v)_{L_2(\Omega)} \tag{11.57}$$

for any $v \in H^1(\Omega; \Gamma_\varepsilon)$.

LEMMA 11.5. *A solution to the boundary value problem* (11.38) *satisfies the estimate*

$$\|u_\varepsilon\|_{H^1(\Omega)} \leq C\|f\|_{L_2(\Omega)}, \tag{11.58}$$

*where the constant $C$ is independent of $\varepsilon$ and $f$.*

PROOF. Substituting $v = u_\varepsilon$ in (11.57), we easily obtain the following *a priori* estimate uniform with respect to $\varepsilon$:

$$\|u_\varepsilon\|_{H^1(\Omega)} \leq C_0 \left(\|f\|_{L_2(\Omega)} + \|u_\varepsilon\|_{L_2(\Omega)}\right). \tag{11.59}$$

Assume that the estimate (11.58) fails to hold. Then there exists a sequence $\varepsilon_n \to 0$ as $n \to \infty$ and a sequence $f_n \in L_2(\Omega)$ such that

$$\|u_{\varepsilon_n}\|_{H^1(\Omega)} > n\|f_n\|_{L_2(\Omega)}. \tag{11.60}$$

From (11.59) and (11.60) we conclude that

$$\|u_{\varepsilon_n}\|_{H^1(\Omega)} < 2C_0\|u_{\varepsilon_n}\|_{L_2(\Omega)}. \tag{11.61}$$

Without loss of generality, we can assume that $\|u_{\varepsilon_n}\|_{L_2(\Omega)} = 1$. By (11.61), there exists a subsequence of $\varepsilon_n$ on which $u_{\varepsilon_n}$ converges to $u_* \in H^1(\Omega)$ weakly in $H^1(\Omega)$

and strongly in $L_2(\Omega)$; moreover, $\|u_*\|_{L_2(\Omega)} = 1$. It is easy to see that $u_* \in \overset{\circ}{H}{}^1(\Omega)$. On the other hand, from (11.60) and (11.61) it follows that $\|f_n\|_{L_2(\Omega)} \to 0$. For $v$ in (11.57) we take an arbitrary function of class $\overset{\circ}{H}{}^1(\Omega)$. Substituting $u_\varepsilon = u_{\varepsilon_n}$ and $f = f_n$ in (11.57) and passing to the limit as $n \to \infty$, we see that $u_*$ is a nontrivial weak solution to the Dirichlet problem

$$\Delta u_* = 0 \quad \text{in } \Omega, \qquad u_* = 0 \quad \text{on } \partial\Omega.$$

We arrive at a contradiction, which proves (11.58). □

Lemma 11.4 and formulas (11.51), (11.58) imply the estimate

$$\|u_\varepsilon - \widetilde{u}_{\varepsilon,N}\|_{H^1(\Omega)} \leq C_N \varepsilon^{N/2}.$$

Since $N$ is arbitrary, the complete asymptotic expansion of the solution to the problem (11.38) is justified.

The method of matching asymptotic expansions is widely used in homogenization theory. In Subsection 15.2 below, this method is applied to boundary value problems with rapidly alternating type of boundary conditions.

## 12. The Bogolyubov method

**12.1. Definitions and the main assertions.** The Bogolyubov method was invented for analyzing dynamical systems in which the variables have essentially different characteristic scales of change in time. Usually, the case of two variables was treated. The parameters of the dynamical system rapidly oscillate with respect to one (*fast*) variable and change slowly relative to the other (*slow*) variable. The simplest example of such a system is a mechanical system subject to an external rapidly oscillating force (vibration). In the study of evolution of such a system during a large time interval, the main role is played by the mean value of the external force over a time period of change, whereas the values of the force at fixed times are not essential. From a formal point of view, this means that if the right-hand side is averaged over time explicitly occurring in it, then the solutions to the simplified system well approximate the solution to the initial system on a large time interval.

In applications, we often encounter complicated systems where functions rapidly oscillating in time do not occur explicitly. Nevertheless, the phase variables can be divided into two groups: the variables evolving slowly and the variables evolving fast, and the fast evolution has a cyclic character. Imagine, for example, a man-made satellite moving along an orbit and simultaneously rotating about the center of mass. In addition to gravity, different weak interactions (for example, the magnetic field of the Earth, the light pressure) affect the satellite. As a result, at sufficiently large time, the characteristics of rotation about the center of mass begin to evolve. For example, the direction of the rotation axis can slowly change. How can we describe such a slow evolution? Of course, we could try to integrate with high accuracy the differential equation of the system over a large time interval. However, in this way, the error of numerical integration will accumulate. Homogenization methods are based on another approach: we construct a simpler "homogenized" system with parameters depending only on the slow variables such that the solution of the simplified system will well approximate the exact solution on a large time interval. The homogenization idea is based on the fact that the evolution of slow variables essentially depends only on the mean value of the fast variables over a characteristic time interval of a cyclic (not necessarily periodic, but close to periodic) change. Note that originally the slow and fast variables are interrelated and the dynamics of the system is determined by their interaction. But, due to homogenization, the system for slow variables becomes separate and can be treated independently. In this case, one usually says that it is possible to realize the asymptotic expansion of the system under consideration.

At first glance, this topic is far from the main subject of our book. However, it was the Bogolyubov method that served as a prototype of the asymptotic homogenization method for spatial characteristics of a medium with periodic microstructure (cf. [**9**] and [**17**]). In fact, we also deal with asymptotic expansions. Indeed, we introduce independent fast variables describing the change of a medium within one small period (cell) and consider equations describing the medium in the fast variables, where the slow (original) variables play the role of parameters. The solvability condition in the space of periodic functions can be written as equations in terms of slow variables only. Such equations are called "homogenized". They describe the macroscopic properties of the medium.

The Bogolyubov averaging method uses the change of variables in order to exclude the dependence on the fast variables in the leading terms (relative to a small parameter). Such a change of variables is possible only if the slow variables satisfy a closed system of equations; moreover, such a system arises from the solvability condition for an auxiliary problem in terms of the fast variables.

We prove the Bogolyubov theorem using a different method (not based on the change of variables) and then formulate the main assertions (in the form of exercises), which suggest an ad hoc change of variables.

We describe the Bogolyubov method (cf. [**20**]) in the simplest situation, restricting ourselves to a system of ordinary differential equations.

Let $Q := \{(x,t) : x \in \Omega,\ 0 \le t < +\infty\}$ be a domain in $\mathbb{R}^{d+1}$, where $\Omega$ is a cylindrical domain in $\mathbb{R}^d$. In $Q$, we consider the nonlinear nonautonomous system

$$\frac{dX_\varepsilon}{dt} = \varepsilon F(t, X_\varepsilon) \tag{12.1}$$

with the initial condition

$$X_\varepsilon(0) = X_0, \tag{12.2}$$

where $\varepsilon$ is a small positive parameter and $F(t,x) \in C([0,\infty); \mathrm{Lip}(\Omega))$ satisfies the estimate

$$\|F(t,\cdot)\| \le M. \tag{12.3}$$

In this section, $\|\cdot\|$ denotes the Lipschitz norm in $\Omega$.

THEOREM 12.1. *Suppose that the finite limit*

$$G(x) = \lim_{T\to\infty} \frac{1}{T}\int_0^T F(t,x)\,dt \in C(\Omega) \tag{12.4}$$

*exists uniformly with respect to $x$, and the problem*

$$\frac{dY_\varepsilon}{dt} = \varepsilon G(Y_\varepsilon), \quad Y_\varepsilon(0) = X_0 \tag{12.5}$$

*has a unique solution $Y_\varepsilon(t)$ for $\varepsilon = 1$ on $[0, T_1]$. Then the sequence $(X_\varepsilon(t) - Y_\varepsilon(t))$ uniformly converges to zero on $0 \le t \le T/\varepsilon$, $0 < T < T_1$ as $\varepsilon \to 0$, where $X_\varepsilon(t)$ is the solution to the Cauchy problem* (12.1), (12.2).

PROOF. Let $\rho > 0$ be the distance from a point $X_0 \in \Omega$ to the boundary of $\Omega$. By (12.3), the solution $X_\varepsilon(t)$ exists, at least on the interval $0 \le t \le \frac{\rho}{M\varepsilon\sqrt{d}}$.

EXERCISE 12.1. Suppose (12.3) holds. Prove that a solution exists for $t \in \left[0, \frac{\rho}{M\varepsilon\sqrt{d}}\right]$ and does not leave the domain $\Omega$.

Introducing the "slow" time variable $\tau = \varepsilon t$, we write equations (12.1) and (12.5) as follows:

$$\frac{dX}{d\tau} = F(\tau/\varepsilon, X) = G_\varepsilon(\tau, X), \tag{12.6}$$

$$\frac{dY}{d\tau} = G(Y). \tag{12.7}$$

We show that the function $G_\varepsilon(\tau, x)$ (defined as $G(x)$ at $\varepsilon = 0$) is *integral-continuous* with respect to $\varepsilon$ for $\varepsilon = 0$, i.e., there exists $\lim_{\varepsilon\to 0} \int_0^t G_\varepsilon(\tau, x)d\tau$ and it is equal to $\int_0^t G(x)d\tau$. Indeed,

$$\begin{aligned}\lim_{\varepsilon\to 0} \int_0^t G_\varepsilon(\tau, x)d\tau &= \lim_{\varepsilon\to 0} \int_0^t F(\tau/\varepsilon, x)d\tau = \lim_{\varepsilon\to 0} \varepsilon \int_0^{t/\varepsilon} F(s, x)ds \\ &= t \lim_{\varepsilon\to 0} \frac{1}{t/\varepsilon} \int_0^{t/\varepsilon} F(s, x)ds = tG(x) = \int_0^t G(x)d\tau.\end{aligned}$$

We write the problem (12.6), (12.7) in the integral form. For any $\tau \in [0, T]$ we have

$$X_\varepsilon(\tau) = X_0 + \int_0^\tau G_\varepsilon(s, X_\varepsilon(s))\, ds, \tag{12.8}$$

$$Y(\tau) = X_0 + \int_0^\tau G(Y(s))\, ds. \tag{12.9}$$

Consider a family of solutions $\{X_\varepsilon(\tau)\}$, $\varepsilon > 0$, $\tau \in [0, T]$, and show that it is uniformly bounded and equicontinuous on $[0, T]$. This follows from the uniform boundedness of the derivatives $\frac{dX_\varepsilon}{d\tau}$. Indeed, taking into account the uniform boundedness of $G_\varepsilon(s, X) \equiv F(s/\varepsilon, X)$, we find from (12.8) that

$$\|X_\varepsilon(\tau)\| \le \|X_0\| + \int_0^\tau \|G_\varepsilon(s, X_\varepsilon(s))\|\, ds \le \|X_0\| + TM.$$

Similarly, (12.6) implies the uniform boundedness of the derivatives $\frac{dX_\varepsilon}{d\tau}$ on $[0, T]$, which yields the equicontinuity of the family.

By the Arzelà theorem, from $X_\varepsilon(\tau)$ we can extract a uniformly converging subsequence on $[0, T]$; i.e., there exists $Y^*(\tau)$ and a sequence $\varepsilon_k$ such that

$$X_{\varepsilon_k}(\tau) \rightrightarrows Y^*(\tau) \text{ uniformly as } \varepsilon_k \to 0. \tag{12.10}$$

Passing to the limit as $\varepsilon_k \to 0$ in the integral identity (12.8), i.e., in the identity

$$X_{\varepsilon_k}(\tau) = X_0 + \int_0^\tau G_{\varepsilon_k}(s, X_{\varepsilon_k}(s))\, ds,$$

we find that

$$Y^*(\tau) = X_0 + \int_0^\tau G(Y^*(s))\,ds, \quad 0 \le \tau \le T. \tag{12.11}$$

Indeed,

$$\left|\int_0^\tau G_{\varepsilon_k}(s, X_{\varepsilon_k})\,ds - \int_0^\tau G(Y^*)\,ds\right| \le \left|\int_0^\tau G_{\varepsilon_k}(s, X_{\varepsilon_k})\,ds - \int_0^\tau G_{\varepsilon_k}(s, Y^*)\,ds\right|$$
$$+ \left|\int_0^\tau G_{\varepsilon_k}(s, Y^*)\,ds - \int_0^\tau G(Y^*)\,ds\right| \to 0 \quad \text{as } \varepsilon \to 0.$$

EXERCISE 12.2. Using the integral continuity of $G_\varepsilon(s, X)$, show that the second integral on the right-hand side of the above inequality converges to zero.

The integral identity (12.11) can be written as

$$\frac{dY^*}{d\tau} = G(Y^*), \quad Y^*(0) = X_0.$$

Returning to the variable $t$, we obtain the problem (12.5). By the uniqueness of the solution, the limit is independent of the choice of a subsequence. □

REMARK 12.1. If $F(t, X)$ in Theorem 12.1 is periodic with respect to $t$, then the following estimate holds on $[0, T/\varepsilon]$:

$$|X_\varepsilon(t) - Y_\varepsilon(t)| \le C(T)\varepsilon,$$

where the constant $C(T) > 0$ depends on $T > 0$, but is independent of $\varepsilon > 0$ (cf. [**20**]).

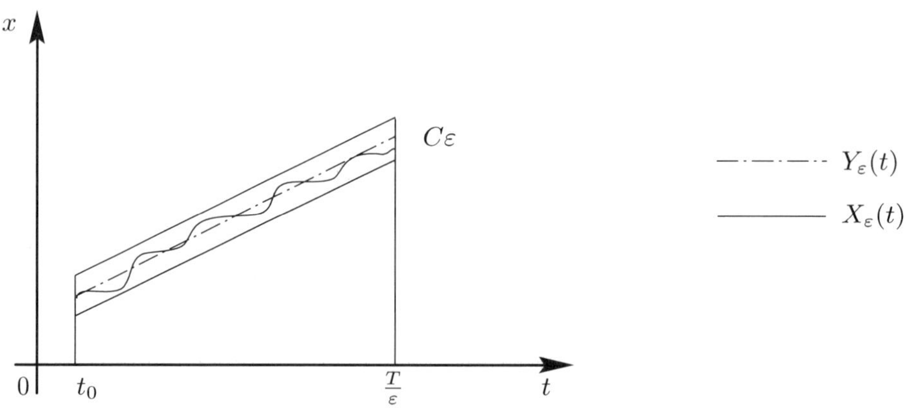


FIGURE 12.1. $|X_\varepsilon(t_0) - X_\varepsilon(T/\varepsilon)| \sim 1$ and $|X_\varepsilon(t) - Y_\varepsilon(t)| \sim C\varepsilon$.

The proof of the following assertions can be found, for example, in [**21**]. Let $F(\theta, X)$ and $F_0(\theta, X)$ be vector-valued functions periodic with respect to $\theta$, and let

$$G(Y) = \frac{1}{\theta_0}\int_0^{\theta_0} F(\theta, Y)\,d\theta, \tag{12.12}$$

where $\theta_0$ is the period of $F(\theta, Y)$.

THEOREM 12.2. *Let $\omega(x)$ be a continuously differentiable function such that $\omega(x) \geq C > 0$, and let vector-valued functions $X_\varepsilon(t)$, $Y_\varepsilon(t)$ be solutions to the problem*

$$\begin{aligned}
\frac{dX_\varepsilon}{dt} &= \varepsilon F(\theta_\varepsilon, X_\varepsilon),\\
\frac{d\theta_\varepsilon}{dt} &= \omega(X_\varepsilon) + \varepsilon F_0(\theta_\varepsilon, X_\varepsilon),\\
X_\varepsilon(t_0) &= X_0, \quad \theta(t_0) = \theta_0,\\
\frac{dY_\varepsilon}{dt} &= \varepsilon G(Y_\varepsilon),\\
Y_\varepsilon(t_0) &= X_0.
\end{aligned}$$

*Then*

$$|X_\varepsilon(t) - Y_\varepsilon(t)| \leq C(T)\varepsilon$$

*on* $[t_0, T/\varepsilon]$, $T = \text{const}$, *where the constant $C(T) > 0$ depends on $T > 0$, but is independent of $\varepsilon > 0$. Here, $G(x)$ is defined by formula* (12.12) (*cf. also Figure* 12.1).

We outline the proof of Theorem 12.2. We use a method (cf., for example, [**11**]) different from the method of the proof of Theorem 12.1.

We try to find a transformation of phase variables $(X, \theta)$ to new phase variables $(Z, \theta)$ such that $Z = X + \varepsilon k(X, \theta)$ and the system takes a simpler form; namely, $\dot{Z} = \varepsilon f(Z) + R(\varepsilon, Z, \theta)$, where $|R| \leq C\varepsilon^2$, i.e., the term of order $\varepsilon$ is independent of $\theta$.

Consider the equation

$$\omega(X)\frac{\partial k}{\partial \theta} = -[F(\theta, X) - G(X)]. \tag{12.13}$$

EXERCISE 12.3. Prove that equation (12.13) for every $X$ has a periodic solution with respect to $\theta$.

EXERCISE 12.4. Let $k(X, \theta)$ define the change of variables. Assume that $k(X, \theta)$ satisfies equation (12.13). Show that $Z$ satisfies the equation $\dot{Z} = \varepsilon G(Z) + R(\varepsilon, Z, \theta)$, where $|R| \leq C\varepsilon^2$.

Note that the function defining the right-hand side of the homogenized system can be found from the solvability condition for equation (12.13).

EXERCISE 12.5. Using the above exercises, complete the proof of Theorem 12.2.

DEFINITION 12.1. Let $x_0$ be an equilibrium state (a stationary solution) of the system $\dot{X} = F(X)$. We say that $x_0$ is *stable at a first approximation* if every characteristic root of the matrix $\left(\frac{\partial F^i}{\partial x_j}(x_0)\right)_{i,j=1}^d$ has a negative real part, i.e., lies in the left half-plane.

We need the notion of an integral manifold in the extended phase space of the system.

DEFINITION 12.2. By an *integral manifold* we mean a smooth surface $I$ in the extended phase space $\mathbb{R}^{d+1} \ni (x_1, \dots, x_d, t)$ of the system $\dot{X} = F(t, X)$ such that each trajectory of the system with an initial condition in some $\sigma$-neighborhood of $I$ tends to $I$ as $t \to +\infty$; moreover, any trajectory $Y(t)$ of the system with the initial condition $Y(t_0) \in I$ entirely lies in $I$ (cf. Figure 12.2).

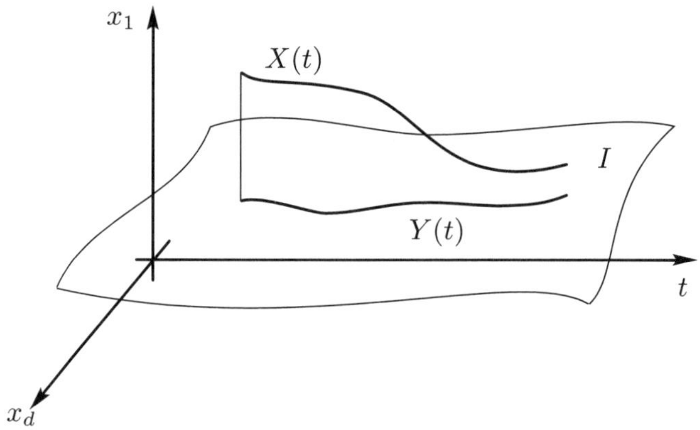


FIGURE 12.2

THEOREM 12.3. *Let $F$ and $G$ be the same as in Theorem 12.2, and let $x_0$ be a stationary point of the system*

$$\frac{dY_\varepsilon}{dt} = \varepsilon G(Y_\varepsilon).$$

*Suppose that $x_0$ is stable at first approximation. Then for each $\sigma$-neighborhood $U_\sigma$ of $x_0$ there exists $\varepsilon_0(\sigma) > 0$ such that for every $\varepsilon < \varepsilon_0(\sigma)$ there exists a unique stable limit cycle in $U_\sigma$ (i.e., asymptotically stable periodic solution $X_\varepsilon^*(t)$) to the system $\frac{dX_\varepsilon}{dt} = \varepsilon F_\varepsilon(t, X_\varepsilon)$). Moreover, for all $t \in (-\infty, +\infty)$,*

$$|X_\varepsilon^*(t) - x_0| \le C\varepsilon,$$

*where the constant $C > 0$ is independent of $t$ and $\varepsilon$ (cf. Figure 12.3).*

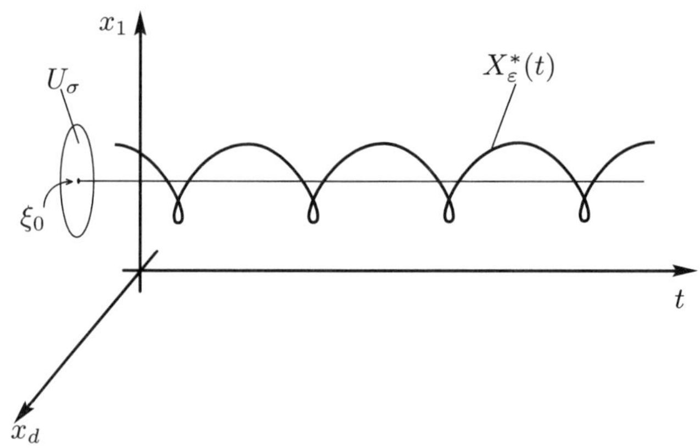


FIGURE 12.3

Consider a linear nonautonomous system $\dot{X} = A(t)X$. Let $X(t)$ be a solution to this system.

DEFINITION 12.3. By the *characteristic exponent* (or *Lyapunov exponent*) of the solution $X(t)$ we mean the number

$$\lambda[x] \equiv \overline{\lim_{t\to\infty}} \frac{\ln |X(t)|}{t}.$$

Let $X_0(t)$ be a periodic solution to some (nonlinear) system $\dot{X} = F(X)$. Consider the linear system

$$\dot{Y} = F'(X_0(t))Y \tag{12.14}$$

which is the linearization of the system $\dot{X} = F(X)$ on the periodic solution $X_0(t)$. Note that the periodic function $Y_0(t) = \dot{X}_0(t)$ is a solution to the linearized system. Indeed, differentiating the equation $\dot{X} = F(X)$ with respect to $t$, we find

$$\ddot{X}_0 = F'(X_0(t))\dot{X}_0(t),$$

which implies

$$\dot{Y}_0 = F'(X_0(t))Y_0.$$

Thus, one of the solutions to the linearized system (in the case $X_0(t) \not\equiv \text{const}$) is a periodic nonzero function, and consequently its characteristic exponent vanishes.

DEFINITION 12.4. A periodic solution $X_0(t)$ is said to be *stable at linear approximation* if the characteristic exponents of the remaining $(d-1)$ linearly independent solutions to the linearized system (12.14) are negative.

THEOREM 12.4. *Let $F$ and $G$ be the same as in Theorem 12.2, and let $Y_\varepsilon = Y(\varepsilon\omega t)$ be a stable limit cycle with period $\frac{2\pi}{\varepsilon\omega}$ of the system*

$$\frac{dY_\varepsilon}{dt} = \varepsilon G(Y_\varepsilon),$$

*where $G(x)$ is defined by (12.12) and $Y(\theta)$ is a $2\pi$-periodic function. Then, in an arbitrary $\sigma$-neighborhood $U_\sigma$ of this limit cycle, there exists $\varepsilon_0(\sigma)$ such that for $\varepsilon < \varepsilon_0(\sigma)$ the system*

$$\frac{dX_\varepsilon}{dt} = \varepsilon F(t, X_\varepsilon)$$

*in $U_\sigma$ has a unique stable two-dimensional integral manifold $I_\varepsilon$ described by the equation $x = f_\varepsilon(t,\theta)$, where $f_\varepsilon(t,\theta)$ is smooth and $2\pi$-periodic with respect to $t$, $\theta$. Moreover, for all $t \in (-\infty, +\infty)$,*

$$|f_\varepsilon(t,\theta) - Y(\theta)| \le C\varepsilon,$$

*where the constant $C > 0$ is independent of $t$ and $\varepsilon$ (cf. Figure 12.4).*

REMARK 12.2. Theorems 12.2–12.4 are proved in [**21**] under the following more general assumptions: first, the dependence of the right-hand side on time $t$ is not necessarily periodic, and, second, the stability of the equilibrium state and of the limit cycle for the homogenized system is not necessary for the existence of a limit cycle or a two-dimensional integral manifold for the original system. We refer the reader to [**21**] for details. Here, we restrict ourselves to simplified formulations of the main theorems.

In [**21**], instead of periodicity of the right-hand side with respect to the explicitly occurring time, the condition of the existence of the uniform (with respect to $T_0$) mean value

$$\lim_{T\to+\infty} \frac{1}{T}\int_{T_0}^{T_0+T} f(t)\,dt$$

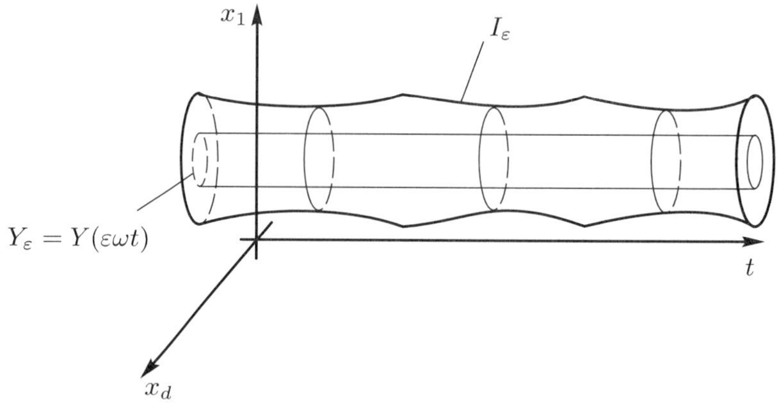


FIGURE 12.4. $I_\varepsilon = \{(x,t) : x = f_\varepsilon(t,\theta), t \geq 0, 0 \leq \theta \leq 2\pi\}$.

was used, and instead of stability at first approximation of the equilibrium state or of the limit cycle of the homogenized system, the following weaker condition was required: the absence of roots of the characteristic equation for the matrix $F'(x_0)$ on the imaginary axis of the complex plane in the case of autonomous systems, or the absence of the zero Lyapunov exponents of the system $\dot{Y} = F'(X_0(\varepsilon\omega t))Y$ (except for one, which is always present), in the case of nonautonomous systems.

**12.2. Examples.**

12.2.1. *Pendulum of varying length.* Consider the *Hill equation* for the motion of a pendulum the length of which varies periodically in accordance with a given law ("swing"):

$$\ddot{x} + \nu^2 x = -\varepsilon f(t) x \tag{12.15}$$

(the change of length is described by the function $\varepsilon f(t)$, where $\varepsilon$ is a small positive parameter). Making the change of variables $(x, \dot{x}) \longmapsto (a, b)$, where

$$x = a \sin \nu t + b \cos \nu t, \quad \dot{x} = a\nu \cos \nu t - b\nu \sin \nu t, \tag{12.16}$$

we reduce equation (12.15) to the system

$$\begin{aligned} \dot{a} &= -\frac{\varepsilon}{\nu} f(t)[a \sin \nu t + b \cos \nu t] \cos \nu t, \\ \dot{b} &= \frac{\varepsilon}{\nu} f(t)[a \sin \nu t + b \cos \nu t] \sin \nu t. \end{aligned} \tag{12.17}$$

EXERCISE 12.6. Show that (12.15) implies (12.17).

We assume that $f(t) = a_0 \sin \lambda t$. For $\lambda = 2\nu$ the average of the right-hand side over time yields the system of equations

$$\begin{aligned} \dot{a} &= -\frac{a_0\varepsilon}{2\nu} a \langle \sin^2 2\nu t \rangle_t \equiv -\frac{a_0\varepsilon}{4\nu} a, \\ \dot{b} &= \frac{a_0\varepsilon}{2\nu} b \langle \sin^2 2\nu t \rangle_t \equiv \frac{a_0\varepsilon}{4\nu} b, \end{aligned}$$

where

$$\langle \sin^2 2\nu t \rangle_t := \lim_{T\to\infty} \frac{1}{T} \int_0^T \sin^2 2\nu t \, dt = \frac{1}{2},$$

which implies the exponential growth of $a$ or $b$. This means that for $\lambda = 2\nu$ the amplitude of oscillations of the pendulum exponentially increases. Thus, in the case $\lambda = 2\nu$, even arbitrarily small periodic changes of the pendulum length could "swing" the pendulum up to an arbitrarily large amplitude.

12.2.2. *Pendulum with vibrating suspension point.* It is well known that, under certain vibration frequencies, an inverted pendulum on a vibrating support (with vibrating suspension point) does not fall down but remains in the vertical position. This means that the unstable upper equilibrium state becomes stable. This phenomenon can be explained by using the Bogolyubov averaging method.

The motion of a pendulum for which the suspension point vibrates in the vertical direction by the law $\delta(t) = a \sin \omega t$, where $\delta(t)$ is the deviation of the suspension point in the vertical direction, is described by the equation

$$\frac{d^2\theta}{dt^2} + \lambda \frac{d\theta}{dt} + \frac{g - a\omega^2 \sin \omega t}{l} \sin \theta = 0,$$

where $\lambda$ is the friction coefficient, $l$ is the suspension length, and $g$ is the gravitation constant. We introduce a new independent variable $\tau = \omega t$ and the constants

$$\omega_0 = \sqrt{\frac{g}{l}}, \quad \alpha = \frac{\lambda}{2\omega_0} k, \quad k^2 = \frac{g}{\omega^2 l} \left( \frac{l}{a} \right)^2.$$

Since

$$\frac{d}{d\tau} = \frac{1}{\omega} \frac{d}{dt}, \qquad \frac{d^2}{d\tau^2} = \frac{1}{\omega^2} \frac{d^2}{dt^2},$$

the equation with respect to the new variable $\tau$ takes the form

$$\frac{d^2\theta}{d\tau^2} + \frac{\lambda}{\omega} \frac{d\theta}{d\tau} + \frac{g - a\omega^2 \sin \tau}{\omega^2 l} \sin \theta = 0.$$

It is obvious that the constants satisfy the relations

$$\frac{g}{\omega^2 l} = \left( \frac{\omega_0}{\omega} \right)^2 = k^2 \left( \frac{a}{l} \right)^2, \quad \frac{\lambda}{\omega} = \frac{\lambda}{\omega_0} \frac{\omega_0}{\omega} = \frac{\lambda}{\omega_0} k \frac{a}{l} = 2\alpha \frac{a}{l}.$$

Therefore, the latter equation can be written in the form

$$\frac{d^2\theta}{d\tau^2} + 2\alpha \frac{a}{l} \frac{d\theta}{d\tau} + \left\{ k^2 \left( \frac{a}{l} \right)^2 - \frac{a}{l} \sin \tau \right\} \sin \theta = 0.$$

Assume that $\varepsilon = a/l$ is a small parameter. Then

$$\frac{d^2\theta}{d\tau^2} + 2\varepsilon\alpha \frac{d\theta}{d\tau} + \left\{ k^2 \varepsilon^2 - \varepsilon \sin \tau \right\} \sin \theta = 0.$$

We cannot directly use the change of variables (12.16) because $\nu = 0$. Making the change of variables $(\theta, \dot{\theta}) \longmapsto (\varphi, \psi)$ by the formulas $\theta = \varphi - \varepsilon \sin \tau \sin \varphi$ and $\dot{\theta} = \varepsilon\psi - \varepsilon \cos \tau \sin \varphi$, we reduce the equation to the following system in the normal form:

$$\begin{aligned} \frac{d\varphi}{d\tau} &= \varepsilon\psi + \varepsilon^2[\cdots], \\ \frac{d\psi}{d\tau} &= \varepsilon[-\sin^2 \tau \sin \varphi \cos \varphi - k^2 \sin \varphi + \psi \cos \tau \cos \varphi - 2\alpha\psi \\ &\qquad + 2\alpha \cos \tau \sin \varphi] + \varepsilon^2[\cdots]. \end{aligned} \tag{12.18}$$

We do not write explicit expressions at higher powers of $\varepsilon$. Averaging the right-hand sides with respect to $\tau$, we obtain the system

$$
\begin{aligned}
\frac{d\widetilde{\varphi}}{d\tau} &= \varepsilon\widetilde{\psi}, \\
\frac{d\widetilde{\psi}}{d\tau} &= -\varepsilon\Big[2\alpha\widetilde{\psi} + \Big(k^2 + \frac{1}{2}\cos\widetilde{\varphi}\Big)\sin\widetilde{\varphi}\Big],
\end{aligned} \tag{12.19}
$$

which is equivalent to the following autonomous second order equation:

$$\frac{d^2\varphi}{d\tau^2} + 2\varepsilon\alpha\frac{d\varphi}{d\tau} + \varepsilon^2\Big(k^2 + \frac{1}{2}\cos\varphi\Big)\sin\varphi = 0.$$

In the case $\frac{1}{2} - k^2 > 0$, the equilibrium state $\varphi = \pi$ is stable at first approximation. Consequently, in a neighborhood of the upper equilibrium state, the original equation has a stable limit cycle (cf. Theorem 12.3) of the form $\theta_\varepsilon(t) = \varphi_\varepsilon(t/\varepsilon) - \varepsilon\sin(t/\varepsilon)\sin\varphi_\varepsilon(t/\varepsilon)$, where $\varphi_\varepsilon(\tau)$ is a periodic solution to the system (12.18) in a small neighborhood of the point $\pi$.

12.2.3. *The van der Pol equation.* Consider nonautonomous second order equation

$$\ddot{x} + \nu^2 x = \varepsilon f(t, x, \dot{x}) \tag{12.20}$$

with a small nonlinear term $\varepsilon f$. We make the change of variables $(x, \dot{x}) \longmapsto (a, \theta)$ in accordance with the formulas

$$x = a\cos\theta, \quad \dot{x} = -\nu a\sin\theta. \tag{12.21}$$

A computation shows that equation (12.20) in new variables takes the form (check it!)

$$
\begin{aligned}
\dot{a} &= -\frac{\varepsilon}{\nu} f(t, a\cos\theta, -a\nu\sin\theta)\sin\theta, \\
\dot{\theta} &= \nu - \frac{\varepsilon}{a\nu} f(t, a\cos\theta, -a\nu\sin\theta)\cos\theta.
\end{aligned} \tag{12.22}
$$

We can apply Theorem 12.2 to the system (12.22). On the other hand, making the change of variables (12.16), we reduce equation (12.20) to the form

$$
\begin{aligned}
\dot{a} &= -\frac{\varepsilon}{\nu} f(t, a\sin\nu t + b\cos\nu t, a\nu\cos\nu t - b\nu\sin\nu t)\cos\nu t, \\
\dot{b} &= -\frac{\varepsilon}{a\nu} f(t, a\sin\nu t + b\cos\nu t, a\nu\cos\nu t - b\nu\sin\nu t)\sin\nu t,
\end{aligned} \tag{12.23}
$$

i.e., we can apply Theorem 12.1. Sometimes, it is convenient to use the change of variables (12.16), whereas in other cases the change (12.21) is more preferable. Consider the van der Pol equation

$$\ddot{x} + \nu^2 x = \varepsilon(1 - bx^2)\dot{x}.$$

Making the change of variables (12.21) and using Theorem 12.2, we reduce the homogenized equation for $a$ to the equation

$$\dot{a} = -\frac{\varepsilon}{\nu}\varphi(a),$$

where

$$\varphi(a) = \int_0^{2\pi} (1 - ba^2\cos^2 y)a\sin^2 y\, dy \equiv \frac{a}{2}\left(\frac{ba^2}{4} - 1\right).$$

The equation $\varphi(a) = 0$ has two nonnegative roots $a_1 = 0$ and $a_2 = 2\sqrt{\frac{1}{b}}$ if $b > 0$ and the single root $a_1 = 0$ if $b < 0$. In the case $b > 0$, since the equilibrium state is stable at the point $a_2$ and unstable at the point $a_1$, we have the following qualitative picture of the behavior of phase trajectories of the van der Pol equation: all the phase trajectories approach a stable periodic solution with amplitude $2\sqrt{\frac{1}{b}}$ either inside or outside the circle (cf. Figure 12.5).

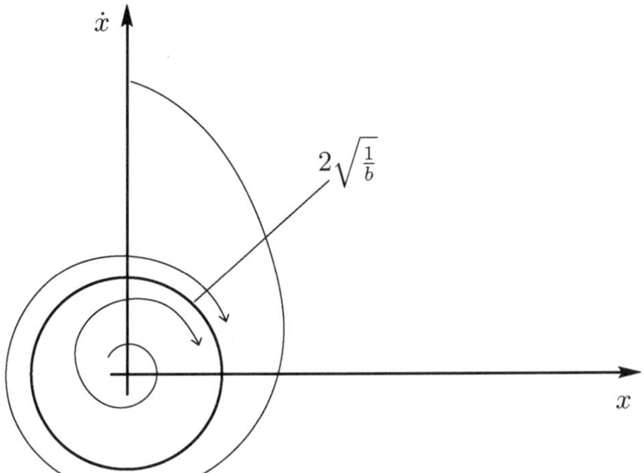


FIGURE 12.5

Theorem 12.2 allows us to assert that this qualitative picture of the behavior of phase trajectories takes place on a time interval of order $1/\varepsilon$.

12.2.4. *Duffing type equations.* The differential equation

$$\ddot{x} + \nu^2 x = \varepsilon f(x, \dot{x}) + \varepsilon g(t) \tag{12.24}$$

is called the *Duffing equation* if $f(x, \dot{x}) = -x^3$. Making the change of variables (12.21), we obtain the system

$$\begin{aligned} \dot{a} &= -\frac{\varepsilon}{\nu} f(a\cos\theta, -a\nu\sin\theta)\sin\theta - \frac{\varepsilon}{\nu} g(t)\sin\theta, \\ \dot{\theta} &= \nu - \frac{\varepsilon}{a\nu} f(a\cos\theta, -a\nu\sin\theta)\cos\theta - \frac{\varepsilon}{\nu a} g(t)\cos\theta. \end{aligned} \tag{12.25}$$

Suppose the external action $g(t)$ has the form $g(t) = a_0 \cos\lambda t$. Denoting $z = \lambda t$, we transform the system (12.25) to the form

$$\begin{aligned} \dot{a} &= -\frac{\varepsilon}{\nu} f(a\cos\theta, -a\nu\sin\theta)\sin\theta - \frac{\varepsilon}{\nu} a_0 \cos z \sin\theta, \\ \dot{\theta} &= \nu - \frac{\varepsilon}{a\nu} f(a\cos\theta, -a\nu\sin\theta)\cos\theta - \frac{\varepsilon a_0}{\nu a} \cos z \cos\theta, \\ \dot{z} &= \lambda. \end{aligned} \tag{12.26}$$

Assume that the frequency $\lambda$ of the "external action" is close to the eigenfrequency $\nu$ of oscillations of the system without perturbations, i.e., $\lambda = \nu + \varepsilon h$, where $h$ is the so-called *detuning*. Introduce the new variable $\delta = z - \theta$ and pass from $(a, \theta, z)$ to $(a, \delta, z)$. Setting $z = (\nu + \varepsilon h)t$ and noting that the system obtained has the normal form (cf. the assumption of Theorem 12.1), we average the right-hand side of the system with respect to $t$. Then we obtain the system

$$(12.27)\qquad \begin{aligned} \dot{a} &= -\frac{\varepsilon}{2\nu\pi}\left(\int_0^{2\pi} f(a\cos y, -a\nu\sin y)\sin y\,dy - a_0\pi\sin\delta\right), \\ \dot{\delta} &= \frac{\varepsilon}{2\pi a\nu}\left(\int_0^{2\pi} f(a\cos y, -a\nu\sin y)\cos y\,dy + a_0\pi\cos\delta\right) + \varepsilon h. \end{aligned}$$

EXERCISE 12.7. Check that the system (12.26) in the variables $(a, \delta, z)$ has the form (12.27).

In the case of the Duffing equation, $f(a\cos y, -a\nu\sin y) = -a^3\sin^3 y$. Therefore, the integrals on the right-hand side of (12.27) can be easily computed, and (12.27) takes the form

$$(12.28)\qquad \begin{aligned} \dot{a} &= -\frac{\varepsilon}{2\nu}a_0\sin\delta, \\ \dot{\delta} &= -\frac{\varepsilon}{2\pi a\nu}\left(\frac{3}{4}a^3 - a_0\pi\cos\delta\right) + \varepsilon h. \end{aligned}$$

The stationary points $(\widetilde{a}, \widetilde{\delta})$ of this system are found from the algebraic equations

$$\begin{aligned} 0 &= \sin\widetilde{\delta}, \\ 0 &= \frac{3}{4}\widetilde{a}^3 - a_0\pi\cos\widetilde{\delta} - 2\pi\nu\widetilde{a}h. \end{aligned}$$

Hence $\widetilde{\delta} = 0$, and the stationary value of the oscillation amplitude $\widetilde{a}$ satisfies the cubic equation

$$(12.29)\qquad \widetilde{a}^3 - \frac{8}{3}\pi\nu h\widetilde{a} - \frac{4}{3}a_0\pi = 0.$$

Our goal is to study the behavior of a weakly nonlinear oscillating system (the nonlinear term $\varepsilon f(x, \dot{x})$ is small for small $\varepsilon$) under the action of a periodic force with frequency close to the eigenfrequency of oscillations of the linear system (the system is linear for $\varepsilon = 0$). Under perturbations of the linear oscillator by a periodic force with period close to the period of eigenoscillations of the oscillator, the well-known resonance phenomenon arises: the amplitude unboundedly increases as the detuning decreases. An analysis of this example shows that the weak nonlinearity essentially changes the qualitative picture of oscillations.

Thus, we have checked that for the Duffing equation

$$(12.30)\qquad \ddot{x} + \nu^2 x = -\varepsilon x^3 + \varepsilon a_0\cos((\nu + \varepsilon h)t),$$

which is a special case of (12.24), the stationary points of the amplitude $\widetilde{x}$ satisfy the cubic equation (12.29). It is of great interest to study the dependence of the roots of equation (12.29), i.e., the amplitudes of stationary oscillating behavior, on the detuning $h$.

Figure 12.6 shows the dependence of positive real roots (the number of which may be 1, 2, or 3) on $h$. The bold line represents the branches of the root of equation (12.29) corresponding to stable cycles and the thin line corresponds to unstable cycles.

By Theorem 12.3, the stability of a stationary point $(\widetilde{a}, \widetilde{\delta})$ of the system (12.28) implies the existence of the limit cycle for the original system in a neighborhood of $(\widetilde{a}, \widetilde{\delta})$. Therefore, the original equation has a periodic solution $x_\varepsilon(t) =$

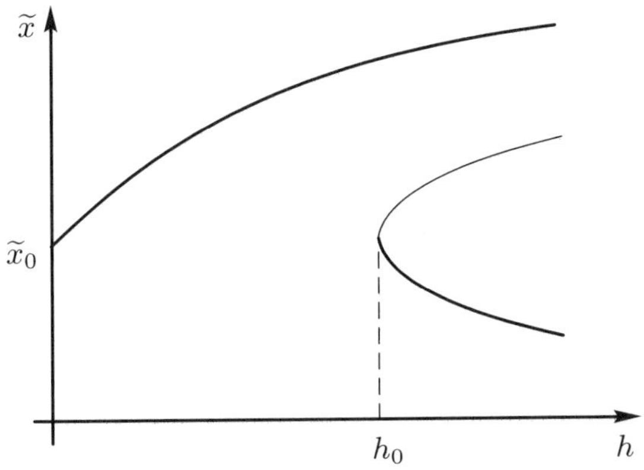


FIGURE 12.6. The bold line corresponds to the amplitudes of stable cycles and the thin line corresponds to the amplitudes of unstable cycles; $h$ represents the detuning.

$a_\varepsilon(t)\cos(\lambda_\varepsilon t - \delta_\varepsilon(t))$, where $\lambda_\varepsilon = \nu + \varepsilon h$, and $a_\varepsilon(t)$ and $\delta_\varepsilon(t)$ are periodic functions with period $\frac{2\pi}{\lambda_\varepsilon}$; moreover, $a_\varepsilon(t) \to \widetilde{a}$ and $\delta_\varepsilon(t) \to \widetilde{\delta}$ as $\varepsilon \to 0$. The limit cycle tends to the cycle $\widetilde{a}\cos(\nu t + \widetilde{\delta})$ as $\varepsilon \to 0$. Note that the diagram of roots presented in Figure 12.6, shows a rather unexpected result: as the "detuning" $h$ tends to zero, the amplitude of stationary stable oscillations is not necessarily increasing (in particular, this happens in linear systems if the frequency of the external force approaches the eigenfrequency of oscillations of the system), and it may also decrease or "jump" when passing through the critical value $h_0$.

CHAPTER 3

# Applications of Homogenization Methods

## 13. Homogenized equations with periodic coefficients. Composite materials

In this section, we show how to apply homogenization methods to elliptic systems. For this purpose, we consider a stationary linear elasticity system with homogeneous Dirichlet boundary condition.

Consider a periodic micro-inhomogeneous material occupying a domain $\Omega$. As usual, $\varepsilon$ denotes the microstructure period and the characteristics of the material are described by the elasticity tensor $a_{ij}^{kl} = a_{ij}^{kl}(\frac{x}{\varepsilon})$, where $a_{ij}^{kl}(\xi)$ is a 1-periodic function (see Section 3 for the definition and properties of the elasticity tensor). The corresponding problem takes the form

$$L_\varepsilon \overline{u}_\varepsilon \equiv \frac{\partial}{\partial x_p}\left(A^{pq}\left(\frac{x}{\varepsilon}\right)\frac{\partial \overline{u}_\varepsilon(x)}{\partial x_q}\right) = f(x), \tag{13.1}$$

where $\overline{u}_\varepsilon(x) \in \overset{\circ}{H}{}^1(\Omega)$, the operator $L_\varepsilon$ is associated with the material under consideration, and $f$ is an arbitrary smooth function.

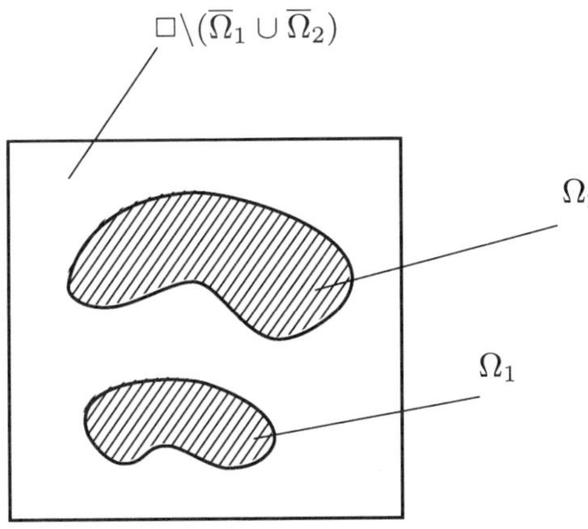


FIGURE 13.1

As an example of a micro-inhomogeneous material we can consider a composite material consisting of two or several different homogeneous materials forming a periodic structure. The mathematical description of composite materials assumes that the periodicity cube $\square$ of functions $a_{ij}^{kl}(\xi)$ contains several disjoint domains $Q_1, Q_2, \ldots, Q_N$ with smooth boundaries at a positive distance from each other. Suppose that the elasticity tensor $a_{ij}^{kl}(\frac{x}{\varepsilon})$ is constant in domains $\Omega_i$ that are the unions of the homothetically contracted and periodically extended domains $Q_i$,

$i = 1, \dots, N$, and also in the domain $\Omega \backslash \big(\bigcup_{i=1}^{N} \Omega_i\big)$, called a *matrix material* or *matrix*. Such a tensor is associated with a material with inclusions of $N$ different materials (cf. Figure 13.1).

Another situation happens if the inclusions form an unbounded connected component and may have common points. For example, consider a reinforced material where the roles of inclusions are played by rods intersecting the reinforcement domain in several different directions. In this case, the periodicity cell $\square$ contains a part of the unbounded periodic set lying in the space $\mathbb{R}^d$. The composite material is described by a tensor $a_{ij}^{kl}\left(\frac{x}{\varepsilon}\right)$ that takes constant values on $\Omega_1$ and $\Omega \backslash \Omega_1$ (cf. Figure 13.2).

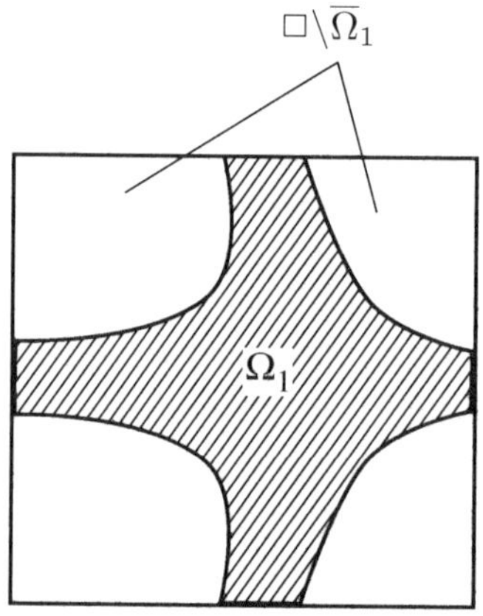


FIGURE 13.2

To construct the homogenized (effective) tensor, we use the following auxiliary problem concerning the periodicity cube $\square$ in the class of periodic functions

$$\frac{\partial}{\partial \xi_k}\left(A^{kj}(\xi)\frac{\partial}{\partial \xi_j}N^q(\xi)\right) = -\frac{\partial}{\partial \xi_k}A^{kq}(\xi) \quad \text{in } \square, \quad q = 1, \dots, d, \tag{13.2}$$

where $A^{kj}(\xi)$ is a $d \times d$-matrix with entries $a_{il}^{kj}(\xi)$, and $N^q(\xi)$ is the unknown vector-valued function with 1-periodic components with respect to $\xi_1, \dots, \xi_d$. The solvability of (13.2) follows from the Lax–Milgram theorem and the Korn inequality for periodic functions.

EXERCISE 13.1. Prove the Korn inequality for periodic functions,

$$\|\overline{u}\|_{H^1(\square)} \leq C\|e(\overline{u})\|_{L_2(\square)},$$

where $\overline{u}(x)$ is 1-periodic with respect to $\xi_1, \dots, \xi_d$ and has zero mean, i.e.,

$$\int_{\square} \overline{u}(x)\, dx = 0;$$

$C > 0$ is a constant independent of $u(x)$.

We define a matrix-valued function $\widehat{A}^{pq}$ by the formula

$$\widehat{A}^{pq} = \int_{\square}\left(A^{pq}(\xi) + A^{pj}(\xi)\frac{\partial N^q(\xi)}{\partial \xi_j}\right) d\xi \tag{13.3}$$

and consider the boundary value problem

$$\widehat{L}\overline{u}_0 := \frac{\partial}{\partial x_p}\left(\widehat{A}^{pq}\frac{\partial \overline{u}_0(x)}{\partial x_q}\right) = f(x), \quad \overline{u}_0 \in \overset{\circ}{H}{}^1(\Omega). \tag{13.4}$$

It turns out that (13.4) is the elasticity system, i.e., $\widehat{A}^{pq}$ is symmetric and positive definite. The proof of this important fact is based on the two assertions formulated in the following exercises.

EXERCISE 13.2. Prove that $\widehat{A}^{pq}(\xi)$ admits the representation

$$\widehat{A}^{pq} = \int_{\square} \frac{\partial}{\partial \xi_k}((N^p)^* + \xi_p E)A^{kj}(\xi)\frac{\partial}{\partial \xi_j}(N^q + \xi_q E)\, d\xi,$$

where $*$ denotes transposition and $E$ is the identity matrix.

EXERCISE 13.3. Prove that

$$\widehat{a}_{ij}^{pq}\eta_{ip}\eta_{jp} := \int_{\square} \frac{\partial}{\partial \xi_k}[(\eta^p)^*((N^p)^* + \xi_p E)]A^{kj}(\xi)\frac{\partial}{\partial \xi_j}[(N^q + \xi_q E)\eta^q]\, d\xi$$

satisfies the inequality $\kappa_0\eta_{ik}\eta_{ik} \le \widehat{a}_{ij}^{pq}\eta_{ip}\eta_{jp}$, where $\widehat{a}_{ij}^{pq}$ are the entries of the matrix $\widehat{A}^{pq}$ and $\eta_{ij}$ are the entries of an arbitrary symmetric $d \times d$-matrix.

Let us verify that the system (13.4) describes the homogenized characteristics of the material under consideration. Consider the vector-valued function

$$\widetilde{u}(x) = \overline{u}_0(x) + \varepsilon N^p\left(\frac{x}{\varepsilon}\right)\frac{\partial \overline{u}_0(x)}{\partial x_p},$$

which will play the role of an approximate solution to the problem (13.1). Indeed, a computation shows that

$$\begin{aligned} L_\varepsilon(\overline{u}_\varepsilon - \widetilde{u}) = & -\varepsilon A^{hk}N^s\frac{\partial^3 \overline{u}_0}{\partial x_h \partial x_k \partial x_s} \\ & + \left[\widehat{A}^{pq} - A^{pq} - A^{pj}\frac{\partial N^q}{\partial \xi_j} - \frac{\partial}{\partial \xi_h}(A^{hp}N^q)\right]\frac{\partial^2 \overline{u}_0}{\partial x_p \partial x_q}. \end{aligned} \tag{13.5}$$

Now we need to prove that $\|\overline{u}_\varepsilon - \widetilde{u}\|_{H^1(\Omega)}$ is small provided that the right-hand side of (13.5) is small and the boundary condition $(\overline{u}_\varepsilon - \widetilde{u})\big|_{\partial\Omega} = \varphi\big|_{\partial\Omega}$ holds, where $\varphi \in H^1(\Omega)$ is a vector-valued function whose boundary values (i.e., the trace of $\varphi$) coincide with the boundary values of the difference $\overline{u}_\varepsilon - \widetilde{u}$, on $\partial\Omega$. In what sense is the right-hand side of (13.5) small? The first term explicitly contains a small parameter. To estimate the second term, we use the following exercise.

EXERCISE 13.4. Suppose that $h(\xi)$ is a piecewise smooth 1-periodic function with respect to $\xi_1, \dots, \xi_d$ with zero mean, i.e., $\int_\square h(\xi)\, d\xi = 0$. Prove that $h(\xi) = \sum_{i=1}^d \frac{\partial h_i(\xi)}{\partial \xi_i}$, where $h_i(\xi)$ are piecewise smooth periodic functions with zero mean.

By Exercise 13.4, we can represent the right-hand side of (13.5) in the form

$$\varepsilon\frac{\partial}{\partial x_i}\left(g_i\left(\frac{x}{\varepsilon}\right)\frac{\partial^2 u_0}{\partial x_p \partial x_q}\right) - \varepsilon g_i\left(\frac{x}{\varepsilon}\right)\frac{\partial^3 u_0}{\partial x_i \partial x_p \partial x_q},$$

where $g_i(\xi) = h_i(\xi) - A^{ip}(\xi)N^q(\xi)$ and

$$\sum_{i=1}^d \frac{\partial h_i}{\partial \xi_i} = \widehat{A}^{pq} - A^{pq} - A^{pj}\frac{\partial N^q}{\partial \xi_j},$$

since

$$\int_{\square}\left[\widehat{A}^{pq}-A^{pq}-A^{pj}\frac{\partial N^q}{\partial \xi_j}\right]d\xi=0$$

by the definition of $\widehat{A}^{pq}$.

Let us estimate the trace of $\overline{u}_\varepsilon-\widetilde{u}$ on $\partial\Omega$ in the $H^{1/2}(\partial\Omega)$-norm.

EXERCISE 13.5. Suppose that $N(\xi)\in H^1(\square)$ is 1-periodic with respect to $\xi_1,\dots,\xi_d$ and $u(x)\in C^\infty(\overline{\Omega})$. Prove that

$$\left\|N\Big(\frac{x}{\varepsilon}\Big)u(x)\right\|_{H^{1/2}(\partial\Omega)}\le\frac{C}{\sqrt{\varepsilon}}.$$

By Exercise 13.5 and the estimate for solutions to the Dirichlet problem for the elasticity system, we have $\|\overline{u}_\varepsilon-\widetilde{u}\|_{H^1(\Omega)}\le C\sqrt{\varepsilon}$. Thus, $\widetilde{u}$ is an approximate solution to the problem (13.1) for small $\varepsilon$.

An important special case of micro-inhomogeneous materials is presented by the so-called laminated composites. A composite material is referred to as a *laminated composite* if all the entries of the elasticity tensor depend on a single variable, i.e., $a_{ij}^{kl}=a_{ij}^{kl}\big(\frac{x_1}{\varepsilon}\big)$ under a suitable choice of coordinates. Assume that a periodic laminated composite is described by the Lamé system. Then the coefficient matrix is symmetric and takes the form

$$\begin{pmatrix}\lambda+2\mu & \lambda & \lambda & 0 & 0 & 0\\ & \lambda+2\mu & \lambda & 0 & 0 & 0\\ & & \lambda+2\mu & 0 & 0 & 0\\ & & & \mu & 0 & 0\\ & & & & \mu & 0\\ & & & & & \mu\end{pmatrix};$$

moreover, each entry is a periodic function of $x_1/\varepsilon$. We construct the matrix of the homogenized elasticity system by formula (13.3) using the fact that all the equations of the system (13.4) become ordinary and can be explicitly solved. A computation yields the homogenized matrix

$$\begin{pmatrix}\frac{1}{\langle\frac{1}{\lambda+2\mu}\rangle} & \langle\frac{\lambda}{\lambda+2\mu}\rangle\langle\frac{1}{\lambda+2\mu}\rangle^{-1} & \langle\frac{\lambda}{\lambda+2\mu}\rangle\langle\frac{1}{\lambda+2\mu}\rangle^{-1} & 0 & 0 & 0\\ & \langle\lambda+2\mu\rangle+\frac{\langle\frac{\lambda}{\lambda+2\mu}\rangle}{\langle\frac{1}{\lambda+2\mu}\rangle}-\langle\frac{\lambda^2}{\lambda+2\mu}\rangle & \langle\lambda\rangle+\frac{\langle\frac{\lambda}{\lambda+2\mu}\rangle}{\langle\frac{1}{\lambda+2\mu}\rangle}-\langle\frac{\lambda^2}{\lambda+2\mu}\rangle & 0 & 0 & 0\\ & & \langle\lambda+2\mu\rangle+\frac{\langle\frac{\lambda}{\lambda+2\mu}\rangle}{\langle\frac{1}{\lambda+2\mu}\rangle}-\langle\frac{\lambda^2}{\lambda+2\mu}\rangle & 0 & 0 & 0\\ & & & \frac{1}{\langle\mu^{-1}\rangle} & 0 & 0\\ & & & & \frac{1}{\langle\mu^{-1}\rangle} & 0\\ & & & & & \frac{1}{\langle\mu^{-1}\rangle}\end{pmatrix}$$

which is also symmetric. As usual, $\langle\cdot\rangle$ denotes the mean value over the period, i.e., $\langle f\rangle=\int_0^1 f(t)\,dt$.

Thus, we have obtained the elasticity system with constant coefficients which has more complicated structure than the Lamé system. The system obtained characterizes an anisotropic or, more exactly, orthotropic material.

## 14. Filtration of viscous incompressible fluid through a porous medium. The Darcy law.

**14.1. The main notions.** Let $\Omega^\varepsilon$ be the perforated domain in $\mathbb{R}^d$ obtained from a bounded domain $\Omega$ with Lipschitz boundary $\partial\Omega$ by eliminating sets of the form $\varepsilon(\Xi_0+\overline{m})$, where $\overline{m}=(m_1,\dots,m_d)\in\mathbb{Z}^d$, $\varepsilon$ is a small positive parameter, and $\Xi_0$ is a closed subset of the unit cube $\square$ with nonempty interior such that the set $\mathbb{R}^d\setminus\bigcup_{\overline{m}\in\mathbb{Z}^d}(\Xi_0+\overline{m})$ is open. Assume that the set $\square\setminus\Xi_0$ can be represented as the finite union of Lipschitz domains $Q_j$, $j=1,\dots,k$, i.e., $\square\setminus\Xi_0=\bigcup_{j=1}^k Q_j$. Note that we eliminate from $\Omega$ only those sets $\varepsilon(\Xi_0+\overline{m})$ for which $\operatorname{dist}(\partial\Omega,\varepsilon(\Xi_0+\overline{m})\geq k_0\varepsilon$, where $k_0>0$ is a constant independent of $\varepsilon>0$ (cf. Figure 14.1).

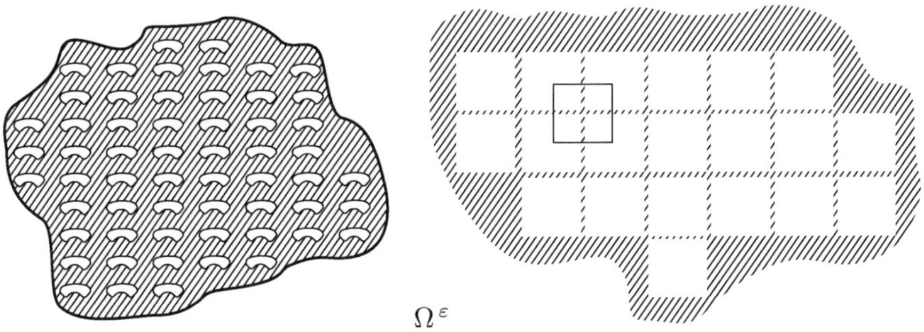


FIGURE 14.1

The obstacles eliminated from $\Omega$ may be disconnected dispersed inclusions or "brick-elements" with cut channels that are adjacent to their neighbors along the faces (cf. Figure 14.2).

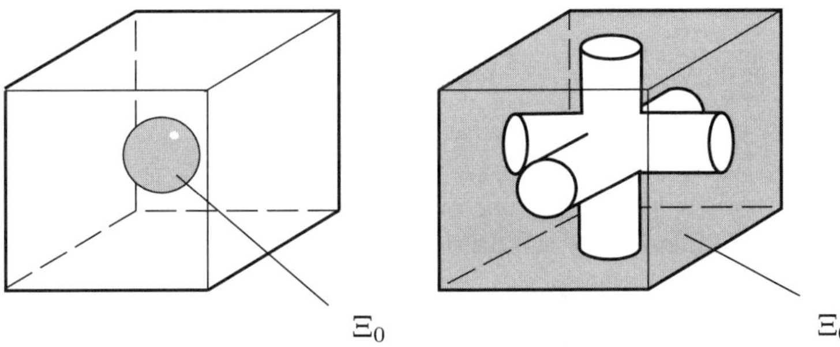


FIGURE 14.2

In the first case, the fluid occupying the domain $\Omega^\varepsilon$ flows around the obstacles. The boundary of obstacles is not necessarily smooth; for example, an obstacle can be a disk with adjoint intervals (cf. Figure 14.3).

In the second case, the fluid filters through channels. We note that, in both cases, the obstacles are at a distance from the boundary of $\Omega$, which allows us to avoid the appearance of non-Lipschitz parts of the boundary of $\Omega^\varepsilon$.

Consider the Stokes problem

$$(14.1)\qquad \begin{aligned} &-\nabla p_\varepsilon+\Delta\overline{u}_\varepsilon+\overline{f}=0 \quad \text{in } \Omega^\varepsilon,\\ &\operatorname{div}\overline{u}_\varepsilon=0 \quad \text{in } \Omega^\varepsilon, \quad \overline{u}_\varepsilon\big|_{\partial\Omega^\varepsilon}=0, \end{aligned}$$

where $\overline{f}\in L_2(\Omega)$, and $(p_\varepsilon,\overline{u}_\varepsilon)$ is a weak solution in the sense of (4.2) or (4.3). Our goal is to study the behavior of the solution $(p_\varepsilon,\overline{u}_\varepsilon)$ to the problem (14.1) as $\varepsilon\to 0$.

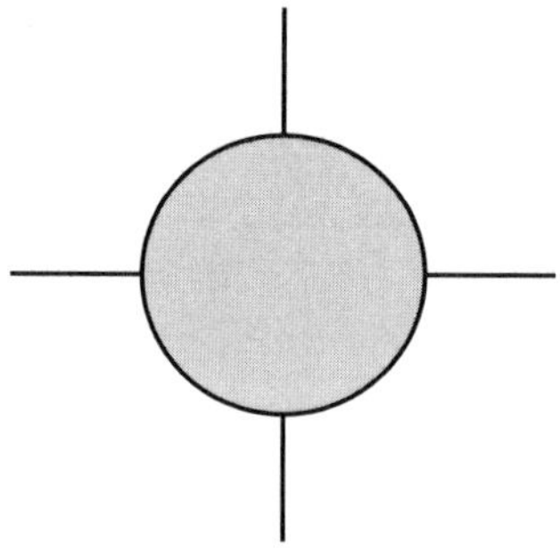

FIGURE 14.3

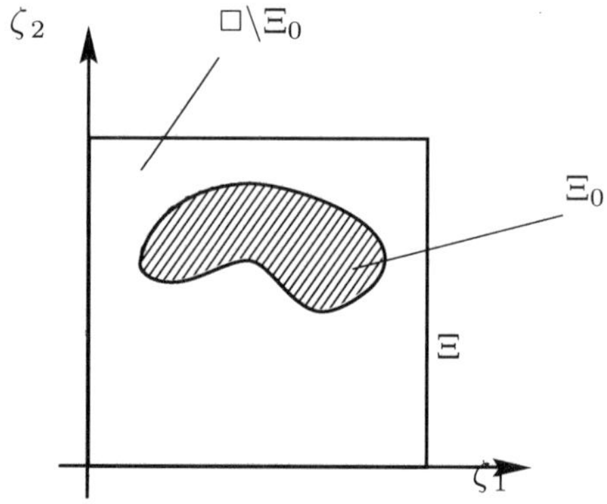


FIGURE 14.4

Let $(q^k, \overline{v}^k(\xi))$, $k = 1, 2$, be a solution to the following auxiliary problem on the periodicity cell $\Xi \equiv \square\backslash\Xi_0$ (cf. Figure 14.4):

$$
\begin{aligned}
& -\nabla_\xi q^k + \Delta_\xi \overline{v}^k + \overline{e}_k = 0 \quad \text{in } \Xi, \\
& \operatorname{div}_\xi \overline{v}^k = 0 \quad \text{in } \Xi, \\
& \overline{v}^k\big|_{\partial\Xi_0} = 0, \\
& \overline{v}^k, q^k \text{ are 1-periodic with respect to } \xi_1, \dots, \xi_d,
\end{aligned} \tag{14.2}
$$

where $\overline{e}_1, \dots, \overline{e}_d$ are the unit vectors of the $\xi_1$-$, \dots, \xi_d$-axes. The existence of a solutions in the class of periodic functions is established in the same way as the existence of a solution to the Stokes equations in an arbitrary domain $\Omega$ in Subsection 4.2 (check it!).

The matrix

$$K_{ij} \equiv \int_\Xi v_j^i(\xi)\, d\xi, \quad \xi = (\xi_1, \dots, \xi_d),$$

where $v_j^i(\xi)$ is the $j$th component of the vector $\overline{v}^i$, is called the *effective filtration matrix* or the *Darcy matrix*.

EXERCISE 14.1. Prove that $K_{ij}$ is a positive definite matrix.

Let $\overline{u}^0(x)$ be a solution to the following boundary value problem in the domain $\Omega$ (without holes):

$$\begin{aligned} &\operatorname{div} \overline{u}^0 = 0 \quad \text{in } \Omega, \\ &u_i^0 = K_{ij}\left(f_j - \frac{\partial p^0}{\partial x_j}\right) \quad \text{in } \Omega, \\ &(\overline{u}^0, \overline{\nu}) = 0 \quad \text{on } \partial\Omega, \end{aligned} \tag{14.3}$$

where $u_i^0$ is the $i$th component of the vector-valued function $\overline{u}^0$, and $\overline{\nu}$ is the outward normal to $\partial\Omega$.

The problem (14.3) is solvable. Indeed, consider the Neumann problem for $p^0(x)$,

$$\begin{aligned} &\frac{\partial}{\partial x_j}(K_{ij} f_i) = \frac{\partial}{\partial x_i}\left(K_{ij}\frac{\partial p^0}{\partial x_j}\right) \quad \text{in } \Omega, \\ &\nu_j K_{ij} f_j = \frac{\partial p^0}{\partial \widetilde{\nu}} \quad \text{on } \partial\Omega, \end{aligned} \tag{14.4}$$

where $\frac{\partial}{\partial \widetilde{\nu}} \equiv \nu_i K_{ij} \frac{\partial}{\partial x_j}$ is the conormal derivative. It is easy to verify the solvability condition for the Neumann boundary value problem. Therefore, there exists a weak solution $p^0 \in H^1(\Omega)$. Further, we find the components of $\overline{u}^0$ using the formula $u_i^0 = K_{ij}\left(f_j - \frac{\partial p^0}{\partial x_j}\right)$. The function $\overline{u}^0$ belongs to the space $H(\Omega, \operatorname{div})$ defined below.

DEFINITION 14.1. We say that $\overline{u}$ belongs to $H(\Omega, \operatorname{div})$ if $\overline{u}(x) \in L_2(\Omega)$ and $\operatorname{div} \overline{u} \in L_2(\Omega)$. The inner product in this (Hilbert) space is introduced by the formula

$$(\overline{u}, \overline{v})_{H(\Omega,\operatorname{div})} = \int_\Omega (\overline{u}, \overline{v})\, dx + \int_\Omega (\operatorname{div} \overline{u})(\operatorname{div} \overline{v})\, dx.$$

Although a function in $H(\Omega, \operatorname{div})$ does not in general have traces on $(d-1)$-dimensional hypersurfaces, the trace of the normal component of $(\overline{u}, \overline{\nu})$ on $\partial\Omega$ has a meaning in the sense of distributions because of the condition $\operatorname{div} \overline{u} \in L_2(\Omega)$.

Indeed, for a smooth vector-valued function $\overline{u}(x)$ in $\Omega$ and a smooth function $\Phi(x)$ in $\Omega$ equal to $\varphi(x)$ on $\partial\Omega$ we have the obvious identity

$$\int_\Omega u_i \frac{\partial \Phi}{\partial x_i}\, dx + \int_\Omega (\operatorname{div} \overline{u}) \Phi\, dx = \int_{\partial\Omega} (u_i, \nu_i)\varphi(x)\, d\sigma.$$

Regarding the integral on the right-hand side of this equality as the action of the distribution $(\overline{u}, \overline{\nu})$ on the smooth test function $\varphi(x)$, we can derive the estimate

$$|\langle (\overline{u}, \overline{\nu}), \varphi \rangle| \le C \|\overline{u}\|_{H(\Omega,\operatorname{div})} \|\Phi\|_{H^1(\Omega)} \le C \|\overline{u}\|_{H(\Omega,\operatorname{div})} \|\varphi\|_{H^{1/2}(\partial\Omega)}, \tag{14.5}$$

where $\Phi$ is an extension of $\varphi$ to the domain $\Omega$ such that

$$\|\Phi\|_{H^1(\Omega)} \le C \|\varphi\|_{H^{1/2}(\partial\Omega)}.$$

Due to this estimate, we can view the distribution $(\overline{u}, \overline{\nu})$ as a linear continuous functional on $H^{1/2}(\partial\Omega)$ not only for smooth functions but also for arbitrary functions $\overline{u}(x) \in H(\Omega, \operatorname{div})$. Indeed, let $\overline{u}_n \to \overline{u}$ in $H(\Omega, \operatorname{div})$, where $\overline{u}_n$ are smooth functions. For $\varphi \in H^{1/2}(\partial\Omega)$ we set

$$\langle (\overline{u}, \overline{\nu}), \varphi \rangle \equiv \lim_{n \to \infty} \langle (\overline{u}_n, \overline{\nu}), \varphi \rangle.$$

Then the normal component of $\overline{u}$ belongs to $H^{-1/2}(\partial\Omega)$. The existence of the limit and the fact that the right-hand side is well defined follow from the inequalities (14.5).

EXERCISE 14.2. Let $\overline{u}_k \rightharpoonup \overline{u}$ weakly in $H(\Omega, \mathrm{div})$ as $k \to \infty$. Prove that $(\overline{u}_k, \overline{\nu}) \to (\overline{u}, \overline{\nu})$ weakly in $[H^{1/2}(\partial\Omega)]'$ as $k \to \infty$.

Thus, the boundary condition on $\overline{u}_k$ of the form $(\overline{u}_k, \overline{\nu})\big|_{\partial\Omega} = 0$ is inherited by the weak limit $\overline{u}$ in the space $H(\Omega, \mathrm{div})$. Note that in the case of weak convergence in $L_2(\Omega)$, the boundary conditions are not necessarily inherited. Indeed, consider a sequence $u_k = \sin^2 kx$ on $[0, 1]$. Then $u_k(0) = 0$ for all $k$, but the weak limit of $u_k$ is a nonzero constant $u = C$ (what is this constant?), and consequently $u(0) \neq 0$.

EXERCISE 14.3. Prove that the component $p^0(x)$ of the solution $(\overline{u}^0, p^0)$ to the problem (14.3), where $\overline{u}^0 \in H(\Omega, \mathrm{div})$ and $p^0(x) \in H^1(\Omega)$, is a weak solution to the Neumann problem (14.4).

**14.2. Homogenization theory.** At the qualitative level, the results in this section on the passage to the limit for the Stokes problem in a perforated domain were known to specialists in fluid filtration theory several decades ago. However, the attempts to rigorously justify the passage to the limit from the Stokes problem in a porous medium to the Darcy law in a homogenous medium faced significant difficulties.

In the appendix to [**115**], Tartar accomplished the passage to the limit in the case of a bounded domain $\Omega$. Earlier, this was done by Volkov [**129**], who used the asymptotic expansion method for unbounded domains. Zhikov [**131**] significantly weakened the smoothness conditions on the boundary of the obstacle $\Xi_0$; namely, he considered $\Xi_0$ with non-Lipschitz boundary $\partial\Xi_0$. Sandrakov [**117**] studied non-stationary filtration in the case where the external force depends on time.

We propose two different approaches. The first approach is based on the passage to the limit technique for integral identities (which is close to the work of Tartar), and the second approach uses the two-scale convergence tools (which is close to the approach suggested by Allaire [**3**]). The second approach leads to a stronger and more complete result. Namely, it makes it possible to prove the strong convergence of pressures and obtain more precise asymptotics of the filtration rate.

We formulate the main results of this section.

THEOREM 14.1. *Let $\widetilde{u}_\varepsilon(x)$ be a solution $\overline{u}_\varepsilon(x)$ extended by zero to $\Omega \setminus \Omega_\varepsilon$. For all $\varepsilon > 0$ there exists an extension $P_\varepsilon(x)$ of the pressure $p_\varepsilon(x)$ to $\Omega \setminus \Omega_\varepsilon$ such that $\frac{\widetilde{u}_\varepsilon}{\varepsilon^2} \rightharpoonup \overline{u}^0$ weakly in $L_2(\Omega)$ and $P_\varepsilon \to p^0$ strongly in $L_2(\Omega)$, where $(\overline{u}^0, p^0)$ is a solution to the problem* (14.3).

PROOF. It is obvious that

$$\int_\Omega \frac{\partial \widetilde{u}^i_\varepsilon}{\partial x_j} \frac{\partial \widetilde{u}^i_\varepsilon}{\partial x_j}\, dx = -\int_\Omega f^i \widetilde{u}^i_\varepsilon \, dx,$$

which implies

$$\int_\Omega |\nabla \widetilde{u}_\varepsilon|^2 \, dx \leq C \|\widetilde{u}_\varepsilon\|_{L_2(\Omega)}. \tag{14.6}$$

Then we use the Friedrichs inequality for $\Omega^\varepsilon$.

EXERCISE 14.4. Prove that for any $u(x) \in \overset{\circ}{H}{}^1(\Omega^\varepsilon)$,

$$\|u\|_{L_2(\Omega^\varepsilon)} \le C\varepsilon \|\nabla u\|_{L_2(\Omega^\varepsilon)}, \tag{14.7}$$

where the constant $C > 0$ is independent of $\varepsilon$ and $u(x)$.

In view of (14.6) and (14.7), we have

$$\int_\Omega |\nabla \widetilde{u}_\varepsilon|^2\, dx \le C\varepsilon \Big( \int_\Omega |\nabla \widetilde{u}_\varepsilon|^2\, dx \Big)^{1/2}.$$

Consequently,

$$\Big( \int_\Omega |\nabla \widetilde{u}_\varepsilon|^2\, dx \Big)^{1/2} \le C\varepsilon.$$

Using the Friedrichs inequality (14.7) again, we find that $\|\widetilde{u}_\varepsilon\|_{L_2(\Omega)} \le C\varepsilon^2$. Thus,

$$\Big\| \frac{\widetilde{u}_\varepsilon}{\varepsilon^2} \Big\|_{L_2(\Omega)} \le C. \tag{14.8}$$

An important step of the proof of this theorem is the following assertion.

LEMMA 14.1. *For all $\varepsilon > 0$ the pressure $p_\varepsilon(x)$ admits an extension (denoted by $P_\varepsilon(x)$) from $\Omega^\varepsilon$ to $\Omega$ such that from any sequence $P_{\varepsilon'}(x)$, $\varepsilon' \to 0$, it is possible to extract a subsequence $P_{\varepsilon''}(x)$ such that $P_{\varepsilon''} \rightharpoonup p^*$ weakly in $L_2(\Omega)$ as $\varepsilon'' \to 0$, $p^* \in L_2(\Omega)$, and for any $\Phi(\xi) \in L_2(\Xi)$, $\varphi(x) \in C_0^\infty(\Omega)$,*

$$\lim_{\varepsilon'' \to 0} \int_\Omega P_{\varepsilon''}(x) \Phi\Big(\frac{x}{\varepsilon''}\Big) \varphi(x)\, dx = \int_\Omega p^*(x)\varphi(x)\, dx \int_\Xi \Phi(\xi)\, d\xi. \tag{14.9}$$

Lemma 14.1 will be proved in Subsection 14.3 below. Now we show how to use it for obtaining a system of equations with respect to $(p^*(x), u^*(x))$, where $p^*(x)$ and $u^*(x)$ are the weak limits of arbitrary converging subsequences $P_{\varepsilon'}(x)$ and $\frac{\widetilde{u}_{\varepsilon'}(x)}{\varepsilon'^2}$ as $\varepsilon' \to 0$. We write the auxiliary system (14.2) in terms of variables $x = \varepsilon\xi$ $(q_\varepsilon^k \equiv q^k(x/\varepsilon), \overline{v}_\varepsilon^k \equiv \overline{v}^k(x/\varepsilon))$:

$$\begin{aligned} & -\varepsilon \nabla_x q_\varepsilon^k + \varepsilon^2 \Delta_x \overline{v}_\varepsilon^k + \overline{e}_k = 0 \quad \text{in } E_\varepsilon, \\ & \operatorname{div}_x \overline{v}_\varepsilon^k = 0 \quad \text{in } E_\varepsilon, \\ & \overline{v}^k \big|_{\partial E_\varepsilon} = 0, \\ & \overline{v}_\varepsilon^k, q_\varepsilon^k \quad \text{are } \varepsilon\text{-periodic with respect to } x_1, \dots, x_d, \end{aligned} \tag{14.10}$$

where $E_\varepsilon = \mathbb{R}^d \backslash \bigcup_{\overline{m} \in \mathbb{Z}^d} \varepsilon(\Xi_0 + \overline{m})$.

Let $\varphi(x) \in C_0^\infty(\Omega)$ be an arbitrary function. Taking the inner product of (14.10) and $\widetilde{u}_\varepsilon \varphi(x)$, integrating by parts, and extending the vectors $\overline{v}_\varepsilon^k$ by zero to $\mathbb{R}^d \backslash E_\varepsilon$, we obtain

$$\int_\Omega \frac{\partial (\overline{v}_\varepsilon^k)_i}{\partial x_j} \frac{\partial ((\widetilde{u}_\varepsilon)_i \varphi)}{\partial x_j}\, dx = \frac{1}{\varepsilon} \int_\Omega q_\varepsilon^k \operatorname{div}(\widetilde{u}_\varepsilon \varphi)\, dx + \frac{1}{\varepsilon^2} \int_\Omega e_k \varphi \widetilde{u}_\varepsilon\, dx.$$

Taking into account the estimate (14.8), using the notation

$$I_1^{\varepsilon,k} := \int_\Omega \frac{\partial (\overline{v}_\varepsilon^k)_i}{\partial x_j} \frac{\partial ((\widetilde{u}_\varepsilon)_i \varphi)}{\partial x_j}\, dx,$$

and passing to the limit as $\varepsilon \to 0$, we find that

$$\lim_{\varepsilon \to 0} I_1^{\varepsilon,k} = \int_\Omega \varphi u_k^*\, dx.$$

Taking the inner product of (14.1) and $\varphi\overline{v}_\varepsilon^k$ and integrating by parts over $\Omega$, we obtain the equality

$$\int_\Omega \frac{\partial(\widetilde{u}_\varepsilon)_i}{\partial x_j}\frac{\partial(\varphi(\overline{v}_\varepsilon^k)_i)}{\partial x_j}\,dx = \int_\Omega f_i\varphi(\overline{v}_\varepsilon^k)_i\,dx + \int_\Omega P_\varepsilon\frac{\partial\varphi}{\partial x_i}(\overline{v}_\varepsilon^k)_i\,dx.$$

Taking into account the definition of the matrix $K_{ij}$, we pass to the limit in the last equality. Denote

$$I_2^{\varepsilon,k} := \int_\Omega \frac{\partial(\widetilde{u}_\varepsilon)_i}{\partial x_j}\frac{\partial(\varphi(\overline{v}_\varepsilon^k)_i)}{\partial x_j}\,dx.$$

By Lemma 14.1, we have

$$\lim_{\varepsilon\to 0} I_2^{\varepsilon,k} = K_{ki}\int_\Omega f_i\varphi\,dx + K_{ki}\int_\Omega p^*\frac{\partial\varphi}{\partial x_i}\,dx.$$

It is easy to show that $\lim_{\varepsilon\to 0}(I_1^{\varepsilon,k} - I_2^{\varepsilon,k}) = 0$ (check it!). Hence

$$\langle u_k^*, \varphi\rangle = K_{ki}\left\langle f_i - \frac{\partial p^*}{\partial x_i}, \varphi\right\rangle$$

or, since $\varphi(x) \in C_0^\infty(\Omega)$ is arbitrary,

$$u_k^* = K_{ki}\left(f_i - \frac{\partial p^*}{\partial x_i}\right).$$

This system of equations coincides with (14.3). Thus, we have proved the weak convergence of $p_\varepsilon(x)$ to $p^0(x)$. It remains to establish the two-scale convergence of $p_\varepsilon(x)$ to $p^0(x)$. This will be done below. □

**14.3. Derivation of the Darcy law using the two-scale convergence method.**

THEOREM 14.2. *Suppose that $f(x) \in L_2(\Omega^\varepsilon)$, a domain $\Xi$ is represented as the finite union of Lipschitz domains $Q_j$, $j = 1,\dots,k$, and $\int_{\Omega^\varepsilon} f(x)\,dx = 0$. Then there exists a vector-valued function $\overline{u}_\varepsilon(x) \in \overset{\circ}{H}{}^1(\Omega^\varepsilon)$ such that $\operatorname{div}\overline{u}_\varepsilon(x) = f(x)$ and*

$$\|\nabla\overline{u}_\varepsilon(x)\|_{L_2(\Omega^\varepsilon)} \le \frac{C}{\varepsilon}\|f(x)\|_{L_2(\Omega^\varepsilon)}, \qquad \|\overline{u}_\varepsilon(x)\|_{L_2(\Omega^\varepsilon)} \le C\|f(x)\|_{L_2(\Omega)},$$

*where $C > 0$ is independent of $f(x)$.*

PROOF. We divide the proof of Theorem 14.2 into several steps and formulate some assertions in the form of exercises.

STEP 1. Assume that $\Xi$ has Lipschitz boundary.

EXERCISE 14.5. Assume that a domain $\Xi$ has Lipschitz boundary. Then the Poincaré inequality holds in $\Omega^\varepsilon$:

$$\left\| u - \int_{\Omega^\varepsilon} u(x)\,dx\right\|_{L_2(\Omega^\varepsilon)} \le K\|\nabla u\|_{L_2(\Omega^\varepsilon)}$$

for any $u \in H^1(\Omega^\varepsilon)$, where the constant $K > 0$ is independent of $\varepsilon > 0$.

*Hint.* Use the result on the extension of an arbitrary function $u \in H^1(\Omega^\varepsilon)$ to the domain $\Omega$ with the corresponding (uniform with respect to $\varepsilon$) estimate for the $H^1(\Omega)$-norm of the extended function. This result is proved, for example, in [**102, 103**]. Although it was already mentioned in the proof of the Friedrichs inequality for a perforated domain (cf. Subsection 7.2), we repeat it here for the sake of convenience.

THEOREM 14.3. *There exists an extension operator* $P_\varepsilon : H^1(\Omega^\varepsilon) \to H^1(\Omega)$ *to the domain* $\Omega$ *(without perforation) such that*

1) $\|P_\varepsilon u\|^2_{L_2(\Omega)} \le C\|u\|^2_{L_2(\Omega^\varepsilon)}$ *for any* $u \in H^1(\Omega^\varepsilon)$, *where the constant* $C > 0$ *is independent of* $\varepsilon > 0$;

2) $\|\nabla(P_\varepsilon u)\|^2_{L_2(\Omega)} \le C\|\nabla u\|^2_{L_2(\Omega_\varepsilon)}$ *for any* $u \in H^1(\Omega^\varepsilon)$, *where the constant* $C > 0$ *is independent of* $\varepsilon > 0$;

3) $u(x) \equiv 1$ *in* $\Omega^\varepsilon$ *implies that* $P_\varepsilon u = 1$ *in* $\Omega$.

EXERCISE 14.6. Consider the Neumann problem

$$\Delta V = f \quad \text{in } \Omega^\varepsilon, \qquad \int_{\Omega^\varepsilon} f(x)dx = 0, \qquad \frac{\partial V}{\partial n} = 0 \quad \text{on } \partial\Omega^\varepsilon,$$

understood in the sense of integral identity. Prove the estimate

$$\|\nabla V\|_{L_2(\Omega^\varepsilon)} \le K\|f\|_{L_2(\Omega^\varepsilon)},$$

where the constant $K > 0$ is independent of $\varepsilon$.

*Hint.* Use Exercise 14.5.

EXERCISE 14.7. Prove that (for sufficiently large $k_0$ in the definition of the perforated domain $\Omega^\varepsilon$) the domain $\Omega^\varepsilon$ is represented as the union of domains $\Omega_i^\varepsilon$, $i = 1, \dots, N(\varepsilon)$: $\Omega^\varepsilon = \bigcup_{i=1}^{N(\varepsilon)} \Omega_i^\varepsilon$, such that for each of the domains $\Omega_i^\varepsilon$ the estimate of Theorem 1.7 holds with a constant $C > 0$ independent of $i = 1, \dots, N(\varepsilon)$ and $\varepsilon > 0$.

Note that under the assumptions of Exercise 14.7, $N(\varepsilon)$ increases to infinity as $\varepsilon$ decreases.

We construct a partition of unity $1 = \sum_{i=1}^{N(\varepsilon)} \psi_i^\varepsilon$ in $\Omega^\varepsilon$ such that

1) $\operatorname{supp} \psi_i^\varepsilon \subset \overline{\Omega}_i^\varepsilon$;

2) $|\nabla \psi_i^\varepsilon| \le M/\varepsilon$, where $M > 0$ is a constant independent of $i = 1, \dots, N(\varepsilon)$;

3) for every $x \in \Omega^\varepsilon$ the sum $\sum_{i=1}^{N(\varepsilon)} \psi_i^\varepsilon(x)$ contains at most $L$ nonzero terms and $L$ is independent of $\varepsilon$.

EXERCISE 14.8. Prove the existence of a partition of unity with the above properties.

We set $f_i = \operatorname{div}(\psi_i^\varepsilon \cdot \overline{F})$, $\overline{F} = \nabla V$, where $V$ is defined in Exercise 14.6. Then $\sum_{i=1}^{N(\varepsilon)} f_i = f$ in $\Omega^\varepsilon$ and, which is important,

$$\int_{\Omega_i^\varepsilon} f_i(x)\, dx = 0.$$

Indeed,

$$\int_{\Omega_i^\varepsilon} f_i\, dx = \int_{\partial\Omega_i^\varepsilon} \psi_i^\varepsilon \left(\overline{F}, \overline{n}\right)\, d\sigma = 0$$

since $(\overline{F}, \overline{n}) = 0$ on $\partial\Omega^\varepsilon$ and $\psi_i^\varepsilon = 0$ on $\partial\Omega_i^\varepsilon \backslash \partial\Omega^\varepsilon$.

By Theorem 1.7, there exist $w_i \in \overset{\circ}{H}{}^1(\Omega_i^\varepsilon)$ such that $\operatorname{div} w_i = f_i$ in $\Omega_i^\varepsilon$ and

$$\|w_i\|_{\overset{\circ}{H}{}^1(\Omega_i^\varepsilon)} \le C\|f_i\|_{L_2(\Omega_i^\varepsilon)},$$

where $C$ is independent of $i = 1, \dots, N(\varepsilon)$ and $\varepsilon > 0$. We set $w = \sum_{i=1}^{N(\varepsilon)} w_i$. Note that for any $x \in \Omega^\varepsilon$ there are at most $L$ nonzero terms $\|f_i\|^2_{L_2(\Omega_i^\varepsilon)}$ such that

$$\|f_i\|^2_{L_2(\Omega_i^\varepsilon)} \le 2\|f\|^2_{L_2(\Omega_i^\varepsilon)} + 2\frac{M^2}{\varepsilon^2} K^2 \|f\|^2_{L_2(\Omega_i^\varepsilon)}.$$

Therefore,

$$\|w\|^2_{\overset{\circ}{H}{}^1(\Omega^\varepsilon)} \leq 2C \sum_{i=1}^{N(\varepsilon)} \|f_i\|^2_{L_2(\Omega_i^\varepsilon)} \leq 2CL\left(2 + \frac{2M}{\varepsilon}K\right)^2 \|f\|^2_{L_2(\Omega^\varepsilon)}.$$

Now it is easy to verify that $\overline{u}_\varepsilon = w$ satisfies all the assumptions of Theorem 14.2.

STEP 2. Assume that $\Xi$ is not a Lipschitz domain, but $\Xi = \bigcup_{j=1}^k Q_j$, where $Q_j$ are domains with Lipschitz boundary.

EXERCISE 14.9. Assume that $f \in L_2(\Omega^\varepsilon)$ and $\int_{\Omega^\varepsilon} f(x)\,dx = 0$. Prove that there exist functions $\widetilde{f}_1, \dots, \widetilde{f}_k$ with the following properties:

1) $\operatorname{supp} \widetilde{f}_j \subset \widetilde{\Omega}_j^\varepsilon$, where $\partial\widetilde{\Omega}_j^\varepsilon$ is Lipschitz and $\widetilde{\Omega}_j^\varepsilon \subset \Omega^\varepsilon$ can be represented (cf. Exercise 14.8) in the form $\widetilde{\Omega}_j^\varepsilon = \bigcup_i^{N(\varepsilon)} \Omega_{i,j}^\varepsilon$; moreover for $\Omega_{i,j}^\varepsilon$ the estimate of Theorem 1.7 holds with the constant $C$ independent of $i$ and $\varepsilon$;

2) $\int_{\widetilde{\Omega}_j^\varepsilon} \widetilde{f}_j\,dx = 0$, $j = 1, \dots, k$;

3) $\int_{\widetilde{\Omega}_j^\varepsilon} \widetilde{f}_j^2\,dx \leq C_1(k) \int_{\Omega^\varepsilon} |f|^2\,dx$, $j = 1, \dots, k$, where $C_1(k)$ depends only on $k$;

4) $f = \sum_{j=1}^k \widetilde{f}_j$.

*Hint.* The representation $f = \sum_j \widetilde{f}_i$ is constructed in the same way as in the proof of Theorem 1.7. To prove the estimate 3), note that $\lim_{\varepsilon\to 0} |\widetilde{\Omega}_i^\varepsilon \cap \widetilde{\Omega}_j^\varepsilon| > 0$, $i, j = 1, \dots, k$.

Using the above arguments, we construct functions $\widetilde{w}_j \in \overset{\circ}{H}{}^1(\widetilde{\Omega}_j^\varepsilon)$ such that $\operatorname{div} \widetilde{w}_j = \widetilde{f}_j$ and $\|\widetilde{w}_j\|_{\overset{\circ}{H}{}^1(\widetilde{\Omega}_j^\varepsilon)} \leq C\|\widetilde{f}_j\|_{L_2(\widetilde{\Omega}_j^\varepsilon)}$. Summarizing the estimates with respect to $j$ and using property 3) in Exercise 14.9, we complete the proof of Theorem 14.2. □

Consider the Stokes problem

$$\begin{aligned} &\nabla p_\varepsilon - \varepsilon^2 \Delta \overline{u}_\varepsilon = \overline{f} \quad \text{in } \Omega^\varepsilon, \\ &\operatorname{div} \overline{u}_\varepsilon = 0 \quad \text{in } \Omega^\varepsilon, \\ &\overline{u}_\varepsilon = 0 \quad \text{on } \partial\Omega^\varepsilon, \\ &\int_{\Omega^\varepsilon} p_\varepsilon(x)\,dx = 0. \end{aligned} \tag{14.11}$$

We introduced a small parameter $\varepsilon^2$ at $\Delta\overline{u}_\varepsilon$ in order to normalize the velocity vector $\overline{u}_\varepsilon(x)$. Then the filtration rate has a finite limit as $\varepsilon \to 0$.

Taking the inner product of the first equation in (14.11) with $\overline{u}_\varepsilon$ and integrating by parts, we find

$$\varepsilon^2 \int_{\Omega^\varepsilon} (\nabla\overline{u}_\varepsilon, \nabla\overline{u}_\varepsilon)\,dx = \int_{\Omega^\varepsilon} (\overline{f}, \overline{u}_\varepsilon)\,dx. \tag{14.12}$$

By the Cauchy–Schwarz–Bunyakovskii inequality and the Friedrichs inequality for $\Omega^\varepsilon$ (cf. Exercise 14.4), we have

$$\begin{aligned} \varepsilon^2 \int_{\Omega^\varepsilon} (\nabla\overline{u}_\varepsilon, \nabla\overline{u}_\varepsilon)\,dx &\leq \Big(\int_{\Omega^\varepsilon} |\overline{f}|^2\,dx\Big)^{1/2} \Big(\int_{\Omega^\varepsilon} |\overline{u}_\varepsilon|^2\,dx\Big)^{1/2} \\ &\leq C\Big(\int_{\Omega^\varepsilon} |\overline{f}|^2\,dx\Big)^{1/2} \Big(\int_{\Omega^\varepsilon} |\varepsilon\nabla\overline{u}_\varepsilon|^2\,dx\Big)^{1/2}, \end{aligned}$$

which implies

$$\int_{\Omega^\varepsilon} |\varepsilon \nabla \overline{u}_\varepsilon|^2 \, dx \le C. \tag{14.13}$$

Using the Friedrichs inequality again, we find

$$\int_{\Omega^\varepsilon} |\overline{u}_\varepsilon|^2 \, dx \le C \int_{\Omega^\varepsilon} |\varepsilon \nabla \overline{u}_\varepsilon|^2 \, dx \le C'. \tag{14.14}$$

Thus, the sequences $\varepsilon \nabla \overline{u}_\varepsilon$ and $\overline{u}_\varepsilon$ are uniformly bounded in $L_2(\Omega^\varepsilon)$. Let us prove that the pressures $p_\varepsilon(x)$ are bounded in $L_2(\Omega^\varepsilon)$. Using Theorem 14.2, we construct a vector-valued function $\overline{v}_\varepsilon \in \overset{\circ}{H}{}^1(\Omega^\varepsilon)$ such that $\operatorname{div} \overline{v}_\varepsilon(x) = p_\varepsilon(x)$ in $\Omega^\varepsilon$ and

$$\|\nabla \overline{v}_\varepsilon\|_{L_2(\Omega^\varepsilon)} \le \frac{C}{\varepsilon} \|p_\varepsilon\|_{L_2(\Omega^\varepsilon)}, \qquad \|\overline{v}_\varepsilon\|_{L_2(\Omega^\varepsilon)} \le C \|p_\varepsilon\|_{L_2(\Omega^\varepsilon)}. \tag{14.15}$$

Multiplying the first equation in (14.11) by $\overline{v}_\varepsilon(x)$ and integrating by parts over $\Omega^\varepsilon$, we find

$$\int_{\Omega^\varepsilon} p_\varepsilon^2 \, dx = \varepsilon^2 \int_{\Omega^\varepsilon} (\nabla \overline{u}_\varepsilon, \nabla \overline{v}_\varepsilon) \, dx - \int_{\Omega^\varepsilon} \overline{f}(x) \overline{v}_\varepsilon(x) \, dx.$$

Next, we use the boundedness of $\varepsilon \nabla \overline{u}_\varepsilon$ in $L_2(\Omega^\varepsilon)$, the estimates (14.15) for $\varepsilon \nabla \overline{v}_\varepsilon$, $\overline{v}_\varepsilon(x)$, and the Cauchy–Schwarz–Bunyakovskii inequality. We have

$$\int_{\Omega^\varepsilon} p_\varepsilon^2(x) \, dx \le C(\|\varepsilon \nabla \overline{v}_\varepsilon\|_{L_2(\Omega^\varepsilon)} + \|\overline{v}_\varepsilon\|_{L_2(\Omega^\varepsilon)}) \le C' \|p_\varepsilon\|_{L_2(\Omega^\varepsilon)}.$$

Hence $\|p_\varepsilon\|_{L_2(\Omega^\varepsilon)} \le C$, where $C > 0$ is independent of $\varepsilon > 0$.

Extend the pressure $p_\varepsilon(x)$ to the entire domain $\Omega$ as follows:

$$P_\varepsilon(x) = \begin{cases} p_\varepsilon(x), & x \in Y_{s,j}^\varepsilon, \\ \frac{1}{|Y_{s,j}^\varepsilon|} \int_{Y_{s,j}^\varepsilon} p_\varepsilon(x) \, dx, & x \in Y_{h,j}^\varepsilon, \end{cases} \tag{14.16}$$

where $Y_{h,j}^\varepsilon$ and $Y_{s,j}^\varepsilon$ are domains of the form $\varepsilon(\Xi_0 + \overline{m})$ and $\varepsilon(\Xi + \overline{m})$ enumerated by a single scalar number $j = 1, \dots, N(\varepsilon)$ such that $\varepsilon\overline{(\Xi_0 + \overline{m})} \subset \Omega$. We extend $\overline{u}_\varepsilon(x)$ by zero to $\Omega$ (preserving the notation for the extended function). Taking into account Theorems 9.1 and 9.3 and the boundedness of $\overline{u}_\varepsilon(x)$ and $P_\varepsilon(x)$, from these sequences we can extract a two-scale converging subsequence such that $\overline{u}_\varepsilon(x) \xrightarrow{2} \overline{u}_0(x,\xi)$ in $\Omega$ and $P_\varepsilon(x) \xrightarrow{2} p(x,\xi)$ in $\Omega$.

LEMMA 14.2. *The two-scale limit $p(x,\xi)$ of pressures is independent of $\xi$, i.e., $p(x,\xi) = p(x)$.*

PROOF. Let $\overline{\psi}(x,\xi)$ be an arbitrary infinitely smooth vector-valued function periodic with respect to $\xi$ and compactly supported in $\Omega$ with respect to $x$. The space of such functions is denoted by $C_0^\infty(\Omega, C_{\text{per}}^\infty(\square))$. We show that

$$\int_\Omega P_\varepsilon(x) \operatorname{div}_\xi \overline{\psi}\Big(x, \frac{x}{\varepsilon}\Big) \, dx \to 0 \quad \text{as } \varepsilon \to 0.$$

Indeed,

$$\begin{aligned} \int_\Omega P_\varepsilon(x) \operatorname{div}_\xi \overline{\psi}\Big(x, \frac{x}{\varepsilon}\Big) \, dx &= \varepsilon \int_\Omega P_\varepsilon(x) \operatorname{div} \Big(\overline{\psi}\Big(x, \frac{x}{\varepsilon}\Big)\Big) \, dx \\ &\quad - \varepsilon \int_\Omega P_\varepsilon(x) \operatorname{div}_x \overline{\psi}\Big(x, \frac{x}{\varepsilon}\Big) \, dx. \end{aligned}$$

Since $P_\varepsilon$ are bounded, the second term on the right-hand side of this equality converges to zero. Therefore, it suffices to prove that the first term also converges to zero. By the definition of $P_\varepsilon$, we have

$$\varepsilon \int_\Omega P_\varepsilon(x)\mathrm{div}\Big(\overline{\psi}\Big(x,\frac{x}{\varepsilon}\Big)\Big)\,dx$$
$$= \varepsilon \sum_j \int_{Y^\varepsilon_{s,j}} p_\varepsilon(x)\Big[\mathrm{div}\,\overline{\psi} + \frac{1}{|Y^\varepsilon_{s,j}|}\int_{Y^\varepsilon_{h,j}} \mathrm{div}\,\overline{\psi}\,dx\Big]\,dx$$
$$+ \varepsilon \int_{\Omega^\varepsilon_{\mathrm{bnd}}} P_\varepsilon(x)\mathrm{div}\Big(\overline{\psi}\Big(x,\frac{x}{\varepsilon}\Big)\Big)\,dx,$$

where

$$\Omega^\varepsilon_{\mathrm{bnd}} \equiv \Omega\setminus \bigcup_{j=1}^{N(\varepsilon)} (Y^\varepsilon_{s,j}\cup Y^\varepsilon_{h,j}) = \Omega\setminus \bigcup_{\{\overline{m}:\mathrm{dist}(\partial\Omega,\varepsilon(\Xi_0+\overline{m}))\ge k_0\varepsilon\}} \varepsilon(\square+\overline{m}).$$

Since $\psi(x,\xi)$ has compact support in $\Omega$, the second integral on the right-hand side vanishes for all sufficiently small $\varepsilon$. Consequently,

$$\varepsilon \int_\Omega P_\varepsilon\ \mathrm{div}\,\overline{\psi}\Big(x,\frac{x}{\varepsilon}\Big)\,dx = \varepsilon \sum_j \int_{Y^\varepsilon_{s,j}} p_\varepsilon(x)\Big[\mathrm{div}\,\overline{\psi} + \frac{1}{|Y^\varepsilon_{s,j}|}\int_{Y^\varepsilon_{h,j}} \mathrm{div}\,\overline{\psi}\,dx\Big]\,dx$$

for sufficiently small $\varepsilon$. Denote by $\kappa^\varepsilon_{s,j}$ the expression in square brackets in the last equality and note that the following relation holds (check it!):

$$\Big|\int_{Y^\varepsilon_{s,j}} \kappa^\varepsilon_{s,j}\,dx\Big| = \Big|\int_{Y^\varepsilon_{s,j}} \Big[\mathrm{div}\,\overline{\psi} + \frac{1}{|Y^\varepsilon_{s,j}|}\int_{Y^\varepsilon_{h,j}} \mathrm{div}\overline{\psi}\,dx\Big]\,dx\Big| \le C\varepsilon^d.$$

By Theorem 14.2, there exists a vector-valued function $\overline{w}_\varepsilon(x)\in \overset{\circ}{H}{}^1(\Omega^\varepsilon)$ such that $\varepsilon\nabla\overline{w}_\varepsilon(x)$ and $\overline{w}_\varepsilon(x)$ are uniformly bounded with respect to $\varepsilon$ in $L_2(\Omega^\varepsilon)$; moreover,

$$\mathrm{div}\,\overline{w}_\varepsilon(x) = \begin{cases} \mathrm{div}\,\overline{\psi} + \frac{1}{|Y^\varepsilon_{s,j}|}\int_{Y^\varepsilon_{h,j}} \mathrm{div}\,\overline{\psi}\,dx - \frac{\kappa^\varepsilon_{s,j}}{|Y^\varepsilon_{s,j}|}, & x\in Y^\varepsilon_{s,j},\ j=1,\dots,N(\varepsilon),\\ 0, & x\in\Omega^\varepsilon\setminus\bigcup_{j=1}^{N(\varepsilon)} Y^\varepsilon_{s,j}. \end{cases}$$

Using the definition of functions $\overline{w}_\varepsilon(x)$ and the Stokes problem (14.11), we find

$$\varepsilon\int_\Omega P_\varepsilon(x)\mathrm{div}\,\Big(\overline{\psi}\Big(x,\frac{x}{\varepsilon}\Big)\Big)\,dx = \varepsilon\int_{\Omega^\varepsilon} p_\varepsilon\,\mathrm{div}\,\overline{w}_\varepsilon\,dx + O(\varepsilon)$$
$$= \varepsilon\int_{\Omega^\varepsilon} (\varepsilon\nabla\overline{u}_\varepsilon,\varepsilon\nabla\overline{w}_\varepsilon)\,dx - \varepsilon\int_{\Omega^\varepsilon} (\overline{f},\overline{w}_\varepsilon)\,dx + O(\varepsilon).$$

By the boundedness of $\varepsilon\nabla\overline{u}_\varepsilon$, $\varepsilon\nabla\overline{w}_\varepsilon$, $\overline{w}_\varepsilon$ in $L_2(\Omega^\varepsilon)$ and the Cauchy–Schwarz–Bunyakovskii inequality, we obtain the limit relation

$$\varepsilon\int_\Omega P_\varepsilon(x)\mathrm{div}\Big(\overline{\psi}\Big(x,\frac{x}{\varepsilon}\Big)\Big)\,dx \to 0 \quad \text{as } \varepsilon\to 0$$

and, consequently,

$$\int_\Omega P_\varepsilon(x)\mathrm{div}_\xi\overline{\psi}\Big(x,\frac{x}{\varepsilon}\Big)\,dx \to 0 \quad \text{as } \varepsilon\to 0.$$

On the other hand, passing to the two-scale limit as $\varepsilon\to 0$ in the same integral, we find

$$\int_{\Omega\times\Xi} p(x,\xi)\mathrm{div}_\xi\overline{\psi}(x,\xi)\,dx\,d\xi = 0.$$

Consequently, $\nabla_\xi p(x,\xi) = 0$ for almost all $x \in \Omega$ in the sense of distributions. Therefore, $p(x,\xi) = p(x)$, i.e., $P_\varepsilon$ is independent of $\xi$. $\square$

Thus, we have proved Lemma 14.1 and thereby almost proved Theorem 14.1. It remains to establish the strong convergence of the extensions $P_\varepsilon(x)$ to the limit pressure $p(x)$.

We give another proof of Theorem 14.1, which is based on the results concerning two-scale convergence. The arguments below are a modification of the proof due to Allaire [**3**].

We write the incompressibility condition in the form

$$\int_\Omega (\overline{u}_\varepsilon, \nabla_x \varphi)\, dx = 0, \qquad \varphi \in C_0^\infty(\Omega).$$

Passing to the two-scale limit, we find

$$\int_{\Omega\times\Xi} (\overline{u}_0(x,\xi), \nabla_x \varphi)\, dx\, d\xi = 0.$$

Since $\varphi(x)$ is arbitrary, we have

$$\operatorname{div}_x \int_\Xi [\overline{u}_0(x,\xi)]\, d\xi = 0. \tag{14.17}$$

Using the estimates (14.14), it is easy to show that $\overline{u}_0(x,\xi) \in L_2(\Omega, [\overset{\circ}{H}{}^1_{\mathrm{per}}(\Xi)]^d)$. We introduce the function class

$$\mathcal{A} = \Big\{ \overline{\psi}(x,\xi) \in C_0^\infty(\Omega, C_{\mathrm{per}}^\infty(\square)) : \overline{\psi}(x,\xi) = 0 \text{ for } \xi \in \Xi_0,$$
$$\operatorname{div}_\xi \overline{\psi}(x,\xi) = 0, \operatorname{div}_x \int_\Xi \overline{\psi}(x,\xi)\, d\xi = 0 \Big\}.$$

Taking the inner product of (14.11) with a vector-valued test function $\overline{\psi}(x,\frac{x}{\varepsilon})$, $\overline{\psi}(x,\xi) \in \mathcal{A}$, and integrating by parts, we find

$$\begin{aligned} \varepsilon^{-1} \int_\Omega P_\varepsilon \operatorname{div}_\xi \overline{\psi}\Big(x,\frac{x}{\varepsilon}\Big)\, dx &+ \int_\Omega P_\varepsilon \operatorname{div}_x \overline{\psi}\Big(x,\frac{x}{\varepsilon}\Big)\, dx \\ &+ \int_\Omega \Big(\varepsilon \nabla \overline{u}_\varepsilon, \nabla_\xi \overline{\psi}\Big(x,\frac{x}{\varepsilon}\Big)\Big)\, dx + \varepsilon^2 \int_\Omega \Big(\nabla \overline{u}_\varepsilon, \nabla_x \overline{\psi}\Big(x,\frac{x}{\varepsilon}\Big)\Big)\, dx \\ &= \int_\Omega \Big(\overline{f}(x), \overline{\psi}\Big(x,\frac{x}{\varepsilon}\Big)\Big)\, dx. \end{aligned}$$

Passing to the two-scale limit, using Theorem 9.3, and taking into account the properties of $\overline{\psi}(x,\xi)$, we obtain the equality

$$\int_{\Omega\times\Xi} (\nabla_\xi \overline{u}_0(x,\xi), \nabla_\xi \overline{\psi}(x,\xi))\, dx\, d\xi = \int_{\Omega\times\Xi} (\overline{f}(x), \overline{\psi}(x,\xi))\, dx\, d\xi. \tag{14.18}$$

We define the vector-valued distribution $\Delta_\xi \overline{u}_0(x,\xi) + \overline{f}(x)$ in $L_2(\Omega, [H^{-1}_{\mathrm{per}}(\Xi)]^d)$ acting on test functions $\overline{\psi}(x,\xi) \in L_2(\Omega, [\overset{\circ}{H}{}^1_{\mathrm{per}}(\Xi)]^d)$ by the rule

$$\langle \Delta_\xi \overline{u}_0 + \overline{f}, \overline{\psi} \rangle \equiv \int_{\Omega\times\Xi} [-(\nabla_\xi \overline{u}_0(x,\xi), \nabla_\xi \overline{\psi}(x,\xi)) + (\overline{f}, \overline{\psi})]\, dx\, d\xi.$$

We also introduce the function space $V$ as the closure of $\mathcal{A}$ in the $L_2(\Omega, [\overset{\circ}{H}{}^1_{\mathrm{per}}(\Xi)]^d)$-norm. The equality (14.18) remains valid for any $\psi(x,\xi) \in V$.

LEMMA 14.3. *Let*

$$\langle \Delta_\xi \overline{u}_0 + \overline{f}, \overline{\psi} \rangle = 0 \quad \forall \overline{\psi}(x,\xi) \in V.$$

*Then there exist functions* $\Phi \in L_2(\Omega)$ *and* $\Phi_1 \in L_2(\Omega, L_2(\Xi))$ *such that*

$$\Delta_\xi \overline{u}_0(x,\xi) + \overline{f}(x) = \nabla_\xi \Phi_1(x,\xi) + \nabla_x \Phi(x).$$

PROOF. Denote by $V_1$ and $V_2$ the closures in the $L_2(\Omega, [\overset{\circ}{H}{}^1_{\text{per}}(\Xi)]^d)$-norm of the classes of smooth functions

$$\mathcal{A}_1 = \{\overline{\varphi}(x,\xi) \in C^\infty_{0,\text{per}}(\Omega \times \Xi) : \operatorname{div}_\xi \overline{\psi}(x,\xi) = 0\},$$

$$\mathcal{A}_2 = \left\{\overline{\varphi}(x,\xi) \in C^\infty_{0,\text{per}}(\Omega \times \Xi) : \operatorname{div}_x \left[\int_\Xi \overline{\psi}(x,\xi)\,d\xi\right] = 0\right\},$$

respectively. Then

$$V_1 = \{\overline{\psi}(x,\xi) \in L_2(\Omega, [\overset{\circ}{H}{}^1_{\text{per}}(\Xi)]^d) : \operatorname{div}_\xi \overline{\psi} = 0 \text{ in } \Omega \times \Xi\},$$

$$V_2 = \left\{\overline{\psi}(x,\xi) \in L_2(\Omega, [\overset{\circ}{H}{}^1_{\text{per}}(\Xi)]^d) : \operatorname{div}_x \left[\int_\Xi \overline{\psi}\,d\xi\right] = 0 \text{ in } \Omega,\ \left(\int_\Xi \overline{\psi}\,d\xi, n(\xi)\right) = 0 \text{ on } \partial\Omega\right\},$$

where $n(\xi)$ is the unit normal to $\partial\Omega$. To derive these equalities, we use the following assertion proved in [**126**]:

*Let* $\mathcal{O}$ *consist of divergence-free* $C_0^\infty(G)$*-functions in a bounded Lipschitz domain* $G$. *Then the closures of* $\mathcal{O}$ *in the spaces* $L_2(G)$ *and* $H^1(G)$ *coincide with the sets* $\{\overline{u} \in (L_2(G))^d : \operatorname{div} \overline{u} = 0$ *in* $G$, $(u,n) = 0$ *on* $\partial G\}$ *and* $\{\overline{u} \in [\overset{\circ}{H}{}^1_{\text{per}}(G)]^d : \operatorname{div} \overline{u} = 0$ *in* $G\}$ *respectively.*

Thus, we can obtain the required assertions about the closure of $\mathcal{A}_1$ and $\mathcal{A}_2$ in the case of Lipschitz domains $\Omega$ and $\Xi$. In our case, the domain $\Xi$ is not necessarily Lipschitz, but it is the finite union of Lipschitz domains. However, an analysis of the proof of Theorem 1.1.6 in [**126**] shows that this theorem remains valid for such sets. The matter is that in the arguments presented in [**126**], the Lipschitz property of the boundary is required only for the use of the de Rham theorem. However, in the present book, the de Rham theorem was proved for sets represented as the union of finitely many Lipschitz domains, and consequently Theorem 1.1.6 in [**126**] remains valid for such sets.

EXERCISE 14.10. Show that a vector-valued distribution $\overline{F}$ in $L_2(\Omega, H^{-1}_{\text{per}}(\Xi))$ such that $\langle \overline{F}, \overline{\psi} \rangle = 0$ for all $\overline{\psi} \in V_1$ has the form $\overline{F} = \nabla_\xi \Phi_1(x,\xi)$, where $\Phi_1 \in L_2(\Omega, L_{2,\text{per}}(\Xi))$.

*Hint.* Use an analog of the de Rham theorem for the domain $\Xi$.

EXERCISE 14.11. Show that a vector-valued distribution $G$ in $L_2(\Omega, H^{-1}_{\text{per}}(\Xi))$ such that $\langle G, \overline{\psi} \rangle = 0$ for all $\psi \in V_2$ has the form $G = \nabla_x \Phi(x)$, where $\Phi \in L_2(\Omega)$.

*Hint.* Prove that $G$, regarded as a distribution, is independent of $\xi$ and then use the de Rham theorem for $\Omega$.

EXERCISE 14.12. Denote by $W^\perp$ the set of linear continuous functionals on $L_2(\Omega, H^{-1}(\Xi))$ such that for $F \in W^\perp$ we have $\langle F, \psi \rangle = 0$ for all $\psi \in W$, where $W$ is a subspace of $L_2(\Omega, H^1_{\text{per}}(\Xi))$. Prove that $(W_1 \cap W_2)^\perp = W_1^\perp \oplus W_2^\perp$.

*Hint.* The assertion of this exercise is not related to any special function space, but is a general result of linear algebra.

It is easy to see that Lemma 14.3 immediately follows from Exercises 14.10–14.12. □

THEOREM 14.4. *The two-scale limit* $(\overline{u}_0(x,\xi), p(x))$ *of the sequence* $(\overline{u}_\varepsilon(x), P_\varepsilon(x))$ *belongs to the space* $L_2(\Omega, H^1_{\text{per}}(\square)) \oplus L_2(\Omega)$ *and is a solution to the problem*

$$(14.19)\qquad \begin{aligned} &\nabla_\xi p_1(x,\xi) + \nabla_x p(x) - \Delta_\xi \overline{u}_0(x,\xi) = \overline{f}(x) \quad \text{in } \Omega \times \Xi, \\ &\operatorname{div}_\xi \overline{u}_0(x,\xi) = 0 \quad \text{in } \Omega \times \Xi, \\ &\operatorname{div}_x \int_\Xi \overline{u}_0(x,\xi)\, d\xi = 0, \\ &\overline{u}_0(x,\xi) = 0 \quad \text{in } \Omega \times \Xi_0, \\ &\Bigl( \int_\Xi \overline{u}_0(x,\xi)\, d\xi, \overline{n} \Bigr) = 0 \quad \text{on } \partial\Omega, \end{aligned}$$

*where* $p_1(x,\xi) \in L_2(\Omega, L_{2,\text{per}}(\Xi))$.

REMARK 14.1. In fact, a solution to the system (14.19) is a triple of functions $(\overline{u}_0(x,\xi), p(x), p_1(x,\xi))$. In this statement, the problem is well posed. However, the function $p_1(x,\xi)$ plays an auxiliary role and contributes nothing to the two-scale limit.

REMARK 14.2. Theorem 14.4 asserts that the two-scale limit $(\overline{u}_0(x,\xi), p(x))$, regarded as a vector-valued function of variables $x$ and $\xi$, satisfies some common boundary value problem with respect to $x$ and $\xi$. It is important to obtain such a two-scale limit problem. Then it will not be difficult to represent functions of variables $x$ and $\xi$ as linear combinations of functions of the variable $x$ and functions of the variable $\xi$, i.e., to realize an asymptotic expansion.

PROOF OF THEOREM 14.4. Since $\varepsilon \nabla u_\varepsilon$ are uniformly bounded in $L_2(\Omega^\varepsilon)$, we make use of Theorem 9.3 to pass to the limit in the equality

$$\varepsilon \int_\Omega \operatorname{div} \overline{u}_\varepsilon(x) \Phi\Bigl(x, \frac{x}{\varepsilon}\Bigr)\, dx = 0$$

as $\varepsilon \longrightarrow 0$. We have

$$\int_{\Omega\times\Xi} \operatorname{div}_\xi \overline{u}_0(x,\xi) \Phi(x,\xi)\, dx\, d\xi = 0,$$

where $\Phi(x,\xi)$ is an arbitrary test function. Since $\Phi(x,\xi)$ is arbitrary, we have

$$(14.20)\qquad \operatorname{div}_\xi \overline{u}_0(x,\xi) = 0.$$

Thus the second equality in (14.19) is proved.

The equality $\bigl(\int_\Xi \overline{u}_0(x,\xi)\, d\xi, \overline{n}\bigr) = 0$ follows from the relation

$$\overline{u}_\varepsilon(x) \rightharpoonup \int_\Xi \overline{u}_0(x,\xi)\, d\xi \quad \text{weakly in } L_2(\Omega) \quad \text{as } \varepsilon \to 0$$

and the condition $(\overline{u}_\varepsilon(x), \overline{n}) = 0$, which is preserved (cf. Exercise 14.2) for the weak limit since $\overline{u}_\varepsilon(x) \in [H(\operatorname{div}, \Omega)]^d$. The second equality in (14.19) can be verified in a similar way. By Lemma 14.3, we have $\nabla_\xi p_1(x,\xi) + \nabla_x \Phi(x) - \Delta_\xi \overline{u}_0(x,\xi) = \overline{f}(x)$ in $\Omega \times \Xi$ with some $p_1(x,\xi) \in L_2(\Omega, L_{2,\text{per}}(\Xi))$ and $\Phi(x) \in L_2(\Omega)$. Substituting a test function $\varphi(x) \in C_0^\infty(\Omega)$ in (14.11) and passing to the two-scale limit, we find that

$\nabla\Phi(x) = \nabla p(x)$ (check it!), i.e., $p(x) = \Phi(x) + \text{const}$, which completes the proof of the theorem. $\square$

EXERCISE 14.13. Is a solution to problem (14.19) unique?

Suppose vector-valued functions $\overline{v}_i(\xi)$ and functions $q_i(\xi)$ are periodic solutions to the problem

$$\begin{aligned} &\nabla_\xi q_i(\xi) - \Delta_\xi \overline{v}_i(\xi) = \overline{e}_i \quad \text{in } \Xi, \\ &\operatorname{div}_\xi \overline{v}_i(\xi) = 0 \quad \text{on } \Xi, \\ &\overline{v}_i(\xi) = 0 \quad \text{on } \partial\Xi_0, \end{aligned} \tag{14.21}$$

where $\overline{e}_i$ is the unit vector of the $\xi_i$-axis. A direct computation shows that the vector-valued function

$$\overline{u}_0(x,\xi) = \sum_{i=1}^d \left(f_i(x) - \frac{\partial p}{\partial x_i}\right) \overline{v}_i(\xi) \tag{14.22}$$

satisfies the first equation in (14.19). The relation (14.22) is called the *Darcy law.* It connects the external forces acting on a fluid and the pressure gradient with the filtration rate of the fluid. Substituting this expression in the equalities

$$\operatorname{div}_x\left[\int_\Xi \overline{u}_0(x,\xi)\,d\xi\right] = 0, \qquad \left(\int_\Xi \overline{u}_0(x,\xi)\,d\xi, \overline{n}\right) = 0,$$

we find that

$$\begin{aligned} &\operatorname{div}_x\left[\sum_{i=1}^d \left(f_i(x) - \frac{\partial p}{\partial x_i}\right) \int_\Xi \overline{v}_i(\xi)\,d\xi\right] = 0, \\ &\left(\sum_{i=1}^d \left(f_i(x) - \frac{\partial p}{\partial x_i}\right) \int_\Xi \overline{v}_i(\xi)\,d\xi, \overline{n}\right) = 0. \end{aligned} \tag{14.23}$$

The first equality in (14.23) is the incompressibility condition, and the second is the condition of impermeability through the boundary $\partial\Omega$.

Thus, we have derived the Darcy law from the Stokes problem by using the theory of two-scale convergence.

REMARK 14.3. We have obtained some additional information on the limit behavior of the velocities $\overline{u}_\varepsilon$ in comparison with Theorem 14.1. Namely, we have found not only the weak limit, as in Theorem 14.1, but also the two-scale limit of the velocities.

To conclude this section, we prove that the two-scale limit is also the strong two-scale limit (cf. Remark 9.2). Furthermore, we show that the pressures $p_\varepsilon(x)$ strongly converge to $p(x)$, i.e., $\|p_\varepsilon(x) - p(x)\|_{L_2(\Omega^\varepsilon)} \to 0$ as $\varepsilon \to 0$. To this end, we assume additionally that the function $f(x)$ and the boundary of $\Omega$ are smooth. The conditions on the boundary $\partial\Xi_0$ of the obstacle remain unchanged.

EXERCISE 14.14. Suppose $u_\varepsilon \xrightarrow{2} u$ strongly, where $u = u(x)$ depends only on $x \in \Omega$. Is it true that $u_\varepsilon(x)$ strongly converges to $u(x)$?

Extending $\overline{u}_\varepsilon$ by zero to $\Omega \setminus \Omega^\varepsilon$ and keeping the same notation for the extended function, we can write the identity (14.12) in the form

$$\int_\Omega (\varepsilon\nabla\overline{u}_\varepsilon, \varepsilon\nabla\overline{u}_\varepsilon)\,dx = \int_\Omega (\overline{f}, \overline{u}_\varepsilon)\,dx.$$

Since $f$ is a smooth function, we can pass to the two-scale limit:

$$\lim_{\varepsilon\to 0}\int_\Omega (\varepsilon\nabla\overline{u}_\varepsilon, \varepsilon\nabla\overline{u}_\varepsilon)\,dx = \int_{\Omega\times\Xi}(\overline{f}(x), \overline{u}_0(x,\xi))\,dx\,d\xi.$$

On the other hand, the function $\overline{u}_0$ belongs to the space $V$ and can be taken for a test function in the integral identity (14.18),

$$\int_{\Omega\times\Xi}(\nabla_\xi\overline{u}_0(x,\xi), \nabla_\xi\overline{u}_0(x,\xi))\,dx\,d\xi = \int_{\Omega\times\Xi}\big(\overline{f}(x), \overline{u}_0(x,\xi)\big)\,dx\,d\xi.$$

Consequently,

$$\begin{aligned}\int_{\Omega\times\Xi}(\nabla_\xi\overline{u}_0(x,\xi), \nabla_\xi\overline{u}_0(x,\xi))\,dx\,d\xi &= \int_{\Omega\times\Xi}(\overline{f}(x), \overline{u}_0(x,\xi))\,dx\,d\xi\\ &= \lim_{\varepsilon\to 0}\int_\Omega(\varepsilon\nabla\overline{u}_\varepsilon, \varepsilon\nabla\overline{u}_\varepsilon)\,dx.\end{aligned}$$

Thus, $\varepsilon\nabla\overline{u}_\varepsilon \xrightarrow{2} \nabla_\xi\overline{u}_0(x,\xi)$ two-scale strongly.

EXERCISE 14.15. Suppose $v_\varepsilon(x)\in L_2(\Omega)$ strongly two-scale converge to $v(x,\xi)$ $\in L_2(\Omega\times\square)$, and let $v(x,\xi)=\sum_{i=1}^k \varphi_i(x)\Phi_i(\xi)$, where the $\varphi_i(x)$ are smooth in $\Omega$ and $\Phi_i(\xi)\in H^1_{\mathrm{per}}(\square)$. Prove that $\|v_\varepsilon(x)-v(x,x/\varepsilon)\|_{L_2(\Omega)}\to 0$ as $\varepsilon\to 0$.

In view of Exercise 14.15, we have

$$\Big\|\varepsilon\nabla\overline{u}_\varepsilon - \nabla_\xi\overline{u}_0\Big(x,\frac{x}{\varepsilon}\Big)\Big\|_{L_2(\Omega)}\to 0.$$

Hence $\varepsilon\nabla\overline{u}_\varepsilon = \nabla_\xi\overline{u}_0\big(x,\frac{x}{\varepsilon}\big)+\delta^\varepsilon(x)$ and $\|\delta^\varepsilon(x)\|_{L_2(\Omega)}\to 0$ as $\varepsilon\to 0$. Then the first equation in the Stokes system can be written in the form

$$\nabla p_\varepsilon(x) - \varepsilon\,\mathrm{div}\Big(\varepsilon\nabla\overline{u}_0\Big(x,\frac{x}{\varepsilon}\Big)+\delta_1^\varepsilon(x)\Big) = \overline{f}(x)\quad\text{in }\Omega^\varepsilon, \tag{14.24}$$

where for the residual $\delta_1^\varepsilon$ we have $\|\delta_1^\varepsilon(x)\|_{L_2(\Omega)}\to 0$ as $\varepsilon\to 0$. On the other hand, it is easy to check directly that the pair $\big(p(x),\overline{u}_0\big(x,\frac{x}{\varepsilon}\big)\big)$ satisfies the equality

$$\nabla p(x) - \varepsilon\,\mathrm{div}\Big(\varepsilon\nabla\overline{u}_0\Big(x,\frac{x}{\varepsilon}\Big)\Big) = \overline{f}(x)+\varepsilon\delta_2^\varepsilon(x)\quad\text{in }\Omega, \tag{14.25}$$

where $\|\delta_2^\varepsilon(x)\|_{H^{-1}(\Omega^\varepsilon)}\to 0$ as $\varepsilon\to 0$.

EXERCISE 14.16. Prove the last assertion.

*Hint.* Define

$$\overline{u}_0^\varepsilon(x) = \Big(f_i(x)-\frac{\partial}{\partial x_i}p(x)\Big)\overline{v}_i\Big(\frac{x}{\varepsilon}\Big),\quad p_1^\varepsilon(x) = p(x)+\varepsilon\Big(f_i(x)-\frac{\partial}{\partial x_i}p(x)\Big)q_i\Big(\frac{x}{\varepsilon}\Big),$$

where $(q_i(\xi),\overline{v}_i(\xi))$ is a periodic solution to the problem (14.21) on the cell $\Xi$ and $\langle q_i(\cdot)\rangle_\Xi = 0$. A direct computation shows that

$$\begin{aligned}\nabla p_1^\varepsilon(x) - \varepsilon^2\Delta\overline{u}_0^\varepsilon(x) &= \varepsilon\Big[\nabla\Big(f_i(x)-\frac{\partial}{\partial x_i}p(x)\Big)\Big]q_i\Big(\frac{x}{\varepsilon}\Big)\\ &\quad-\varepsilon^2\Big[\Delta\Big(f_i(x)-\frac{\partial}{\partial x_i}p(x)\Big)\Big]\overline{v}_i\Big(\frac{x}{\varepsilon}\Big)\\ &\quad-2\varepsilon\Big(\nabla\Big(f_i(x)-\frac{\partial}{\partial x_i}p(x)\Big),\nabla_\xi\overline{v}_i\Big(\frac{x}{\varepsilon}\Big)\Big)+f(x).\end{aligned}$$

Show that the $H^{-1}(\Omega^\varepsilon)$-norm of the first three terms on the right-hand side of the last equality is of order $\varepsilon^2$.

Subtracting (14.25) from (14.24), we find

$$\nabla(p_\varepsilon - p) = \varepsilon \operatorname{div} \delta_1^\varepsilon(x) - \varepsilon \delta_2^\varepsilon(x) \quad \text{in } \Omega^\varepsilon. \tag{14.26}$$

Denote by $w_\varepsilon(x)$ the right-hand side of (14.26) and note that it is an element of $H^{-1}(\Omega^\varepsilon)$. Then

$$\nabla(p_\varepsilon - p) = w_\varepsilon \quad \text{in } \Omega^\varepsilon, \qquad \frac{\|w_\varepsilon(x)\|_{H^{-1}(\Omega^\varepsilon)}}{\varepsilon} \to 0 \quad \text{as } \varepsilon \to 0. \tag{14.27}$$

To estimate $(p_\varepsilon - p)$, we use the inequality

$$\|q(x)\|_{L_2(\Omega^\varepsilon)} \le \frac{C}{\varepsilon}\|\nabla q(x)\|_{H^{-1}(\Omega^\varepsilon)}, \qquad \int_{\Omega^\varepsilon} q(x)\,dx = 0, \tag{14.28}$$

where $C > 0$ is independent of $\varepsilon > 0$. The inequality (14.28) immediately follows from the assertion on the solvability of the problem

$$\operatorname{div} Q(x) = q(x) \quad \text{in } \Omega^\varepsilon, \qquad \int_{\Omega^\varepsilon} q(x)\,dx = 0, \quad Q(x) \in \overset{\circ}{H}{}^1(\Omega^\varepsilon),$$

with the estimates

$$\|Q(x)\|_{H^1(\Omega^\varepsilon)} \le \frac{C}{\varepsilon}\|q(x)\|_{L_2(\Omega^\varepsilon)}, \qquad \|Q(x)\|_{L_2(\Omega^\varepsilon)} \le C\|q(x)\|_{L_2(\Omega^\varepsilon)},$$

where the constant $C > 0$ is independent of $\varepsilon > 0$.

EXERCISE 14.17. Derive (14.28) from the assertion on the solvability of the boundary value problem.

Applying the estimate (14.28) to (14.27), we find that for every $\varepsilon > 0$ there exists a constant $k_\varepsilon$ such that $\|p - p_\varepsilon - k_\varepsilon\|_{L_2(\Omega^\varepsilon)} \to 0$ as $\varepsilon \to 0$.

Thus, we have proved that

$$\|p_\varepsilon(x) - p(x)\|_{L_2(\Omega^\varepsilon)} \to 0 \tag{14.29}$$

provided that the additive constant in the pressure $p_\varepsilon$ is chosen in a suitable way.

Now we prove that

$$\|P_\varepsilon(x) - p(x)\|_{L_2(\Omega)} \to 0, \tag{14.30}$$

where $P_\varepsilon(x)$ is the extension of $p_\varepsilon(x)$ to the domain $\Omega$. We define

$$p_\varepsilon^0(x) = \begin{cases} p(x), & x \in Y_{s,i}^\varepsilon, \ i = 1, \dots, N(\varepsilon), \\ \frac{1}{|Y_{s,i}^\varepsilon|}\int_{Y_{s,i}^\varepsilon} p(x)\,dx, & x \in Y_{h,i}^\varepsilon, \ i = 1, \dots, N(\varepsilon). \end{cases}$$

EXERCISE 14.18. Prove that $\|p_\varepsilon^0(x) - p(x)\|_{L_2(\Omega)} \to 0$ as $\varepsilon \to 0$.

*Hint.* Use the Poincaré inequality for each of the cells $Y_{h,i}^\varepsilon$ and pay attention to the value of the constant in the Poincaré inequality.

Further, we have

$$\begin{aligned} \|P_\varepsilon(x) - p(x)\|_{L_2(\Omega)} &\le \|P_\varepsilon(x) - p_\varepsilon^0(x) + p_\varepsilon^0(x) - p(x)\|_{L_2(\Omega)} \\ &\le \|p_\varepsilon(x) - p(x)\|_{L_2(\Omega^\varepsilon)} + \|P_\varepsilon(x) - p_\varepsilon^0(x)\|_{L_2(\Omega_h^\varepsilon)} + \|p_\varepsilon^0(x) - p(x)\|_{L_2(\Omega)}. \end{aligned}$$

The first and third terms on the right-hand side of the last inequality converge to zero as $\varepsilon \to 0$ in view of (14.29) and Exercise 14.18. It remains to prove that the second term converges to zero as $\varepsilon \to 0$. This can be done using the inequality

$$[\langle f \rangle_G - \langle g \rangle_G]^2 \le \frac{1}{|G|}\int_G (f - g)^2\,dx$$

for any $f, g \in L_2(G)$, where $G$ is an arbitrary domain and

$$\langle f\rangle_G \equiv \frac{1}{|G|}\int_G f(\xi)d\xi.$$

EXERCISE 14.19. Prove the above inequality.

By the last inequality, we have

$$\|P_\varepsilon(x) - p_\varepsilon^0(x)\|^2_{L_2(\Omega_h^\varepsilon)} \le \frac{|\Omega_h^\varepsilon|}{|\Omega^\varepsilon|}\|p_\varepsilon(x) - p(x)\|_{L_2(\Omega^\varepsilon)}.$$

Hence (14.30) is proved.

Thus, Theorem 14.1 is completely proved. We have established the strong convergence of the extensions of pressures $P_\varepsilon(x)$ to the limit pressure $p(x)$ in $L_2(\Omega)$. We have also specified the assertion of Theorem 14.1 concerning the velocity convergence. Namely, not only have we established the weak convergence but also proved that the norm $\|\nabla_\xi \overline{u}_0(x, x/\varepsilon) - \varepsilon\nabla\overline{u}_\varepsilon(x)\|_{L_2(\Omega)}$ tends to 0 as $\varepsilon \to 0$. This means that the oscillating function given explicitly by formula (14.22) approaches the solutions $\overline{u}_\varepsilon$ in the space $L_2(\Omega)$.

EXERCISE 14.20. Prove that $\|\overline{u}_0(x, x/\varepsilon) - \overline{u}_\varepsilon(x)\|_{L_2(\Omega)} \to 0$ as $\varepsilon \to 0$.

Thus, we have proved the following assertion on the convergence of sequences of velocities and pressures.

THEOREM 14.5. *Suppose that a function $f(x)$ and the boundary of a domain $\Omega$ are smooth and $\Xi$ is a generalized Lipschitz domain, i.e., $\Xi$ can be represented as the union of finitely many Lipschitz domains. Then for the extended pressures $P_\varepsilon$ (defined by (14.16)) and velocities $\overline{u}_\varepsilon$ (extended by zero to $\Omega \setminus \Omega^\varepsilon$) the following equalities hold:*

$$\lim_{\varepsilon\to 0}\|P_\varepsilon(x) - p(x)\|_{L_2(\Omega)} = 0,$$
$$\lim_{\varepsilon\to 0}\left\|\overline{u}_\varepsilon(x) - \overline{u}_0\left(x, \frac{x}{\varepsilon}\right)\right\|_{L_2(\Omega)} = 0,$$
$$\lim_{\varepsilon\to 0}\varepsilon\left\|\nabla\overline{u}_\varepsilon(x) - \nabla\overline{u}_0\left(x, \frac{x}{\varepsilon}\right)\right\|_{L_2(\Omega)} = 0,$$

*where $\overline{u}_0(x,\xi)$ is defined by formula (14.22), and $p(x)$ is a solution to the Neumann problem*

$$\sum_{i,j=1}^d K_{ij}\frac{\partial^2 p(x)}{\partial x_i \partial x_j} = \sum_{i,j=1}^d K_{ij}\frac{\partial f_i(x)}{\partial x_j} \quad \text{in } \Omega,$$
$$\sum_{i,j=1}^d K_{ij}\frac{\partial p(x)}{\partial x_i}n_j = \sum_{i,j=1}^d K_{ij} f_i(x) n_j \quad \text{on } \partial\Omega;$$

*moreover, $K_{ij} = \int_\Xi (\overline{v}_i)_j\, d\xi$ and the vector-valued functions $\overline{v}_i(\xi)$ are solutions to the Stokes problem in $\Xi$ :*

$$\nabla_\xi q_i(\xi) - \Delta_\xi \overline{v}_i(\xi) = \overline{e}_i \quad \text{in } \Xi,$$
$$\operatorname{div}\overline{v}_i = 0, \quad \overline{v}_i \in H^1_{\text{per}}(\Xi),$$
$$\overline{v}_i(\xi) = 0 \quad \text{on } \partial\Xi_0.$$

## 15. Boundary value problems with rapidly alternating type of boundary conditions

### 15.1. Homogenization theorem.

15.1.1. *Notation and statement of the problem.* Many papers devoted to problems with rapidly alternating type of boundary conditions appeared in the mid-1980s (cf. [**51, 35, 82**]). A complete analysis of problems with periodic rapidly alternating type of boundary conditions can be found in [**36**]. The nonperiodic case (including the random case) was treated in [**104, 16, 40**].

In this section, we give an example of asymptotic analysis of a problem with rapidly alternating type of boundary conditions. The arguments mainly follow [**35**].

Consider a membrane with partially clamped boundary. Assume that the clamped parts of the boundary have length of order $\varepsilon_1$ and the free parts have length of order $\varepsilon_2$. The following question arises: What relation between the parameters $\varepsilon_1$ and $\varepsilon_2$ guarantees that the equilibrium state of such a membrane with small $\varepsilon_1$ and $\varepsilon_2$ is close to the equilibrium state of the membrane clamped along the entire boundary?

We show that both membranes have the same asymptotic behavior if $\varepsilon_1$ and $\varepsilon_2$ are of the same order. Furthermore, if $\varepsilon_1$ is small and $\varepsilon_2$ is of order $|\ln \varepsilon_1|^{\delta-1}$, $\delta \in (0,1)$, then the equilibrium state of the membrane under consideration is close to the equilibrium state of the free membrane.

Consider the boundary value problem for the second order elliptic equation,

$$(15.1)\qquad \begin{aligned} &L[u_\varepsilon] \equiv \frac{\partial}{\partial x_j}\left(a^{ij}(x)\frac{\partial u_\varepsilon}{\partial x_i}\right) = f(x) \quad \text{in } \Omega \subset \mathbb{R}^2, \\ &u_\varepsilon = 0 \quad \text{on } \Gamma_D^\varepsilon, \\ &\frac{\partial u_\varepsilon}{\partial \gamma} \equiv a^{ij}(x)\frac{\partial u_\varepsilon}{\partial x_i}\nu_j = 0 \quad \text{on } \Gamma_N^\varepsilon. \end{aligned}$$

Suppose that $\lambda_1|\xi|^2 \geq a^{ij}(x)\xi_i\xi_j \geq \lambda_2|\xi|^2$, where $\lambda_1 > 0$, $\lambda_2 > 0$, $\Omega \subset \mathbf{R}^2$ is a domain with smooth boundary $\partial\Omega$, and $a^{ij}(x)$ are bounded measurable functions in $\Omega$.

Let $\Gamma_D^\varepsilon = \bigcup_i \Gamma_D^i$, $\Gamma_N^\varepsilon = \bigcup_j \Gamma_N^j$, $\Gamma_D^\varepsilon \cup \Gamma_N^\varepsilon = \partial\Omega$, $|\Gamma_D^i| = \varepsilon_1$, and $|\Gamma_N^j| = \varepsilon_2$ for all $i$, $j$, where $|l|$ denotes the length of a curve $l$ (cf. Figure 15.1).

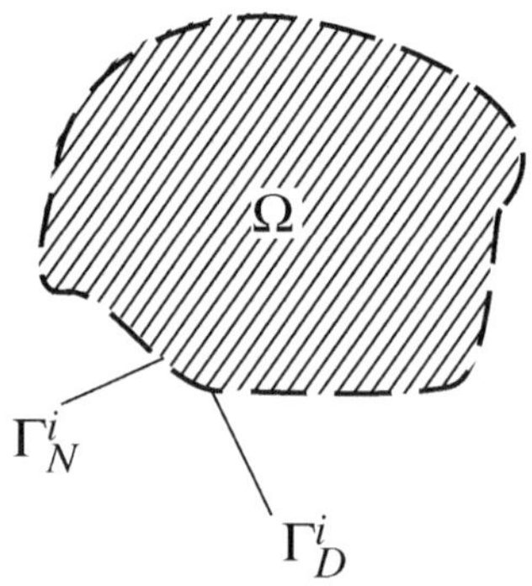


FIGURE 15.1

Consider two cases.

CASE (a). $\varepsilon_1 = O(\varepsilon)$ and $\varepsilon_2 = O(\varepsilon)$.
CASE (b). $\varepsilon_1 = O(\varepsilon)$ and $\varepsilon_2 = O(|\ln \varepsilon|^{\delta-1})$, where $\delta \in (0,1)$.
We look for a solution to the problem (15.1) in the space

$$H^1(\Omega, \Gamma_D^\varepsilon) = \{v(x) : v(x) \in H^1(\Omega), v(x)\big|_{\Gamma_D^\varepsilon} = 0\}$$

equipped with the norm

$$\|v\|_{H^1(\Omega,\Gamma_D^\varepsilon)} = \Big(\int_\Omega (v^2 + |\nabla v|^2)\, dx\Big)^{1/2}.$$

DEFINITION 15.1. A function $u_\varepsilon \in H^1(\Omega, \Gamma_D^\varepsilon)$ is called a *weak solution* to the problem (15.1) if for any $\varphi \in H^1(\Omega, \Gamma_D^\varepsilon)$,

$$\int_\Omega a^{ij}(x) \frac{\partial u_\varepsilon}{\partial x_i} \frac{\partial \varphi}{\partial x_j}\, dx = -\int_\Omega f(x)\, \varphi(x)\, dx. \tag{15.2}$$

We study the limit behavior of $u_\varepsilon$ as $\varepsilon \to 0$.

15.1.2. *The limit Dirichlet problem.* Consider case (a). For the sake of definiteness, we set $\varepsilon_1 = k_1\varepsilon$ and $\varepsilon_2 = k_2\varepsilon$, where $k_1, k_2 > 0$. We need an analog of the classical Friedrichs inequality (cf. Exercise 15.1 below and [**86**] for the proof).

EXERCISE 15.1. Suppose that $\Omega$ is a bounded Lipschitz domain and $u \in H^1(\Omega)$ vanishes on $\Gamma \subset \partial\Omega$. Let $|\Gamma| > 0$, where $|\Gamma|$ is the surface Lebesgue measure of $\Gamma$. Prove the inequality

$$\int_\Omega u^2(x)\, dx \le C \int_\Omega |\nabla u(x)|^2\, dx, \tag{15.3}$$

where the constant $C$ depends only on $\Omega$ and $|\Gamma|$ is independent of $u$:

$$C \le C_1(\Omega) + \frac{C_2(\Omega)}{|\Gamma|}.$$

By (15.3), it is possible to introduce the new inner product in $H^1(\Omega, \Gamma_D^\varepsilon)$:

$$[u, \varphi] = \int_\Omega (\nabla u, \nabla \varphi)\, dx$$

and the corresponding norm $\|u\| = [u, u]^{1/2}$.

THEOREM 15.1. *A weak solution $u_\varepsilon$ to the problem* (15.1) *exists and is unique.*

The existence and uniqueness of a weak solution to the problem (15.1) can be proved in a standard way using the Lax–Milgram theorem (cf., for example, [**130, 88**]).

Now we estimate $u_\varepsilon$ uniformly with respect to $\varepsilon$. In case (a), the estimate (15.3) for the solutions $u_\varepsilon$ to the problem (15.1) is uniform with respect to $\varepsilon$ since $|\Gamma_D^\varepsilon| \geq \text{const}\,|\partial\Omega|$, where const is independent of $\varepsilon$.

Using the Cauchy–Schwarz–Bunyakovskii inequality, the inequality (15.3), and the ellipticity of the operator $L$, we find that

$$\lambda_2 \int_\Omega |\nabla u_\varepsilon|^2\,dx \leq \left|\int_\Omega a^{ij}(x)\frac{\partial u_\varepsilon}{\partial x_i}\frac{\partial u_\varepsilon}{\partial x_j}\,dx\right| = \left|\int_\Omega f(x)\,u_\varepsilon(x)\,dx\right|$$
$$\leq \left(\int_\Omega f^2(x)\,dx\right)^{1/2}\left(\int_\Omega u_\varepsilon^2(x)\,dx\right)^{1/2} \leq M\left(\int_\Omega |\nabla u_\varepsilon(x)|^2\,dx\right)^{1/2},$$

where $M = C\big(\int_\Omega f^2(x)\,dx\big)^{1/2}$. Thus,

$$\int_\Omega |\nabla u_\varepsilon|^2\,dx \leq M_1,$$

where $M_1$ is independent of $\varepsilon$.

By the Rellich theorem, we can choose a subsequence $\varepsilon^k$ converging to zero such that

(i) $u_\varepsilon \to U \in H^1(\Omega)$ in the $L_2(\Omega)$-norm as $\varepsilon^k \to 0$;

(ii) $\frac{\partial u_\varepsilon}{\partial x_j} \rightharpoonup \frac{\partial U}{\partial x_j}$, $j = 1, 2$, weakly in $L_2(\Omega)$ as $\varepsilon^k \to 0$.

Therefore, passing to the limit in the integral identity (15.2) as $\varepsilon^k \to 0$, we find that

$$\int_\Omega a^{ij}(x)\frac{\partial U}{\partial x_i}\frac{\partial v}{\partial x_j}\,dx = -\int_\Omega f(x)v(x)\,dx \quad \forall v \in \overset{\circ}{H}{}^1(\Omega).$$

Now we show that $U$ vanishes on $\partial\Omega$. We make a local change of variables $(x_1, x_2) \to (y_1, y_2)$ in a neighborhood of the endpoint $p_i$ of $\Gamma_D^i$ so that in this neighborhood, $\partial\Omega$ will be transformed to the line $y_2 = 0$, and the point $p_i$, to the origin. Consider a strip of width $\omega$ in a neighborhood of $p_i$ (cf. Figure 15.2).

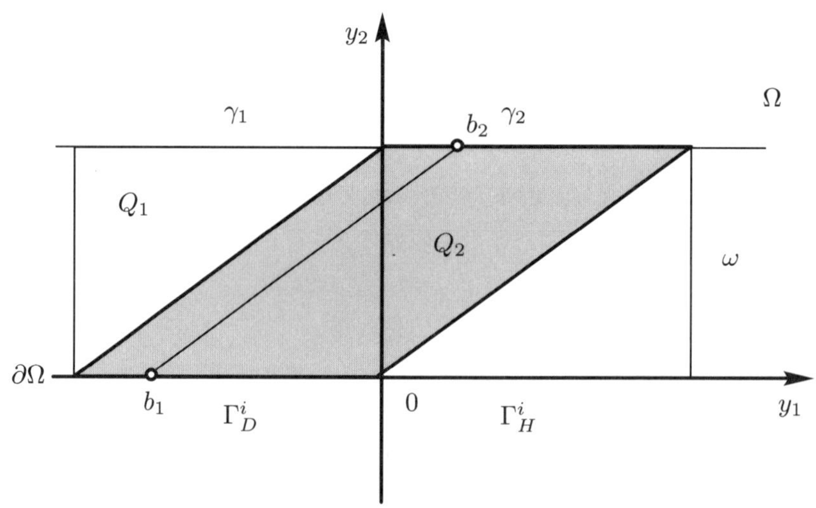


FIGURE 15.2

We set

$$\begin{aligned}\gamma_1 &= \{(y_1, y_2) : y_2 = \omega, \ -k_1\varepsilon < y_1 < 0\},\\ \gamma_2 &= \{(y_1, y_2) : y_2 = \omega, \ 0 < y_1 < k_2\varepsilon\},\\ Q_1 &= \{(y_1, y_2) : 0 < y_2 < \omega, \ -k_1\varepsilon < y_1 < 0\},\\ Q_2 &= \left\{(y_1, y_2) : 0 < y_2 < \omega, \ -k_1\varepsilon + \frac{k_1}{\omega}y_2 < y_1 < \frac{k_1}{\omega}y_2\right\},\end{aligned}$$

where $k_1$, $k_2$ are the constants independent of $\varepsilon$ introduced at the beginning of this section. It suffices to consider the case where $|\gamma_1| = |\gamma_2|$. Other cases can be treated in a similar way and are left to the reader.

We assume that $u_\varepsilon$ belongs to $C^1(\overline{\Omega})$ and the integrals of the derivatives of $u_\varepsilon(x)$ along one-dimensional curves can be understood in the classical sense. However, the required estimate (15.5) remains valid for $u_\varepsilon(x) \in H^1(\Omega, \Gamma_\varepsilon^D)$, which easily follows from the fact that the $C^1(\overline{\Omega})$ functions vanishing on $\Gamma_\varepsilon^D$ are dense in the space $H^1(\Omega, \Gamma_\varepsilon^D)$.

EXERCISE 15.2. How can the above arguments be completed to obtain (15.5) for $u_\varepsilon(x) \in H^1(\Omega, \Gamma_\varepsilon^D)$?

We have

$$u_\varepsilon(y_1, \omega) = \int_0^\omega \frac{\partial u_\varepsilon}{\partial y_2}\, dy_2$$

if $y_1 \in \gamma_1$. Therefore,

$$u_\varepsilon^2(y_1, \omega) \le \omega \int_0^\omega \left(\frac{\partial u_\varepsilon}{\partial y_2}\right)^2 dy_2. \tag{15.4}$$

Integrating the inequality (15.4) over $\gamma_1$ and using the Cauchy–Schwarz–Bunyakovskii inequality, we find

$$\int_{\gamma_1} u_\varepsilon^2(y_1, \omega)\, dy_1 \le \omega \int_{\gamma_1} \int_0^\omega \left(\frac{\partial u_\varepsilon}{\partial y_2}\right)^2 dy_2\, dy_1 \le \omega \int_{Q_1} |\nabla u_\varepsilon|^2\, dy_1\, dy_2.$$

Now we estimate $u_\varepsilon(y_1, \omega)$, where $y_1 \in \gamma_2$. We have

$$u_\varepsilon(y_1, \omega) = \int_{b_1}^{b_2} \frac{\partial u_\varepsilon}{\partial l}\, dl,$$

where $b_1$ and $b_2$ are points on $\Gamma_D^i$ and $\gamma_2$, respectively, that at the same time belong to the line $l$ parallel to the side of the parallelogram $Q_2$ with vertices at the endpoints of $\Gamma_D^i$ and $\gamma_2$. Then

$$u_\varepsilon^2(y_1, \omega) \le \sqrt{\omega^2 + \varepsilon^2 k_1^2} \int_{b_1}^{b_2} \left(\frac{\partial u_\varepsilon}{\partial l}\right)^2.$$

Integrating this inequality over $\gamma_2$, we find that

$$\begin{aligned}\int_{\gamma_2} u_\varepsilon^2(y_1, \omega)\, dy_1 &\le \sqrt{\omega^2 + \varepsilon^2 k_1^2} \int_{\gamma_2} \int_{b_1}^{b_2} \left(\frac{\partial u_\varepsilon}{\partial l}\right)^2 dl\, dy_1\\ &= \frac{\omega^2 + \varepsilon^2 k_1^2}{\omega} \int_{Q_2} \left(\frac{\partial u_\varepsilon}{\partial l}\right)^2 dy_1\, dy_2\\ &\le \frac{\omega^2 + \varepsilon^2 k_1^2}{\omega} \int_{Q_2} |\nabla u_\varepsilon|^2\, dy_1\, dy_2.\end{aligned}$$

Making the inverse coordinate transformation $(y_1, y_2) \to (x_1, x_2)$ and summing over all neighborhoods covering $\partial\Omega$, we obtain

$$\int_{\widetilde{\gamma}} u_\varepsilon^2(x)\, ds \le C_4 \frac{\omega^2 + \varepsilon^2 k_1^2}{\omega} \int_Q |\nabla u_\varepsilon|^2\, dx \le C_4 M_1 \frac{\omega^2 + \varepsilon^2 k_1^2}{\omega}, \tag{15.5}$$

where $\widetilde{\gamma}$ is a curve at a distance of order $\omega$ from $\partial\Omega$, and $Q$ is the layer between $\widetilde{\gamma}$ and $\partial\Omega$; the constants $C_4$ and $M_1$ are independent of $\varepsilon$.

Taking into account the continuity of the trace operator from $H^1(\Omega)$ to $L_2(\widetilde{\gamma})$ (cf. [**120, 121**]), we pass to the limit in (15.5) as $\varepsilon^k \to 0$. We have

$$\int_{\widetilde{\gamma}} U^2\, ds \le \omega M_1 C_4. \tag{15.6}$$

Since $\omega$ is an arbitrarily small positive number and $U \in H^1(\Omega)$, from (15.6) it follows that $U = 0$ on $\partial\Omega$. Here we have used the continuity of the trace of functions in $H^1(\Omega)$ if the sequences of the surfaces (where the trace is considered) converge to some limit surface (we recommend that the reader will formulate and rigorously prove this assertion).

Thus, we have proved the following.

THEOREM 15.2. *In case* (a), *as* $\varepsilon \to 0$, *the sequence of solutions* $u_\varepsilon$ *of problem* (15.1) *converges in the* $L_2(\Omega)$*-norm and weakly in* $H^1(\Omega)$ *to the solution of the Dirichlet problem*

$$L[U] = f(x) \quad \text{in } \Omega, \qquad U = 0 \quad \text{on } \partial\Omega. \tag{15.7}$$

This assertion has already been proved for $\varepsilon^k$. The convergence of $u_\varepsilon$ as $\varepsilon \to 0$ follows from the uniqueness of the solution to the problem (15.7).

15.1.3. *The limit Neumann problem.* We estimate the solution $u_\varepsilon$ to the problem (15.1) in case (b). Assume that

$$\int_\Omega f(x)\, dx = 0. \tag{15.8}$$

Note that (15.8) is the solvability condition for a Neumann problem with a homogeneous boundary condition. If the solvability condition is not satisfied, the $L_2(\Omega)$-norms of $u_\varepsilon$ can unboundedly increase as $\varepsilon \to 0$. Introduce the notation

$$\langle u_\varepsilon \rangle = \frac{1}{|\Omega|} \int_\Omega u_\varepsilon(x)\, dx.$$

Setting $\varphi = u_\varepsilon$ in (15.2), we find that

$$\begin{aligned}
\lambda_2 \int_\Omega |\nabla u_\varepsilon|^2\, dx &\le \left| \int_\Omega a^{ij}(x) \frac{\partial u_\varepsilon}{\partial x_i} \frac{\partial u_\varepsilon}{\partial x_j}\, dx \right| = \left| \int_\Omega f(x) u_\varepsilon(x)\, dx \right| \\
&= \left| \int_\Omega f(x) u_\varepsilon(x) dx - \int_\Omega f(x) \langle u_\varepsilon \rangle dx \right| \\
&= \left| \int_\Omega f(x)(u_\varepsilon(x) - \langle u_\varepsilon \rangle) dx \right| \\
&\le \left( \int_\Omega f^2(x)\, dx \right)^{1/2} \left( \int_\Omega (u_\varepsilon(x) - \langle u_\varepsilon \rangle)^2\, dx \right)^{1/2} \\
&\le K \left( \int_\Omega f^2(x)\, dx \right)^{1/2} \left( \int_\Omega |\nabla u_\varepsilon|^2\, dx \right)^{1/2}.
\end{aligned}$$

Thus,

$$\int_\Omega |\nabla u_\varepsilon|^2 dx \le \frac{K}{\lambda_2}\int_\Omega f^2\,dx, \tag{15.9}$$

where $K$ is independent of $\varepsilon$.

Consider the sequence $(u_\varepsilon(x) - \langle u_\varepsilon\rangle)$. From (15.9) and the Poincaré inequality we obtain the estimate $\|u_\varepsilon - \langle u_\varepsilon\rangle\|_{H^1(\Omega)} \le C_0$, where $C_0$ is independent of $\varepsilon$. By the Rellich theorem, there exists $U \in H^1(\Omega)$ such that the sequence $u_\varepsilon - \langle u_\varepsilon\rangle$ converges to $U$ in the $L_2(\Omega)$-norm and weakly in $H^1(\Omega)$ on a subsequence $\varepsilon^k \to 0$. By construction, $\langle U\rangle = 0$. We show that $U$ is a weak solution to the problem

$$\begin{gathered} L[U] = f(x) \quad \text{in } \Omega, \\ \frac{\partial U}{\partial \nu} \equiv a^{ij}(x)\frac{\partial U}{\partial x_i}\nu_j = 0 \quad \text{on } \partial\Omega, \end{gathered} \tag{15.10}$$

i.e., $U$ satisfies the integral identity

$$\int_\Omega a^{ij}(x)\frac{\partial U}{\partial x_i}\frac{\partial \varphi}{\partial x_j}\,dx = -\int_\Omega f(x)\varphi(x)\,dx$$

for any $\varphi \in H^1(\Omega)$. Let $p_i$ be the midpoint of the curve $\Gamma_D^i$. We make a local change of variables $(x_1, x_2) \to (y_1, y_2)$ in a neighborhood of $\Gamma_D^i$ such that in a neighborhood of $p_i$, the boundary $\partial\Omega$ transformes to the line $y_2 = 0$ and $p_i$ goes to the origin (cf. Figure 15.3).

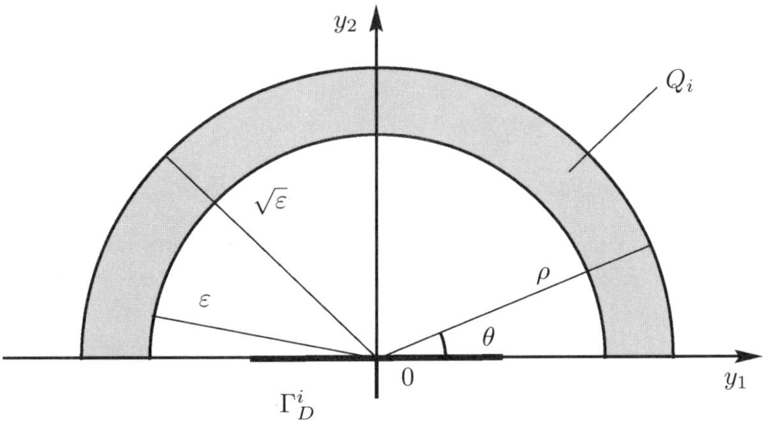


FIGURE 15.3

Without loss of generality, we can assume that the length of the image of $\Gamma_D^i$ is equal to $\varepsilon$.

We introduce polar coordinates $(\rho, \theta)$, $\rho = (y_1^2 + y_2^2)^{1/2}$. Consider $\psi \in C^\infty(\mathbf{R})$ such that $\psi(s) = 0$ for $|s| \le 1$ and $\psi(s) = 1$ for $|s| > 2$, $0 \le \psi \le 1$. We set $\psi_\varepsilon^i = \psi\big(\frac{\ln\varepsilon}{\ln\rho}\big)$ and $\zeta = \varphi\psi_\varepsilon$, where $\varphi \in H^1(\Omega)$ and $\psi_\varepsilon = \prod_i \psi_\varepsilon^i$. Note that, by definition, $\psi_\varepsilon^i(\rho) = 1$ if $\rho \ge \sqrt{\varepsilon}$ (cf. Figure 15.3). We also note that the functions $\psi_\varepsilon^i$ written in the coordinates $(y_1, y_2)$ are independent of $i$. However, if we return to the original coordinates $(x_1, x_2)$, the dependence on $i$ is not trivial.

Taking $\zeta$ as a test function, we have

$$\int_\Omega a^{ij}(x)\frac{\partial u_\varepsilon}{\partial x_i}\frac{\partial \zeta}{\partial x_j}\,dx = -\int_\Omega f(x)\zeta(x)\,dx.$$

Taking into account the fact that $\zeta = \varphi\psi_\varepsilon$, we find that

$$\int_\Omega \left[a^{ij}(x)\frac{\partial u_\varepsilon}{\partial x_i}\frac{\partial \psi_\varepsilon}{\partial x_j}\varphi + a^{ij}(x)\frac{\partial u_\varepsilon}{\partial x_i}\frac{\partial \varphi}{\partial x_j}\psi_\varepsilon\right] dx = -\int_\Omega f(x)\zeta(x)\,dx.$$

Since the sequence $\psi_\varepsilon$ converges to the function $\equiv 1$ in $L_2(\Omega)$ as $\varepsilon \to 0$, we have

$$\int_\Omega f(x)\varphi(x)\psi_\varepsilon(x)\,dx \to \int_\Omega f(x)\varphi(x)\,dx \quad \text{as} \quad \varepsilon \to 0.$$

Recalling that the sequence $u_\varepsilon - \langle u_\varepsilon\rangle$ weakly converges in $H^1(\Omega)$ to some function $U \in H^1(\Omega)$ as $\varepsilon^k \to 0$, we arrive at the limit relation

$$\int_\Omega a^{ij}(x)\frac{\partial u_\varepsilon}{\partial x_i}\frac{\partial \varphi}{\partial x_j}\psi_\varepsilon\,dx \to \int_\Omega a^{ij}(x)\frac{\partial U}{\partial x_i}\frac{\partial \varphi}{\partial x_j}\,dx \quad \text{as} \quad \varepsilon \to 0.$$

It remains to show that $\int_\Omega a^{ij}(x)\frac{\partial u_\varepsilon}{\partial x_i}\frac{\partial \psi_\varepsilon}{\partial x_j}\varphi\,dx \to 0$ as $\varepsilon \to 0$. Using the Cauchy–Schwarz–Bunyakovskii inequality, we obtain

$$\begin{aligned}\left|\int_\Omega a^{ij}(x)\frac{\partial u_\varepsilon}{\partial x_i}\frac{\partial \psi_\varepsilon}{\partial x_j}\varphi\,dx\right| &\le \left(\int_\Omega \varphi^2\sum_{j=1}^{2}\left(a^{ij}(x)\frac{\partial u_\varepsilon}{\partial x_i}\right)^2 dx\right)^{1/2}\left(\int_\Omega |\nabla\psi_\varepsilon|^2\,dx\right)^{1/2}\\ &\le C_5\left(\int_\Omega |\nabla u_\varepsilon|^2\,dx\right)^{1/2}\left(\int_\Omega |\nabla\psi_\varepsilon|^2\,dx\right)^{1/2}.\end{aligned}$$

Denote $Q^\varepsilon = \{(\rho,\theta) : \rho \le \sqrt{\varepsilon}\}$. It is easy to see that

$$\int_\Omega |\nabla\psi_\varepsilon|^2\,dx \le C_6\sum_i\int_{Q_i}\left(\frac{\partial\psi^i_\varepsilon}{\partial\rho}\right)^2\rho\,d\rho\,d\theta.$$

Let $s = (\ln\varepsilon/\ln\rho)$. Then

$$\begin{aligned}\int_{Q_i}\left(\frac{\partial\psi^i_\varepsilon}{\partial\rho}\right)^2\rho\,d\rho\,d\theta &= \int_{Q_i}\left(\frac{\partial\psi}{\partial s}\right)^2\left(\frac{\partial s}{\partial\rho}\right)^2\rho\,d\rho\,d\theta\\ &= \ln^2\varepsilon\int_{Q_i}\left(\frac{\partial\psi}{\partial s}\right)^2\left(\frac{1}{\ln\rho}\right)^4\frac{1}{\rho^2}\rho\,d\rho\,d\theta\\ &\le C_7\ln^2\varepsilon\int_{Q_i}\left(\frac{1}{\ln\rho}\right)^4 d(\ln\rho)\,d\theta \le \frac{C_8}{|\ln\varepsilon|}.\end{aligned}$$

Since the number of $\Gamma^i_D$'s is of order $|\ln\varepsilon|^{1-\delta}$, we have

$$\left|\int_\Omega a^{ij}(x)\frac{\partial u_\varepsilon}{\partial x_i}\frac{\partial \psi_\varepsilon}{\partial x_j}\varphi\,dx\right| \le \frac{C_9}{|\ln\varepsilon|^{\delta/2}}.$$

Therefore, $\int_\Omega a^{ij}(x)\frac{\partial u_\varepsilon}{\partial x_i}\frac{\partial \psi_\varepsilon}{\partial x_j}\varphi\,dx$ converges to zero as $\varepsilon \to 0$. Thus, we have proved the following assertion.

THEOREM 15.3. *In case* (b), *the sequence* $u_\varepsilon - \langle u_\varepsilon\rangle$, *where* $u_\varepsilon$ *is the weak solution to the problem* (15.1) *and* $\langle u_\varepsilon\rangle = \frac{1}{|\Omega|}\int_\Omega u_\varepsilon x dx$, *converges in the* $L_2(\Omega)$-*norm and weakly in* $H^1(\Omega)$ *to the solution of the Neumann problem* (15.10) *as* $\varepsilon \to 0$.

This assertion has already been proved for a subsequence $\varepsilon^k$. The convergence of the entire sequence $u_\varepsilon - \langle u_\varepsilon\rangle$ follows from the uniqueness of the solution to the problem (15.10) and from $\int_\Omega U\,dx = 0$.

**15.2. Matching asymptotic expansions.** In a number of papers published in the mid1990s, the method of matching asymptotic expansions was applied to problems with rapidly alternating type of boundary conditions (cf., for example, [**64**] and the bibliography therein). Consider one of these problems.

Introduce the notation: $N \gg 1$ is an integer, $\varepsilon = 2N^{-1}$, $(r,\theta)$ are polar coordinates, $\delta_\varepsilon = \{(r,\theta) : r = 1, -\varepsilon a(\varepsilon) < \theta < \varepsilon a(\varepsilon)\}$, $0 < a(\varepsilon) < \pi/2$, $\Gamma_\varepsilon$ is the union of arcs obtained from $\delta_\varepsilon$ by rotation about the origin through angles multiple of $\varepsilon\pi$, $\Omega$ is the unit circle centered at the origin, and $\gamma_\varepsilon = \partial D \backslash \overline{\Gamma}_\varepsilon$. Our goal is to construct asymptotics of solutions to the following boundary value problem as $\varepsilon \to 0$:

$$
\begin{aligned}
&\Delta u_\varepsilon = f \quad \text{for } x \in \Omega,\\
&u_\varepsilon = 0 \quad \text{for } x \in \Gamma_\varepsilon,\\
&\frac{\partial u_\varepsilon}{\partial r} = 0 \quad \text{for } x \in \gamma_\varepsilon,
\end{aligned}
\tag{15.11}
$$

under the condition that $\varepsilon \ln a(\varepsilon) \to 0$ as $\varepsilon \to 0$.

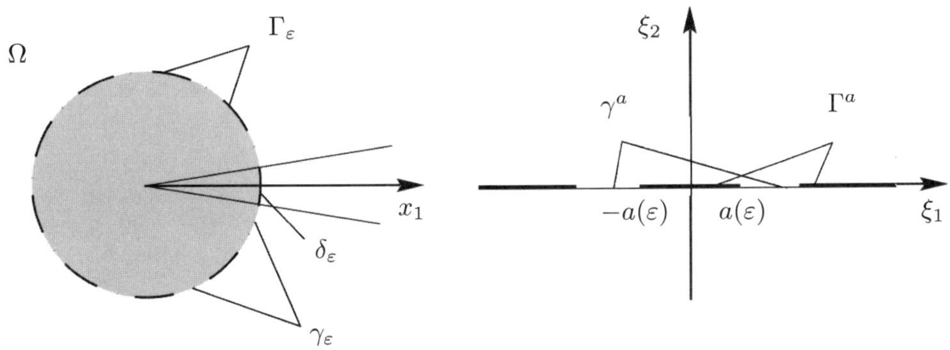


FIGURE 15.4

The same argument as before shows (cf. [**36**]) that the sequence of solutions to problem (15.11) converges strongly in $L_2(\Omega)$ and weakly in $H^1(\Omega)$ to the solution $u_0$ of the Dirichlet problem in a ball. Therefore, it is natural to look for the leading term of the asymptotics in the form $u_0$ outside a small neighborhood of $\partial\Omega$. For the sake of simplicity, we assume that $f$ depends only on $r$ and vanishes in some neighborhood of $\partial\Omega$. Then there exists a constant $a_1^{(0)}$ such that

$$
u_0(x) = a_1^{(0)} \sum_{i=1}^{\infty} \frac{1}{i}(1-r)^i \quad \text{as } r \to 1.
\tag{15.12}
$$

EXERCISE 15.3. Under the above assumptions, prove the representation (15.12) in a neighborhood of the boundary of the unit disk $\Omega$.

Since $u_0$ does not satisfy the boundary condition in the problem (15.11), we try to look for the asymptotics of $u_\varepsilon$ in a neighborhood of $\partial\Omega$ in some other form. Taking into account the fact that the size of the periodicity cell is of order $O(\varepsilon)$, we represent the solution in a neighborhood of $\partial\Omega$ as a series $v_\varepsilon(\xi_1,\xi_2)$ with coefficients depending on the inner variables $\xi = (\xi_1,\xi_2)$, $\xi_1 = \varepsilon^{-1}\theta$, $\xi_2 = \varepsilon^{-1}(1-r)$. Passing to the variables $\xi$ in (15.12), for $\varepsilon^{1/2} < (1-r) < 2\varepsilon^{1/2}$ (or, which is the same, for $\varepsilon^{-1/2} < \xi_2 < 2\varepsilon^{-1/2}$) we obtain

$$u_0(x) = a_1^{(0)} \sum_{i=1}^{\infty} \varepsilon^i \frac{1}{i} \xi_2^i. \tag{15.13}$$

From the matching condition for asymptotic expansions (cf. Subsections 11.1–11.3), we see that the coefficients of the series

$$v_\varepsilon(\xi) = \sum_{i=1}^{\infty} \varepsilon^i v_i(\xi) \tag{15.14}$$

satisfy the relation

$$v_i(\xi) \sim \frac{a_1^{(0)}}{i} \xi_2^i \quad \text{as } \xi_2 \to \infty. \tag{15.15}$$

Substituting (15.14) in (15.11), passing to the variables $\xi$, and collecting terms with equal power of $\varepsilon$, we arrive at the following family of boundary value problems:

$$\begin{aligned} &\Delta v_i = \sum_{j=1}^{i-1} \left( \xi_2^{j-1} \frac{\partial}{\partial \xi_2} - (j+1) \xi_2^j \frac{\partial^2}{\partial \xi_1^2} \right) v_{i-j} \quad \text{for } \xi_2 > 0, \\ &v_i = 0 \quad \text{for } \xi \in \Gamma^a, \\ &\frac{\partial v_i}{\partial \xi_2} = 0 \quad \text{for } \xi \in \gamma^a, \end{aligned} \tag{15.16}$$

where $\Gamma^a$ is the union of intervals on the line $\xi_2 = 0$ obtained from the interval $(-a, a)$ by shifts multiple of $\pi$, and $\gamma^a$ is the complement of $\overline{\Gamma^a}$ to the $\xi_1$-axis. By (15.16), the equations for the first two terms of the series (15.14) have the form

$$\Delta v_1 = 0 \quad \text{for } \xi_2 > 0, \tag{15.17}$$

$$\Delta v_2 = \left( \frac{\partial}{\partial \xi_2} - 2\xi_2^j \frac{\partial^2}{\partial \xi_1^2} \right) v_1 \quad \text{for } \xi_2 > 0. \tag{15.18}$$

We introduce the function $Y(\xi) = \operatorname{Re} \ln(\sin z + (\sin^2 z - \sin^2 a)^{1/2}) - \ln \sin a$, where $z = \xi_1 + i\xi_2$. It can be shown that $Y$ is $\pi$-periodic with respect to $\xi_1$, satisfies the homogeneous Neumann boundary condition on $\gamma^a$, vanishes on $\Gamma^a$, and admits the asymptotic representation

$$Y(\xi) = \xi_2 - \ln \sin a + O(e^{-\xi_2}) \quad \text{as } \xi_2 \to \infty. \tag{15.19}$$

EXERCISE 15.4. Prove the last assertion.

Thus, the function

$$v_1(\xi) = a_1^{(0)} Y(\xi) \tag{15.20}$$

satisfies (15.15)–(15.17). Since $Y$ is harmonic, it is easy to verify that the function

$$v_2(\xi) = \frac{a_1^{(0)}}{2} \xi_2^2 \frac{\partial Y(\xi)}{\partial \xi_2} + A_1^{(1)} Y(\xi) \tag{15.21}$$

satisfies (15.15), (15.16), and (15.18) for any constant $A_1^{(1)}$ (check it!). From (15.19)–(15.21) it follows that

$$\begin{aligned}\varepsilon v_1(\xi)+\varepsilon^2 v_2(\xi) = a_1^{(0)}\Big((1-r)+\frac{1}{2}(1-r)^2\Big)+\varepsilon(A_1^{(1)}(1-r)\\ - a_1^{(0)}\ln\sin a) - \varepsilon^2 A_1^{(1)}\ln\sin a + O((1-r)e^{-\frac{1-r}{\varepsilon}})\end{aligned} \tag{15.22}$$

for $\varepsilon^{1/2} < (1-r) < 2\varepsilon^{1/2}$ (or, which is the same, for $\varepsilon^{-1/2} < \xi_2 < 2\varepsilon^{-1/2}$). From (15.22) and the matching condition for asymptotic expansions it follows that the terms of the series of outer expansion

$$u_\varepsilon(x) = u_0(x) + \varepsilon u_1(x) + \varepsilon^2 u_2(x) + \cdots \tag{15.23}$$

are to satisfy the relation

$$u_1(x) \sim -a_1^{(0)}\ln\sin a + A_1^{(1)}(1-r), \quad u_2(x) \sim A_1^{(1)} \quad \text{as } r \to 1. \tag{15.24}$$

Substituting (15.23) in (15.11), we obtain the following equation for $u_1$ and $u_2$:

$$\Delta u_k = 0 \quad \text{for } x \in \Omega. \tag{15.25}$$

To match the series (15.14) and (15.23), we need to prove the existence of functions $u_k$, $k = 1, 2$, satisfying (15.24) and (15.25). It is obvious that $u_1(x) = -a_1^{(0)}\ln\sin a$ and $u_2(x) = 0$ are solutions to (15.25) and satisfy (15.24) for $A_1^{(1)} = 0$. Thus, we have constructed the first terms $v_1$, $v_2$, $u_0$, $u_1$, and $u_2$ of the asymptotic expansions (15.14) and (15.23). The further construction and justification of the asymptotic expansions can be done in the same way as in Subsection 11.3. In particular, it is easy to check that

$$v_3(\xi) = a_1^{(0)}\Big(\frac{1}{8}\xi_2^4\frac{\partial^2}{\partial\xi_2^2} + \frac{1}{6}\xi_2^3\frac{\partial}{\partial\xi_2}\Big)Y(\xi)$$

and, consequently, $u_3 = 0$.

EXERCISE 15.5. Unlike Subsections 11.2 and 11.3, we do not formulate an assertion that the $u_\varepsilon(x)$ are close to partial sums of the corresponding asymptotic series. Formulate and rigorously prove the results analogous to those in Subsections 11.2 and 11.3.

REMARK 15.1. As was shown in [**62**], in the case $\varepsilon\ln a(\varepsilon) \to 0$, the asymptotic expansion of the solution to the problem (15.11) can be constructed by a simple method, namely, by the method of boundary functions of exponential type [**128**]. However, if the limit problem is the second or third boundary value problem, i.e.,

$$a(\varepsilon) = \exp\Big(-\frac{1}{\varepsilon^2 A}\Big), \quad a(\varepsilon) = \exp\Big(-\frac{1}{\varepsilon A}\Big), \quad A > 0,$$

respectively (cf. [**36**]), then the method of matching asymptotic expansions (together with the method of boundary functions of exponential type and the two-scale convergence method) should be used (cf. [**63, 64, 22**]).

## 16. Appearance of a potential (or étrange term) in the homogenized equation

**16.1. Homogenization of solutions to the Dirichlet problem in a perforated domain with small total concentration of holes.** Suppose that $\Omega$ is a domain in $\mathbb{R}^3$ with smooth boundary and $G_0 = \{x : |x| < 1\}$ is the unit ball in $\mathbb{R}^3$. We set

$$\Omega_\varepsilon = \Omega \setminus \bigcup_{z\in\mathbb{Z}^3} (\varepsilon^3 G_0 + 2\varepsilon z),$$

i.e., $\Omega_\varepsilon$ is a domain with sphere-shaped cavities of radius $\varepsilon^3$ located at the vertices of the cubic lattice with period $2\varepsilon$.

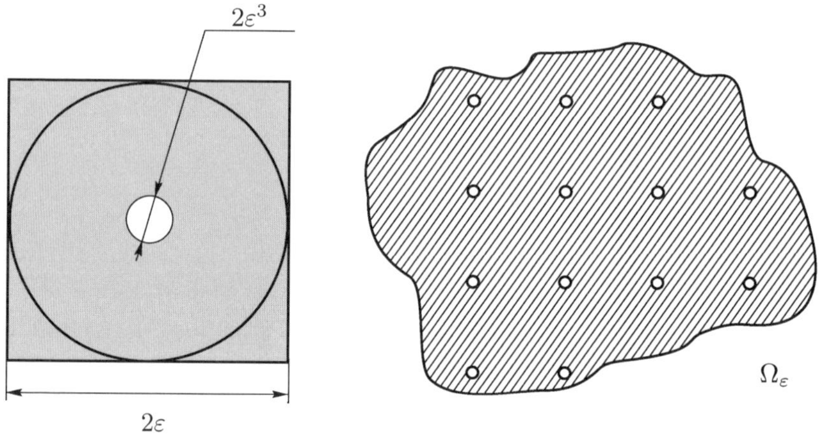


FIGURE 16.1

Consider the boundary value problems

$$\Delta u_\varepsilon = f(x) \quad \text{in } \Omega_\varepsilon, \qquad u_\varepsilon \in \overset{\circ}{H}{}^1(\Omega_\varepsilon), \tag{16.1}$$

$$(\Delta + \mu)u = f(x) \quad \text{in } \Omega, \qquad u \in \overset{\circ}{H}{}^1(\Omega), \quad \mu = -\pi/2. \tag{16.2}$$

We show that the solution $u_\varepsilon$ of the problem (16.1) is close to the solution $u$ of the problem (16.2) for small $\varepsilon$, i.e., the problem (16.2) can be regarded as the homogenized problem (16.1). In such a situation, an interesting effect is observed: the leading terms of the homogenized operator remains unchanged, whereas an additional potential $\mu I$ appears, where $I$ is the identity operator. This effect was first observed and studied in detail in [**85**]. Methods of [**85**], allow one to study models with arbitrary (nonperiodic) location of holes. The arguments of [**85**] use mainly the method of potentials for solutions to boundary value problems. In this section, we consider a special case where the holes are located periodically and each hole is of spherical shape. In this situation, the proof of the homogenization theorem is considerably simpler and we can derive sharp estimates for the convergence rate. At present, there are effective methods, not using the potential technique, that can be applied in the case of a periodic or nonperiodic location of holes (cf., for example, [**45, 105**]).

THEOREM 16.1. *Let $f \in L_2(\Omega)$. Then the solutions to the problems* (16.1) *and* (16.2) *satisfy the limit relations*

$$\lim_{\varepsilon\to 0} \|u_\varepsilon - w_\varepsilon u\|_{H^1(\Omega_\varepsilon)} = 0, \quad \lim_{\varepsilon\to 0} \|u_\varepsilon - u\|_{L_2(\Omega_\varepsilon)} = 0,$$

*where $w_\varepsilon(x)$ is a $2\varepsilon$-periodic function defined explicitly by the formula*

$$w_\varepsilon = \begin{cases} 0, & x \in \bigcup_{z\in\mathbb{Z}^3}(\varepsilon^3 G_0 + 2\varepsilon z), \\ 1, & x \in \mathbb{R}^3 \setminus \bigcup_{z\in\mathbb{Z}^3} \varepsilon(G_0 + 2z), \\ \dfrac{\left(\frac{1}{r} - \frac{1}{\varepsilon^3}\right)}{\left(\frac{1}{\varepsilon} - \frac{1}{\varepsilon^3}\right)}, & x \in \bigcup_{z\in\mathbb{Z}^3}(\varepsilon G_0 \setminus \varepsilon^3 G_0 + 2\varepsilon z). \end{cases} \tag{16.3}$$

*If $f \in C^\alpha(\overline{\Omega})$ with some $\alpha > 0$, then*

$$\|u_\varepsilon - w_\varepsilon u\|_{H^1(\Omega_\varepsilon)} \le C\varepsilon \|f\|_{C^\alpha(\overline{\Omega})}, \quad \|u_\varepsilon - u\|_{L_2(\Omega_\varepsilon)} \le C\varepsilon \|f\|_{C^\alpha(\overline{\Omega})},$$

*where $C$ is independent of $\varepsilon$ and $f(x)$.*

PROOF. It suffices to consider a Hölder continuous function $f$. In the general case, the required assertions can be obtained by approximating $f$ and using *a priori* estimates. The proof is based on the following exercises.

EXERCISE 16.1. Prove that $\|w_\varepsilon - 1\|_{L_2(\Omega_\varepsilon)} \to 0$ as $\varepsilon \to 0$ and $\|w_\varepsilon - 1\|_{L_2(\Omega_\varepsilon)} \le C\varepsilon^2$, where $C$ is independent of $\varepsilon$.

EXERCISE 16.2. Prove that for any $u \in \overset{\circ}{H}{}^1(\Omega)$,

$$\|(\Delta w_\varepsilon - \mu)u\|_{H^{-1}(\Omega_\varepsilon)} \le C\varepsilon \|u\|_{\overset{\circ}{H}{}^1(\Omega)},$$

where the constant $C > 0$ is independent of $\varepsilon$ and $u(x)$.

*Hint.* Use the explicit form of $w_\varepsilon(x)$ and the definition of the $H^{-1}(\Omega_\varepsilon)$-norm.

Consider the identity

$$\begin{aligned} \Delta(u_\varepsilon - w_\varepsilon u) &= f - \Delta u w_\varepsilon - 2(\nabla u, \nabla w_\varepsilon) - u\Delta w_\varepsilon \\ &= f(1 - w_\varepsilon) - (\Delta w_\varepsilon - \mu)u + \mu(w_\varepsilon - 1)u - 2(\nabla u, \nabla w_\varepsilon). \end{aligned}$$

The right-hand side consists of four terms. Therefore, it suffices to estimate each term in the $H^{-1}(\Omega_\varepsilon)$-norm via the norm $\varepsilon C\|f\|_{C^\alpha(\overline{\Omega})}$. This can be easily done using the estimates (16.1), (16.2) and the Schauder estimates $\|u\|_{C^{\alpha+2}(\overline{\Omega})} \le K\|f\|_{C^\alpha(\overline{\Omega})}$ for solutions to the boundary value problem (16.2) (cf., for example, [**65**]). Here, $\alpha > 0$, $K > 0$ are constants and $K$ is independent of $f(x)$. A detailed proof is left to the reader.

The second inequality follows from the first and the definition of $w_\varepsilon$. □

**16.2. Homogenization of a boundary value problem in a perforated domain with oscillating third boundary condition on the boundary of cavities.** Consider the elliptic boundary value problem in a perforated domain with the Fourier boundary condition (the condition of the third type) on the boundary (surface) of perforation. Under this choice of the boundary condition, a nonzero potential may appear in the homogenized equation.

We emphasize that the $(d-1)$-dimensional volume of the perforation surface unboundedly increases under homothetic contractions of a perforated medium, i.e., if the period converges to zero. Therefore, the problems with the third boundary condition on the boundary of perforation may show degeneration or unbounded growth of the solutions (depending on the sign of the coefficient in the Fourier condition). However, this phenomenon does not occur if the coefficient of the boundary operator asymptotically vanishes as the small parameter tends to zero, or has zero average over the perforation surface.

We consider both cases. In the first case, the potential appearing in the limit operator depends only on the average of the coefficient over the perforation surface, whereas the oscillation of this coefficient about this average does not contribute to the limit operator (cf. [**13, 39**]). In the second case (cf. [**14**]), the well-posedness of the boundary value problem does not follow directly from the positive definiteness of the differential operator. However, the problem will be well posed under certain additional assumptions. By [**111**], for a sufficiently small microstructure period, the well-posedness of the problem follows from the positive definiteness of the formally homogenized operator. In particular, estimates for the potential in the homogenized equation play an important role. Here, we propose a method for obtaining such estimates based on an auxiliary Steklov type problem.

16.2.1. *Statement of the problem.* Let $\Omega$ be a smooth bounded domain in $\mathbb{R}^d$, $d \geq 2$. We set

$$J^\varepsilon = \{j \in \mathbb{Z}^d : \operatorname{dist}(\varepsilon j, \partial\Omega) \geq \varepsilon\sqrt{d}\},$$
$$\square \equiv \{\xi : -1/2 < \xi_j < 1/2,\ j = 1, \dots, d\}.$$

Let $F(\xi)$ be a smooth 1-periodic function such that $F(\xi)\big|_{\xi\in\partial\square} \geq \text{const} > 0$, $F(0) = -1$, and $\nabla_\xi F \neq 0$ for $\xi \in \square\backslash\{0\}$. We set

$$Q_j^\varepsilon = \{x \in \varepsilon(\square + j) : F(x/\varepsilon) \leq 0\}$$

and introduce the perforated domain $\Omega^\varepsilon = \Omega\backslash\bigcup_{j\in J^\varepsilon} Q_j^\varepsilon$. By construction, $\partial\Omega^\varepsilon$ consists of $\partial\Omega$ and the boundary of the set $S_\varepsilon \subset \Omega$, $S_\varepsilon = (\partial\Omega^\varepsilon) \cap \Omega$.

We introduce the following notation: $Q = \{\xi : -1/2 < \xi_j < 1/2,\ j = 1, \dots, d,$ $F(\xi) \leq 0\}$ is an inclusion, $S = \{\xi : F(\xi) = 0\}$ is the boundary of $Q$, and $\nu$ is the unit outward normal to $S$ in the slow coordinates. Consider the problem

$$\begin{aligned}
&-\mathcal{L}_\varepsilon u_\varepsilon := \frac{\partial}{\partial x_k}\left(a_{kj}\left(\frac{x}{\varepsilon}\right)\frac{\partial u_\varepsilon}{\partial x_j}\right) = f(x) \quad \text{in } \Omega^\varepsilon,\\
&\frac{\partial u_\varepsilon}{\partial \gamma} + p\left(\frac{x}{\varepsilon}\right)u_\varepsilon + \varepsilon q\left(\frac{x}{\varepsilon}\right)u_\varepsilon = 0 \quad \text{on } S_\varepsilon,\\
&u_\varepsilon = 0 \quad \text{on } \partial\Omega,
\end{aligned} \tag{16.4}$$

where

$$\frac{\partial u_\varepsilon}{\partial \gamma} := \nu_k^\varepsilon a_{kj}\left(\frac{x}{\varepsilon}\right)\frac{\partial u_\varepsilon}{\partial x_j},$$

and $\nu^\varepsilon = (\nu_1^\varepsilon, \dots, \nu_d^\varepsilon)$ is the unit outward normal to the boundary of the inclusion. Assume that the matrix $\{a_{kj}(\xi)\}$ is symmetric and positive definite, i.e., $\kappa_1\eta^2 \leq a_{kj}\eta_k\eta_j \leq \kappa_2\eta^2$ for any vector $\eta \in \mathbb{R}^d$, where $\kappa_1$ and $\kappa_2$ are positive constants and $a_{kj}(\xi)$, $p(\xi)$, and $q(\xi)$ are 1-periodic functions with respect to $\xi \in \mathbb{R}^d$. Let

$$\langle p(\xi)\rangle_S = 0, \tag{16.5}$$

where $\langle\cdot\rangle_S := \int_S \cdot\, d\sigma$ and $\sigma$ is the $(d-1)$-dimensional volume element on the surface $S$. The operator of the problem (16.4) is denoted by $A^\varepsilon$.

16.2.2. *Formal asymptotic analysis.* We look for a solution in the form of an asymptotic series

$$u_\varepsilon(x) \sim u_0(x) + \varepsilon u_1(x,\xi) + \varepsilon^2 u_2(x,\xi) + \cdots, \qquad \xi = x/\varepsilon, \tag{16.6}$$

where the functions $u_i(x,\xi)$ are periodic with respect to $\xi$. We set

$$-\mathcal{L}_{xx}\varphi(x,\xi) := \frac{\partial}{\partial x_k}\left(a_{kj}(\xi)\frac{\partial\varphi(x,\xi)}{\partial x_j}\right), \quad \frac{\partial\varphi(x,\xi)}{\partial\gamma_x} := a_{kj}(\xi)\frac{\partial\varphi(x,\xi)}{\partial x_j}\nu_k,$$

$$-\mathcal{L}_{x\xi}\varphi(x,\xi) := \frac{\partial}{\partial x_k}\left(a_{kj}(\xi)\frac{\partial\varphi(x,\xi)}{\partial\xi_j}\right), \quad -\mathcal{L}_{\xi x}\varphi(x,\xi) := \frac{\partial}{\partial\xi_k}\left(a_{kj}(\xi)\frac{\partial\varphi(x,\xi)}{\partial x_j}\right),$$

$$-\mathcal{L}_{\xi\xi}\varphi(x,\xi) := \frac{\partial}{\partial\xi_k}\left(a_{kj}(\xi)\frac{\partial\varphi(x,\xi)}{\partial\xi_j}\right), \quad \frac{\partial\varphi(x,\xi)}{\partial\gamma_\xi} := a_{kj}(\xi)\frac{\partial\varphi(x,\xi)}{\partial\xi_j}\nu_k.$$

Substituting (16.6) in (16.4) and collecting terms with equal power of $\varepsilon$, we obtain a recurrent sequence of problems, the first of which has the form

$$\begin{aligned} &\mathcal{L}_{\xi\xi}u_1 + \mathcal{L}_{\xi x}u_0 = 0 \quad \text{in } \square\backslash Q, \\ &\frac{\partial u_1}{\partial\gamma_\xi} + \frac{\partial u_0}{\partial\gamma_x} + p(\xi)u_0 = 0 \quad \text{on } S, \end{aligned} \tag{16.7}$$

where $x \in \Omega$ plays the role of a parameter. We write the integral identity corresponding to problem (16.7):

$$\int_{\square\backslash Q} a_{kj}\frac{\partial u_1}{\partial\xi_j}\frac{\partial v}{\partial\xi_k}d\xi + \int_{\square\backslash Q} a_{kj}\frac{\partial u_0}{\partial x_j}\frac{\partial v}{\partial\xi_k}d\xi + \int_S p(\xi)u_0 v\, d\sigma = 0, \tag{16.8}$$

where $v \in H^1_{\text{per}}(\square\backslash Q)$. The integral identity suggests the structure of $u_1(x,\xi)$:

$$u_1(x,\xi) = M(\xi)u_0(x) + N_i(\xi)\frac{\partial u_0(x)}{\partial x_i}. \tag{16.9}$$

Substituting (16.9) in (16.8) and combining the corresponding terms, we obtain the following problem for $N_i(\xi)$:

$$\int_{\square\backslash Q} a_{kj}\frac{\partial N_i}{\partial\xi_j}\frac{\partial v}{\partial\xi_k}d\xi + \int_{\square\backslash Q} a_{ki}\frac{\partial v}{\partial\xi_k}d\xi = 0 \tag{16.10}$$

or, in the classical formulation,

$$\begin{aligned} &\mathcal{L}_{\xi\xi}\left(N_i(\xi) + \xi_i\right) = 0 \quad \text{in } \square\backslash Q, \\ &\frac{\partial N_i(\xi)}{\partial\gamma_\xi} = -a_{ki}(\xi)\nu_k \quad \text{on } S, \end{aligned}$$

where $i = 1,\dots,d$, and the problem for $M(\xi)$,

$$\int_{\square\backslash Q} a_{kj}\frac{\partial M}{\partial\xi_j}\frac{\partial v}{\partial\xi_k}d\xi + \int_S p(\xi)v\, d\sigma = 0, \tag{16.11}$$

or, in the classical formulation,

$$\begin{aligned} &\mathcal{L}_{\xi\xi}M(\xi) = 0 \quad \text{in } \square\backslash Q, \\ &\frac{\partial M(\xi)}{\partial\gamma_\xi} = -p(\xi) \quad \text{on } S. \end{aligned}$$

The compatibility condition can be easily verified by integrating by parts for the problem (16.10) and directly follows from (16.5) for the problem (16.11). We note that $M(\xi)$ and $N_i(\xi)$ are determined up to an additive constant. Hence it is natural to use the equalities $\langle M\rangle_{\square\backslash Q} = \langle N_i\rangle_{\square\backslash Q} = 0$, $i = 1, \dots, d$, for normalization. We assume that these conditions are satisfied.

The next problem of the recurrent chain has the form

$$\begin{aligned} &\mathcal{L}_{\xi\xi}u_2 + \mathcal{L}_{x\xi}u_1 + \mathcal{L}_{\xi x}u_1 + \mathcal{L}_{xx}u_0 = -f \quad \text{in } \square\backslash Q, \\ &\frac{\partial u_2}{\partial \gamma_\xi} + \frac{\partial u_1}{\partial \gamma_x} + p(\xi)u_1 + q(\xi)u_0 = 0 \quad \text{on } S. \end{aligned} \tag{16.12}$$

We need the following assertion.

LEMMA 16.1. *The functions $M(\xi)$ and $N_k(\xi)$ are connected by the integral equalities*

$$\int_{\square\backslash Q} a_{kj}\frac{\partial M}{\partial \xi_j}\,d\xi - \int_S pN_k\,d\sigma = 0, \quad k = 1, 2, \dots, d.$$

PROOF. Substituting $N_i(\xi)$ for a test function in (16.11), we find

$$\int_{\square\backslash Q} a_{kj}\frac{\partial M}{\partial \xi_j}\frac{\partial N_i}{\partial \xi_k}d\xi + \int_S p(\xi)N_i\,d\sigma = 0.$$

Similarly, substituting $M(\xi)$ for a test function in (16.10), we obtain

$$\int_{\square\backslash Q} a_{kj}\frac{\partial N_i}{\partial \xi_j}\frac{\partial M}{\partial \xi_k}d\xi + \int_{\square\backslash Q} a_{ki}\frac{\partial M}{\partial \xi_k}d\xi = 0.$$

Since the matrix $\{a_{kj}\}$ is symmetric, we have

$$\int_{\square\backslash Q} a_{kj}\frac{\partial M}{\partial \xi_j}\,d\xi = \int_S pN_k\,d\sigma.$$

The lemma is proved. $\square$

We write the integral identity corresponding to the problem (16.12):

$$\begin{aligned} &\int_{\square\backslash Q} a_{kj}\frac{\partial u_2}{\partial \xi_j}\frac{\partial v}{\partial \xi_k}d\xi + \int_{\square\backslash Q} a_{kj}\frac{\partial u_1}{\partial x_j}\frac{\partial v}{\partial \xi_k}d\xi + \int_S p(\xi)u_1 v\,d\sigma + u_0(x)\int_S q(\xi)v\,d\sigma \\ &- \int_{\square\backslash Q} a_{kj}\frac{\partial M}{\partial \xi_j}v\,d\xi\frac{\partial u_0}{\partial x_k} - \int_{\square\backslash Q}\left(a_{ij}\frac{\partial N_k}{\partial \xi_j} + a_{ik}\right)v\,d\xi\frac{\partial^2 u_0}{\partial x_i \partial x_k} + |\square\backslash Q|f = 0. \end{aligned}$$

The solvability condition for this problem is an equation for $u_0(x)$. It is the required homogenized (limit) equation. Using Lemma 16.1, we write it in the form

$$\widehat{a}_{kj}\frac{\partial^2 u_0(x)}{\partial x_k \partial x_j} - u_0(x)\left(\int_S q(\xi)\,d\sigma + \int_S p(\xi)M(\xi)\,d\sigma\right) = |\square\backslash Q|f(x), \tag{16.13}$$

where

$$\widehat{a}_{ik} := \int_{\square\backslash Q}\left(a_{ij}(\xi)\frac{\partial N_k(\xi)}{\partial \xi_j} + a_{ik}(\xi)\right)d\xi.$$

Thus, the homogenized problem has the form

$$\begin{aligned} &\widehat{a}_{kj}\frac{\partial^2 u_0(x)}{\partial x_k \partial x_j} - \langle q\rangle_S\, u_0(x) + m\,u_0(x) = |\square\backslash Q|f(x) \quad \text{in } \Omega, \\ &u_0(x) = 0 \quad \text{on } \partial\Omega, \end{aligned} \tag{16.14}$$

where $m := -\langle pM\rangle_S$. Denote by $\widehat{A}$ the operator of problem (16.14).

REMARK 16.1. The potential $m$ in the limit equation is always positive and thereby the presence of the potential $m$ can cause the loss of the coerciveness of the problem (16.14). It is easy to see that the problem (16.14) is well posed if $m - \langle q\rangle_S < \lambda_0$, where $\lambda_0$ is the first eigenvalue of the operator $\widehat{a}_{ij}\frac{\partial}{\partial x_i}\frac{\partial}{\partial x_j}$ in the space $\overset{\circ}{H}{}^1(\Omega)$.

16.2.3. *The main assertions and estimates.* We begin with an estimate for the homogenized potential in (16.14) and then formulate the main result of this section.

Consider the auxiliary spectral Steklov type problem

$$
\begin{aligned}
&\frac{\partial}{\partial \xi_k}\Big(a_{kj}(\xi)\frac{\partial \theta}{\partial \xi_j}\Big) = 0 \quad \text{in } \square\backslash Q, \\
&\frac{\partial \theta}{\partial \gamma} = \Upsilon\theta \quad \text{on } S, \quad \langle\theta\rangle_S = 0,
\end{aligned}
\tag{16.15}
$$

in the space of 1-periodic functions. Denote by $\Upsilon_1$ the first eigenvalue of this problem. It can be found as the minimum of the variational problem

$$
\Upsilon_1 = \inf \frac{a(\psi,\psi)}{\langle\psi^2\rangle_S},
$$

where the infimum is taken over $\psi \in H^1_{\rm per}(\square)\backslash\{0\}$ such that $\langle\psi\rangle_S = 0$ and

$$
a(u,v) := \int_{\square\backslash Q} a_{kj}\frac{\partial u}{\partial \xi_j}\frac{\partial v}{\partial \xi_k}\, d\xi.
$$

LEMMA 16.2. *The constant $m$ is positive. The following inequalities hold*:

$$
\langle p^2\rangle_S\frac{\langle p^2\rangle_S}{a(p,p)} \le m \le \frac{\langle p^2\rangle_S}{\Upsilon_1},
\tag{16.16}
$$

*where $p(\xi)$ is the coefficient in the boundary condition of the original problem.*

REMARK 16.2. Both inequalities in (16.16) become equalities for functions $p(\xi)$ in the eigenspace of problem (16.15) corresponding to the first eigenvalue $\Upsilon_1$.

PROOF OF LEMMA 16.2. Taking $M(\xi)$ for a test function in (16.11), we have

$$
\int_{\square\backslash Q} a_{kj}\frac{\partial M}{\partial \xi_j}\frac{\partial M}{\partial \xi_k}\,d\xi + \int_S p(\xi)M\,d\sigma = 0.
$$

Thus,

$$
m = -\langle pM\rangle_S = \Big\langle a_{kj}\frac{\partial M}{\partial \xi_j}\frac{\partial M}{\partial \xi_k}\Big\rangle_{\square\backslash Q} > 0
$$

if $M \neq 0$. Note that $M$ is identically equal to zero if and only if $p$ vanishes.

Consider the variational problem

$$
\inf_{\psi\in H^1_{\rm per}(\square)} H(\psi) \equiv \inf_{\psi\in H^1_{\rm per}(\square)}\{a(\psi,\psi) + 2\langle p\psi\rangle_S\}.
\tag{16.17}
$$

By (16.11), the minimum (16.17) is attained at the function $M$. Consequently,

$$
-\inf_{\psi\in H^1_{\rm per}(\square)}\{a(\psi,\psi) + 2\langle p\psi\rangle_S\} = -a(M,M) - 2\langle pM\rangle_S = -\langle pM\rangle_S = m.
$$

Substituting $\psi = -tp$ in $H(\psi)$, we find $H(-tp) = t^2a(p,p) - 2t\langle p^2\rangle_S$. To find the minimum of $H(-tp)$ over $t$, we equate the derivative of $H(-tp)$ with respect to

$t$ to zero. Then $0 = H_t'(-t_0p) = 2t_0a(p,p) - 2\langle p^2\rangle_S$. Further, $t_0 = \frac{\langle p^2\rangle_S}{a(p,p)}$ and, consequently,

$$H(-t_0p) = \frac{(\langle p^2\rangle_S)^2}{a(p,p)} - 2\frac{(\langle p^2\rangle_S)^2}{a(p,p)} = -\frac{(\langle p^2\rangle_S)^2}{a(p,p)}.$$

Thus, the first inequality in (16.16) is proved

Substituting $\psi = -t\varphi$ and performing a similar procedure, we find

$$H(-t_0\varphi) = -\frac{(\langle p\varphi\rangle_S)^2}{a(\varphi,\varphi)}.$$

Since

$$m = \sup_{\varphi\in H^1_{\mathrm{per}}(\square)} \frac{(\langle p\varphi\rangle_S)^2}{a(\varphi,\varphi)} \tag{16.18}$$

and the maximum is attained at $\varphi = M$, for any $\varphi$ we have $m \geq \frac{(\langle p\varphi\rangle_S)^2}{a(\varphi,\varphi)}$. From (16.18) it follows that

$$\frac{1}{m} = \inf_{\varphi\in H^1_{\mathrm{per}}(\square)\setminus\{0\}} \frac{a(\varphi,\varphi)}{(\langle p\varphi\rangle_S)^2} \geq \inf_{\substack{\varphi\in H^1_{\mathrm{per}}(\square)\setminus\{0\},\\ \langle\varphi\rangle_S=0}} \frac{a(\varphi,\varphi)}{\langle p^2\rangle_S\langle\varphi^2\rangle_S} = \frac{\Upsilon_1}{\langle p^2\rangle_S}.$$

Finally, $m \leq \frac{\langle p^2\rangle_S}{\Upsilon_1}$, where $\Upsilon_1$ is the first eigenvalue of the problem (16.15). Thus, we have proved the second inequality in (16.16). □

The following assertion justifies the constructed asymptotics of the solution to the problem (16.4) and the estimate for the remainder.

THEOREM 16.2. *Suppose that $f(x) \in H^1(\Omega)$, $p(\xi)$, $q(\xi)$ are 1-periodic $C^1$-functions, $u_0$ is the solution to the problem (16.14), $u_1$ has the form (16.9), and $M(\xi)$ and $N_i(\xi)$ are the above-constructed functions. Assume that*

$$m < \lambda_0 + \langle q\rangle_S, \tag{16.19}$$

*where $\lambda_0$ is defined in Remark 16.1. Then for all sufficiently small $\varepsilon$ the problem (16.4) has a unique solution $u_\varepsilon$ and $\|u_\varepsilon - u_0\|_{L_2(\Omega^\varepsilon)} \to 0$ as $\varepsilon \to 0$. Furthermore,*

$$\|u_0 + \varepsilon u_1 - u_\varepsilon\|_{H^1(\Omega^\varepsilon)} \leq K_1\sqrt{\varepsilon}, \tag{16.20}$$

*where the constant $K_1 > 0$ is independent of $\varepsilon$.*

16.2.4. *Preliminary lemmas.* The technical assertions below will be used in the further consideration. The proof of the first two lemmas can be found in [**13, 41**]. Therefore, we restrict ourselves to their formulations.

LEMMA 16.3. *Let $\zeta(x,\xi)$ be a $C^1(\Omega,S)$-function, 1-periodic with respect to $\xi$, and let*

$$Z(x) \equiv \int_S \zeta(x,\xi)\, d\sigma. \tag{16.21}$$

*Then*

$$\left|\frac{1}{|\square\setminus Q|}\int_{\Omega^\varepsilon} Z(x)\, u(x)v(x)\, dx - \varepsilon\int_{S_\varepsilon} \zeta\Big(x, \frac{x}{\varepsilon}\Big)\, u(x)v(x)\, d\sigma^\varepsilon\right| \leq C_2\varepsilon\|u\|_{H^1(\Omega^\varepsilon)}\|v\|_{H^1(\Omega^\varepsilon)} \tag{16.22}$$

*for any $u(x), v(x) \in H^1(\Omega^\varepsilon)$, where the constant $C_2$ is independent of $\varepsilon$.*

REMARK 16.3. Similar estimates were obtained in [**109**]–[**111**].

The following assertion is a slight modification of Lemma 1.1.

LEMMA 16.4. *Let $\zeta(x,\xi)$ be a function of class $C^1(\Omega, L_\infty(\square\backslash Q))$, 1-periodic with respect to $\xi$, and let*

$$\int_{\square\backslash Q} \zeta(x,\xi)\, d\xi \equiv 0. \tag{16.23}$$

*Then*

$$\left|\frac{1}{|\square\backslash Q|}\int_{\Omega^\varepsilon} \zeta\Big(x, \frac{x}{\varepsilon}\Big)\, u(x)v(x)\, dx\right| \le C_3\varepsilon\|u\|_{H^1(\Omega^\varepsilon)}\|v\|_{H^1(\Omega^\varepsilon)}$$

*for any $u(x), v(x) \in H^1(\Omega^\varepsilon)$, where the constant $C_3$ is independent of $\varepsilon$.*

Let $\lambda_0$ be the same as in Remark 16.1.

LEMMA 16.5. *If $m < \lambda_0 + \langle q\rangle_S$, then the problem* (16.14) *is coercive.*

PROOF. Using the variational properties of the eigenvalues of the operator $\widehat{A}$, we find that

$$\begin{aligned} \inf_{\substack{v\in \overset{\circ}{H}{}^1(\Omega),\\ \|v\|_{L_2(\Omega)}=1}} (-\widehat{A}v, v)_{L_2(\Omega)} &= \inf_{\substack{v\in \overset{\circ}{H}{}^1(\Omega),\\ \|v\|_{L_2(\Omega)}=1}} \int_\Omega \widehat{a}_{ij}\frac{\partial v}{\partial x_i}\frac{\partial v}{\partial x_j} + (\langle q\rangle_S - m)\, v^2\, dx \\ &= \inf_{\substack{v\in \overset{\circ}{H}{}^1(\Omega),\\ \|v\|_{L_2(\Omega)}=1}} \int_\Omega \widehat{a}_{ij}\frac{\partial v}{\partial x_i}\frac{\partial v}{\partial x_j}\, dx + (\langle q\rangle_S - m) = \lambda_0 + \langle q\rangle_S - m. \end{aligned}$$

Thus, $(-\widehat{A}v, v)_{L_2(\Omega)} \ge C_4\|v\|^2_{L_2(\Omega)}$, where $C_4 > 0$, which was to be proved. $\square$

LEMMA 16.6. *If $p \in C^1(S)$ and $\langle p\rangle_S = 0$, then for any $u(\xi), v(\xi) \in H^1(\square)$ (without the periodicity condition),*

$$\left|\int_S p(\xi)\, u(\xi)v(\xi)\, d\sigma\right| \le C_5(\|\nabla u\|_{L_2(\square)}\|v\|_{L_2(\square)} + \|u\|_{L_2(\square)}\|\nabla v\|_{L_2(\square)}). \tag{16.24}$$

PROOF. By assumption, the problem

$$\begin{aligned} &\Delta_\xi\Psi(\xi) = 0 \quad \text{in } \square\backslash Q, \\ &\frac{\partial\Psi}{\partial n} = p(\xi) \quad \text{on } S, \qquad \frac{\partial\Psi}{\partial n} = 0 \quad \text{on } \partial\square \end{aligned} \tag{16.25}$$

is solvable. Furthermore, the solution is unique up to an additive constant.

We multiply the equation in (16.25) by $uv$, where $u(\xi), v(\xi) \in H^1(\square)$, and integrate over $\square\backslash Q$. Integrating by parts, we obtain

$$\begin{aligned} &\left|\int_S p(\xi)\, u(\xi)v(\xi)\, d\sigma\right| \\ &\quad = \left|\int_{\square\backslash Q} \Delta_\xi\Psi(\xi)u(\xi)v(\xi)\, d\xi - \int_S p(\xi)\, u(\xi)v(\xi)\, d\sigma\right| \\ &\quad \le \left|\int_{\square\backslash Q} (\nabla_\xi\Psi(\xi), \nabla_\xi(u(\xi)v(\xi)))\, d\xi\right| \\ &\quad \le C_5(\|\nabla u\|_{L_2(\square)}\|v\|_{L_2(\square)} + \|u\|_{L_2(\square)}\|\nabla v\|_{L_2(\square)}). \end{aligned} \tag{16.26}$$

The lemma is proved. $\square$

LEMMA 16.7. *The coerciveness of the homogenized problem* (16.14) *implies the coerciveness of the problem* (16.4) *for a sufficiently small* $\varepsilon$.

PROOF. We show that

$$\int_{S_\varepsilon} p\Big(\frac{x}{\varepsilon}\Big)u^2(x)\,ds \le C_6\Big(\alpha\int_{\Omega^\varepsilon}|\nabla u|^2\,dx + \frac{1}{\alpha}\int_{\Omega^\varepsilon}u^2\,dx\Big) \tag{16.27}$$

for any $\alpha > 0$. Indeed, by Lemma 16.6,

$$\left|\int_S p(\xi)\,u^2(\xi)\,d\sigma\right| \le 2C_5\|\nabla u\|_{L_2(\square)}\|u\|_{L_2(\square)} \le C_7\Big(\frac{\alpha}{\varepsilon}\int_\square|\nabla u|^2\,d\xi + \frac{\varepsilon}{\alpha}\int_\square u^2\,d\xi\Big).$$

Passing to the coordinates $x = \varepsilon\xi$ and taking the sum over all cells, we obtain the required inequality (cf. also [**109, 110**]).

We prove that there is a sufficiently large number $\Lambda$, independent of $\varepsilon$, such that the boundary value problem

$$\begin{aligned} &\mathcal{L}_\varepsilon u_\varepsilon + \Lambda u_\varepsilon = -f(x) \quad \text{in } \Omega^\varepsilon,\\ &\frac{\partial u_\varepsilon}{\partial\gamma} + p\Big(\frac{x}{\varepsilon}\Big)u_\varepsilon + \varepsilon q\Big(\frac{x}{\varepsilon}\Big)u_\varepsilon = 0 \quad \text{on } S_\varepsilon,\\ &u_\varepsilon = 0 \quad \text{on } \partial\Omega \end{aligned} \tag{16.28}$$

is well posed for any $\varepsilon > 0$.

Using Lemmas 16.3 and 16.6, we find that

$$\begin{aligned} &\int_{\Omega^\varepsilon} a_{ik}\Big(\frac{x}{\varepsilon}\Big)\frac{\partial v}{\partial x_k}\frac{\partial v}{\partial x_i}dx + \int_{S_\varepsilon} p\Big(\frac{x}{\varepsilon}\Big)v^2 ds + \varepsilon\int_{S_\varepsilon} q\Big(\frac{x}{\varepsilon}\Big)v^2 ds + \int_{\Omega^\varepsilon}\Lambda v^2\,dx\\ &\quad\ge \kappa_1\|\nabla v\|^2_{L_2(\Omega^\varepsilon)} - (C_6\alpha + O(\varepsilon))\|\nabla u\|^2_{L_2(\Omega^\varepsilon)} + \Big(\langle q\rangle_S - \frac{C_6}{\alpha} + \Lambda\Big)\|v\|^2_{L_2(\Omega^\varepsilon)}. \end{aligned} \tag{16.29}$$

Choosing a sufficiently small $\alpha$ and, after that, sufficiently large $\Lambda$, we achieve the positive definiteness of the quadratic form on the right-hand side of (16.29). Hence we arrive at the required estimate.

Consider the spectral problem

$$(-A^\varepsilon + \Lambda)^{-1}u_\varepsilon^k = \lambda_\varepsilon^k u_\varepsilon^k, \tag{16.30}$$

$$(-\widehat{A} + \Lambda)^{-1}u^k = \lambda^k u^k, \tag{16.31}$$

where $A^\varepsilon$ is the operator of the problem (16.4) and $\widehat{A}$ is the operator of the homogenized problem (16.14). By $(-A^\varepsilon + \Lambda)^{-1}$ we understand an operator sending an arbitrary function $f$ in $L_2(\Omega^\varepsilon)$ to the solution $u_\varepsilon \in H^1(\Omega^\varepsilon)$ to the problem (16.28). The operator $(-\widehat{A}+\Lambda)^{-1}$ is defined in a similar way. Taking into account the coerciveness property, it is easy to verify that the operators $(-A^\varepsilon + \Lambda)^{-1}$ and $(-\widehat{A} + \Lambda)^{-1}$ satisfy conditions C1–C4 of Theorem 1.9 in [**103**, Chapter III] concerning the convergence of the spectra of a family of operators defined in different spaces. In particular, this theorem implies that $\lambda_\varepsilon^0 \to \lambda^0$ as $\varepsilon \to 0$. Denote by $\mu_\varepsilon^0$ and $\mu^0$ the first eigenvalues of the operators $-A^\varepsilon$ and $-\widehat{A}$ respectively. Then $\mu_\varepsilon^0 \equiv -\Lambda + \frac{1}{\lambda_\varepsilon^0} \to \mu^0 \equiv -\Lambda + \frac{1}{\lambda^0}$ as $\varepsilon \to 0$. By the variational principle, the positive definiteness of $-\widehat{A}$ implies the positive definiteness of $-A^\varepsilon$ for all sufficiently small $\varepsilon$. $\square$

REMARK 16.4. The trick involving the passage to an auxiliary problem by displacement along the spectrum is used in [**111**, Lemma 3], and the coerciveness of the problem (16.28) is proved by a method similar to the method of [**111**].

16.2.5. *Justification of asymptotics.*

PROOF OF THEOREM 16.2. In order to estimate the $H^1$-norm of the remainder $\|u_0+\varepsilon u_1-u_\varepsilon\|_{H^1(\Omega^\varepsilon)}$, we substitute the expression $z_\varepsilon(x,\frac{x}{\varepsilon})=u_0(x)+\varepsilon u_1(x,\frac{x}{\varepsilon})-u_\varepsilon(x)$ in (16.4). This yields

$$\begin{aligned}\mathcal{L}_\varepsilon\Big(z_\varepsilon\Big(x,\frac{x}{\varepsilon}\Big)\Big)&=\frac{1}{\varepsilon}\mathcal{L}_{\xi x}u_0(x)\big|_{\xi=x/\varepsilon}+\mathcal{L}_{xx}u_0(x)+\varepsilon\mathcal{L}_{xx}u_1(x,\xi)\big|_{\xi=x/\varepsilon}\\&\quad+\mathcal{L}_{\xi x}u_1(x,\xi)\big|_{\xi=x/\varepsilon}+\mathcal{L}_{x\xi}u_1(x,\xi)\big|_{\xi=x/\varepsilon}\\&\quad+\frac{1}{\varepsilon}\mathcal{L}_{\xi\xi}u_1(x,\xi)\big|_{\xi=x/\varepsilon}-\mathcal{L}_\varepsilon u_\varepsilon(x).\end{aligned}\tag{16.32}$$

Taking into account the relations

$$\begin{aligned}&\mathcal{L}_{\xi\xi}u_1(x,\xi)=-\mathcal{L}_{\xi x}u_0(x),\quad \mathcal{L}_\varepsilon u_\varepsilon(x)=-f(x),\\&-\mathcal{L}_{\xi x}u_1(x,\xi)=\frac{\partial}{\partial\xi_i}\Big(a_{ij}(\xi)\frac{\partial u_0(x)}{\partial x_j}M(\xi)\Big)+\frac{\partial}{\partial\xi_i}\Big(a_{ij}(\xi)\frac{\partial^2u_0(x)}{\partial x_j\partial x_k}N_k(\xi)\Big),\\&-\mathcal{L}_{x\xi}u_1(x,\xi)=a_{ij}(\xi)\frac{\partial u_0(x)}{\partial x_i}\frac{\partial M(\xi)}{\partial\xi_j}+a_{ij}(\xi)\frac{\partial^2u_0(x)}{\partial x_i\,\partial x_k}\frac{\partial N_k(\xi)}{\partial\xi_j},\end{aligned}\tag{16.33}$$

and

$$\widehat{a}_{kj}\frac{\partial^2u_0(x)}{\partial x_k\partial x_j}-\langle q\rangle_S\,u_0(x)+m\,u_0(x)=|\square\backslash Q|f(x)\quad\text{in }\Omega,\tag{16.34}$$

we can write (16.32) in $\Omega^\varepsilon$ as follows:

$$\begin{aligned}-\mathcal{L}_\varepsilon\Big(z_\varepsilon\Big(x,\frac{x}{\varepsilon}\Big)\Big)&=-\varepsilon\mathcal{L}_{xx}u_1(x,\xi)\big|_{\xi=x/\varepsilon}\\&\quad+\frac{\partial}{\partial\xi_i}\Big(a_{ij}(\xi)\frac{\partial u_0(x)}{\partial x_j}M(\xi)\Big)\Big|_{\xi=x/\varepsilon}\\&\quad+\frac{\partial}{\partial\xi_i}\Big(a_{ij}(\xi)\frac{\partial^2u_0(x)}{\partial x_j\,\partial x_k}N_k(\xi)\Big)\Big|_{\xi=x/\varepsilon}\\&\quad+a_{ij}(\xi)\frac{\partial u_0(x)}{\partial x_i}\frac{\partial M(\xi)}{\partial\xi_j}\Big|_{\xi=x/\varepsilon}\\&\quad+a_{ij}(\xi)\frac{\partial^2u_0(x)}{\partial x_i\,\partial x_k}\frac{\partial N_k(\xi)}{\partial\xi_j}\Big|_{\xi=x/\varepsilon}\\&\quad-\mathcal{L}_{xx}u_0(x)-\frac{1}{|\square\backslash Q|}\widehat{a}_{kj}\frac{\partial^2u_0(x)}{\partial x_k\partial x_j}+\frac{(\langle q\rangle_S-m)}{|\square\backslash Q|}\,u_0(x).\end{aligned}\tag{16.35}$$

Similarly, setting

$$\begin{aligned}\frac{\partial}{\partial\gamma_x}g\Big(x,\frac{x}{\varepsilon}\Big)&=\nu^\varepsilon a\Big(\frac{x}{\varepsilon}\Big)\nabla_x g(x,\xi)\Big|_{\xi=x/\varepsilon},\\\frac{\partial}{\partial\gamma_\xi}g\Big(x,\frac{x}{\varepsilon}\Big)&=\nu^\varepsilon a\Big(\frac{x}{\varepsilon}\Big)\nabla_\xi g(x,\xi)\Big|_{\xi=x/\varepsilon},\end{aligned}$$

on $S_\varepsilon$, we have

$$\begin{aligned}
&\frac{\partial z_\varepsilon(x, \frac{x}{\varepsilon})}{\partial \gamma} - p\Big(\frac{x}{\varepsilon}\Big) z_\varepsilon - \varepsilon q\Big(\frac{x}{\varepsilon}\Big) z_\varepsilon \\
&\quad = -\frac{\partial u_\varepsilon(x)}{\partial \gamma} + \frac{\partial u_0(x)}{\partial \gamma} + \varepsilon \frac{\partial}{\partial \gamma_x} u_1\big(x, \frac{x}{\varepsilon}\big) + \frac{\partial}{\partial \gamma_\xi} u_1\Big(x, \frac{x}{\varepsilon}\Big) \\
&\quad = p\Big(\frac{x}{\varepsilon}\Big) u_\varepsilon(x) + \varepsilon q\Big(\frac{x}{\varepsilon}\Big) u_\varepsilon(x) + \frac{\partial u_0(x)}{\partial \gamma} + \varepsilon \frac{\partial}{\partial \gamma_x} u_1\Big(x, \frac{x}{\varepsilon}\Big) \\
&\qquad + u_0(x) \frac{\partial}{\partial \gamma_\xi} M\Big(\frac{x}{\varepsilon}\Big) + \frac{\partial u_0(x)}{\partial x_i} \frac{\partial}{\partial \gamma_\xi} N_i\Big(\frac{x}{\varepsilon}\Big).
\end{aligned}$$

Multiplying (16.35) by $v(x) \in \overset{\circ}{H}{}^1(\Omega)$ and integrating over $\Omega^\varepsilon$, we find that

$$\begin{aligned}
&-\int_{\Omega^\varepsilon} \mathcal{L}_\varepsilon\Big(z_\varepsilon\Big(x, \frac{x}{\varepsilon}\Big)\Big) v(x)\, dx \\
&\quad = -\varepsilon \int_{\Omega^\varepsilon} \mathcal{L}_{xx} u_1(x, \xi)\Big|_{\xi = x/\varepsilon} v(x)\, dx \\
&\qquad + \int_{\Omega^\varepsilon} \frac{\partial}{\partial \xi_i}\Big(a_{ij}(\xi) \frac{\partial u_0(x)}{\partial x_j} M(\xi)\Big)\Big|_{\xi = x/\varepsilon} v(x)\, dx \\
&\qquad + \int_{\Omega^\varepsilon} \frac{\partial}{\partial \xi_i}\Big(a_{ij}(\xi) \frac{\partial^2 u_0(x)}{\partial x_j\, \partial x_k} N_k(\xi)\Big)\Big|_{\xi = x/\varepsilon} v(x)\, dx \\
&\qquad + \int_{\Omega^\varepsilon} a_{ij}(\xi) \frac{\partial u_0(x)}{\partial x_i} \frac{\partial M(\xi)}{\partial \xi_j}\Big|_{\xi = x/\varepsilon} v(x)\, dx \\
&\qquad + \int_{\Omega^\varepsilon} a_{ij}(\xi) \frac{\partial^2 u_0(x)}{\partial x_i\, \partial x_k} \frac{\partial N_k(\xi)}{\partial \xi_j}\Big|_{\xi = x/\varepsilon} v(x)\, dx \\
&\qquad - \int_{\Omega^\varepsilon} \mathcal{L}_\varepsilon u_0(x) v(x)\, dx - \frac{1}{|\square \backslash Q|} \int_{\Omega^\varepsilon} \widehat{a}_{kj} \frac{\partial^2 u_0(x)}{\partial x_k \partial x_j} v(x)\, dx \\
&\qquad + \frac{1}{|\square \backslash Q|} \int_{\Omega^\varepsilon} t(\langle q \rangle_S - m)\, u_0(x)\, v(x)\, dx.
\end{aligned} \tag{16.36}$$

On the other hand, using the Green formula, it is possible to transform the left-hand side of (16.36) as follows:

$$\begin{aligned}
&-\int_{\Omega^\varepsilon} \mathcal{L}_\varepsilon\Big(z_\varepsilon\Big(x, \frac{x}{\varepsilon}\Big)\Big) v(x)\, dx \\
&\quad = \int_{S_\varepsilon} \frac{\partial z_\varepsilon}{\partial \gamma} v(x)\, ds - \int_{\Omega^\varepsilon} \nabla z_\varepsilon \nabla v(x)\, dx \\
&\quad = \int_{S_\varepsilon} p\Big(\frac{x}{\varepsilon}\Big) u_\varepsilon(x)\, v(x)\, ds + \varepsilon \int_{S_\varepsilon} q\Big(\frac{x}{\varepsilon}\Big) u_\varepsilon(x)\, v(x)\, ds \\
&\qquad + \int_{S_\varepsilon} \frac{\partial u_0(x)}{\partial \gamma_x} v(x)\, ds + \varepsilon \int_{S_\varepsilon} \frac{\partial u_1(x, \xi)}{\partial \gamma_x}\Big|_{\xi = x/\varepsilon} v(x)\, ds \\
&\qquad + \int_{S_\varepsilon} \Big(u_0(x) \frac{\partial M(\xi)}{\partial \gamma_\xi} + \frac{\partial u_0(x)}{\partial x_i} \frac{\partial N_i(\xi)}{\partial \gamma_\xi}\Big)\Big|_{\xi = x/\varepsilon} v(x)\, ds \\
&\qquad - \int_{\Omega^\varepsilon} \nabla z_\varepsilon\Big(x, \frac{x}{\varepsilon}\Big) \nabla v(x)\, dx.
\end{aligned} \tag{16.37}$$

From (16.36) and (16.37) we find that

$$
\begin{aligned}
&\int_{\Omega^\varepsilon} \nabla z_\varepsilon\Big(x, \frac{x}{\varepsilon}\Big)\, \nabla v(x)\, dx \\
&\quad = \int_{S_\varepsilon} p\Big(\frac{x}{\varepsilon}\Big) u_\varepsilon(x)\, v(x)\, ds + \varepsilon \int_{S_\varepsilon} q\Big(\frac{x}{\varepsilon}\Big) u_\varepsilon(x)\, v(x)\, ds \\
&\quad + \int_{S_\varepsilon} \frac{\partial u_0(x)}{\partial \gamma}\, v(x)\, ds + \varepsilon \int_{S_\varepsilon} \frac{\partial}{\partial \gamma_x} u_1\Big(x, \frac{x}{\varepsilon}\Big) v(x)\, ds \\
&\quad + \int_{S_\varepsilon} \Big(u_0(x) \frac{\partial}{\partial \gamma_\xi} M\Big(\frac{x}{\varepsilon}\Big) + \frac{\partial u_0(x)}{\partial x_i}\, \frac{\partial}{\partial \gamma_\xi} N_i\Big(\frac{x}{\varepsilon}\Big)\Big) v(x)\, ds \\
&\quad - \int_{\Omega^\varepsilon} \frac{\partial}{\partial \xi_i}\Big(a_{ij}(\xi) \frac{\partial u_0(x)}{\partial x_j} M(\xi)\Big)\Big|_{\xi = x/\varepsilon} v(x)\, dx \\
&\quad - \int_{\Omega^\varepsilon} \frac{\partial}{\partial \xi_i}\Big(a_{ij}(\xi) \frac{\partial^2 u_0(x)}{\partial x_j\, \partial x_k} N_k(\xi)\Big)\Big|_{\xi = x/\varepsilon} v(x)\, dx \\
&\quad - \int_{\Omega^\varepsilon} a_{ij}(\xi) \frac{\partial u_0(x)}{\partial x_i}\, \frac{\partial M(\xi)}{\partial \xi_j}\Big|_{\xi = x/\varepsilon} v(x)\, dx \\
&\quad + \varepsilon \int_{\Omega^\varepsilon} \mathcal{L}_{xx} u_1(x, \xi)\Big|_{\xi = x/\varepsilon} v(x)\, dx \\
&\quad - \int_{\Omega^\varepsilon} a_{ij}(\xi) \frac{\partial^2 u_0(x)}{\partial x_i \partial x_k}\, \frac{\partial N_k(\xi)}{\partial \xi_j}\Big|_{\xi = x/\varepsilon} v(x)\, dx \\
&\quad + \int_{\Omega^\varepsilon} \mathcal{L}_{xx} u_0(x) v(x)\, dx \\
&\quad + \frac{1}{|\square \backslash Q|} \int_{\Omega^\varepsilon} \widehat{a}_{kj} \frac{\partial^2 u_0(x)}{\partial x_k \partial x_j}\, v(x)\, dx \\
&\quad - \frac{1}{|\square \backslash Q|} \int_{\Omega^\varepsilon} (\langle q \rangle_S - m) u_0(x)\, v(x)\, dx.
\end{aligned} \tag{16.38}
$$

By the oblivious relation

$$
\begin{aligned}
\frac{\partial}{\partial \xi_i}\Big(a_{ij}(\xi) \frac{\partial^2 u_0(x)}{\partial x_j \partial x_k} N_k(\xi)\Big)\Big|_{\xi = x/\varepsilon} &= \varepsilon \frac{\partial}{\partial x_i}\Big(a_{ij}(\xi) \frac{\partial^2 u_0(x)}{\partial x_j \partial x_k} N_k(\xi)\Big|_{\xi = x/\varepsilon}\Big) \\
&\quad - \varepsilon \frac{\partial}{\partial x_i}\Big(a_{ij}(\xi) \frac{\partial^2 u_0(x)}{\partial x_j \partial x_k} N_k(\xi)\Big)\Big|_{\xi = x/\varepsilon}
\end{aligned}
$$

and the Stokes formula, we have

$$
\begin{aligned}
&\int_{\Omega^\varepsilon} \frac{\partial}{\partial \xi_i}\Big(a_{ij}(\xi) \frac{\partial^2 u_0(x)}{\partial x_j\, \partial x_k} N_k(\xi)\Big)\Big|_{\xi = x/\varepsilon} v(x)\, dx \\
&\qquad + \int_{\Omega^\varepsilon} \frac{\partial}{\partial \xi_i}\Big(a_{ij}(\xi) \frac{\partial u_0(x)}{\partial x_j} M(\xi)\Big)\Big|_{\xi = x/\varepsilon} v(x)\, dx \\
&\quad = \varepsilon \int_{S_\varepsilon} \frac{\partial}{\partial \gamma_x} u_1\Big(x, \frac{x}{\varepsilon}\Big) v(x)\, ds + O(\varepsilon) \|v\|_{H^1(\Omega^\varepsilon)}.
\end{aligned} \tag{16.39}
$$

Using (16.38) and the boundary conditions in (16.7) and (16.14), we estimate the expression

(16.40)

$$\begin{aligned}
&\Big| \int_{\Omega^\varepsilon} \nabla z_\varepsilon\Big(x, \frac{x}{\varepsilon}\Big) \nabla v(x)\, dx + \int_{S_\varepsilon} \Big(p\Big(\frac{x}{\varepsilon}\Big) + \varepsilon q\Big(\frac{x}{\varepsilon}\Big)\Big) z_\varepsilon\Big(x, \frac{x}{\varepsilon}\Big) v(x)\, ds \Big| \\
&\leq \varepsilon \Big| \varepsilon \int_{S_\varepsilon} q\Big(\frac{x}{\varepsilon}\Big) u_1\big(x, \frac{x}{\varepsilon}\big) v(x)\, ds\Big| + \Big| \int_{S_\varepsilon} u_0 \frac{\partial}{\partial \gamma_\xi} M\Big(\frac{x}{\varepsilon}\Big) v\, ds + \int_{S_\varepsilon} p\Big(\frac{x}{\varepsilon}\Big) u_0 v\, ds \Big| \\
&\quad + \Big| \varepsilon \int_{S_\varepsilon} q\Big(\frac{x}{\varepsilon}\Big) u_0(x)\, v(x)\, ds - \frac{1}{|\square \backslash Q|} \int_{\Omega^\varepsilon} \langle q \rangle_S u_0(x)\, v(x)\, dx \Big| \\
&\quad + \Big| \varepsilon \int_{\Omega^\varepsilon} \mathcal{L}_{xx} u_1(x, \xi) \Big|_{\xi = x/\varepsilon} v(x)\, dx \Big| + O(\varepsilon) \|v\|_{H^1(\Omega^\varepsilon)} \\
&\quad + \Big| \int_{S_\varepsilon} \Big( \frac{\partial u_0(x)}{\partial \gamma} + \frac{\partial u_0(x)}{\partial x_i} \frac{\partial}{\partial \gamma_\xi} N_i\Big(x, \frac{x}{\varepsilon}\Big)\Big) v(x)\, ds \Big| \\
&\quad + \Big| \int_{S_\varepsilon} \varepsilon M\Big(\frac{x}{\varepsilon}\Big) p\Big(\frac{x}{\varepsilon}\Big) u_0(x)\, v(x)\, ds + \frac{1}{|\square \backslash Q|} \int_{\Omega^\varepsilon} m u_0(x) v(x)\, dx \Big| \\
&\quad + \Big| \int_{S_\varepsilon} \varepsilon p\Big(\frac{x}{\varepsilon}\Big) \frac{\partial u_0}{\partial x_k} N_k\Big(\frac{x}{\varepsilon}\Big) v(x)\, ds - \int_{\Omega^\varepsilon} a_{ij}\Big(\frac{x}{\varepsilon}\Big) \frac{\partial u_0}{\partial x_i} \frac{\partial M(\xi)}{\partial \xi_j} \Big|_{\xi = x/\varepsilon} v(x)\, dx \Big| \\
&\quad + \Big| \int_{\Omega^\varepsilon} \Big( \frac{\widehat{a}_{kj}}{|\square \backslash Q|} \frac{\partial^2 u_0(x)}{\partial x_k \partial x_j} v(x) \\
&\qquad\qquad - a_{ij}(\xi) \frac{\partial^2 u_0(x)}{\partial x_i \partial x_k} \frac{\partial N_k(\xi)}{\partial \xi_j} \Big|_{\xi = x/\varepsilon} v(x) + \mathcal{L}_{xx} u_0(x) v(x) \Big)\, dx \Big| \\
&= I_1 + I_2 + I_3 + I_4 + I_5 + I_6 + I_7 + I_8.
\end{aligned}$$

Let us estimate $I_3$. By Lemma 16.3, we have

$$\begin{aligned}
I_3 &= \Big| \varepsilon \int_{S_\varepsilon} q\Big(\frac{x}{\varepsilon}\Big) u_0(x)\, v(x)\, ds - \frac{1}{|\square \backslash Q|} \int_{\Omega_\varepsilon} \langle q \rangle_S u_0(x)\, v(x)\, dx \Big| \\
&\leq C_2 \varepsilon \|u_0\|_{H^1(\Omega^\varepsilon)} \|v\|_{H^1(\Omega^\varepsilon)}.
\end{aligned}$$

Similarly, using Lemma 16.3, we can estimate $I_6$ as follows:

$$I_6 \leq C_8 \varepsilon \|v\|_{H^1(\Omega^\varepsilon)}.$$

Using Lemma 16.4, for $I_8$ we have

$$I_8 \leq C_{10} \varepsilon \|v\|_{H^1(\Omega^\varepsilon)}.$$

The terms $I_1$ and $I_4$ admit the following estimate:

$$|I_1| + |I_4| \leq C_{11} \varepsilon \|v\|_{H^1(\Omega^\varepsilon)}.$$

EXERCISE 16.3. Prove the last estimate.

The identities $I_2 \equiv 0$ and $I_5 \equiv 0$ follow from (16.10)–(16.11). Lemma 16.1 implies that $|I_7| \leq C\varepsilon$. The function $z_\varepsilon$ does not vanish on $\partial\Omega$ since the function $u_1$ (corrector) does not vanish on $\partial\Omega$. Let $\chi_\varepsilon(x)$ be a cut-off function, in the $\varepsilon$-neighborhood of the external boundary, such that $\chi_\varepsilon(x) \equiv \chi\big(\frac{\rho(x)}{\varepsilon}\big)$, where $\chi(t) = 0$ for $t < 0$ and $\chi(t) = 1$ for $t > 1$, $\chi \in C^\infty(\mathbb{R})$, and $\rho(x)$ is the distance from a point $x \in \Omega$ to $\partial\Omega$. Take the test function $v = u_0 + \varepsilon\chi_\varepsilon(x) u_1 - u_\varepsilon$. It is easy to check that $\|\varepsilon u_1(1 - \chi_\varepsilon)\|_{H^1(\Omega^\varepsilon)} \leq C_{12}\sqrt{\varepsilon}$ (check it!). Substituting $v$ in (16.40) and taking into account the previous estimates, we obtain the desired inequality (16.20). $\square$

## 17. Parabolic equations in media with rapidly pulsating perforation

**17.1. Statement of the problem and examples.** To simplify the presentation, we consider the heat equation on a set with periodic perforation with rapidly alternating geometry. Under the assumption that the geometry of the microstructure alternates periodically in time and the alternations are automodel, i.e., agree with the automodel change of variables $x \to \varepsilon x$, $t \to \varepsilon^2 t$, we construct the homogenized model, establish the convergence, and study properties of the homogenized problem. In particular, we show that time-oscillations (also called pulsations) of perforation may cause an unboundedly increasing convection in the effective dynamics. In this case, the effective diffusion is observed against the background of large convections, and the homogenization theorem holds in the moving coordinates. This problem was studied in [**46**].

We define a perforated domain with pulsating boundary. Let $F_t : \mathbb{R}^d \to \mathbb{R}^d$ be a family of diffeomorphisms that are 1-periodic with respect to $t$ and $F_t(x+k) = F_t(x) + k$ for all $x \in \mathbb{R}^d$, $k \in \mathbb{Z}^d$, and $-\infty < t < \infty$. Let $G_{\text{ref}}$ be a $\square$-periodic connected open set in $\mathbb{R}^d$ with Lipschitz boundary, where $\square = [0,1]^d$. We emphasize that the complement of $G_{\text{ref}}$ may be a dispersed set or may have an unbounded connected component.

Let $G(t) = F_t(G_{\text{ref}})$. The set

$$Q_T^\varepsilon = \left\{(x,t) \in \mathbb{R}^{d+1} : 0 \le t \le T;\ \frac{x}{\varepsilon} \in G(t/\varepsilon^2)\right\}, \quad \varepsilon > 0,$$

is referred to as a *perforated domain with pulsating boundary.*

EXAMPLE 17.1. A typical example of dispersed perforations is the set $G_{\text{ref}} = \mathbb{R}^d \setminus \bigcup_{k \in \mathbb{Z}^d} (B_0 + k)$, where $B_0$ is the ball of radius $r < 1/2$ centered at the origin.

If the diffeomorphisms $F_t$ are independent of $t$, we obtain an ordinary homogenization problem in a perforated medium.

Consider the problem

$$\begin{aligned} &\frac{\partial}{\partial t} u^\varepsilon(x,t) - \Delta u^\varepsilon(x,t) = f(x,t) \quad \text{in } Q_T^\varepsilon, \\ &\frac{\partial}{\partial n^\varepsilon(t)} u^\varepsilon(x,t) = 0 \quad \text{on } \partial Q_T^\varepsilon \cap \{0 < t < T\}, \\ &u^\varepsilon\big|_{t=0} = g(x) \quad \text{on } \varepsilon G(0), \end{aligned} \tag{17.1}$$

where $n^\varepsilon(t) = n^\varepsilon(x,t)$ is the unit normal to the boundary of the section of $G_T^\varepsilon$ by the plane $t = \text{const}$, i.e., $n^\varepsilon(t)$ is normal to the boundary of $\varepsilon G(t/\varepsilon^2)$ in $\mathbb{R}^d$.

Assume that

$$f \in L_2(0,T; L_2(\mathbb{R}^d)), \quad g \in L_2(\mathbb{R}^d). \tag{17.2}$$

Hereinafter, we use the notation $G^\varepsilon(t) = \varepsilon G(t/\varepsilon^2)$.

PROPOSITION 17.1. *Under the above assumptions, for every $\varepsilon > 0$ the problem (17.1) is uniquely solvable in the space*

$$V = L_2(0,T; H^1(G^\varepsilon(t))) \cap L_\infty(0,T; L_2(G^\varepsilon(t))).$$

PROOF. Making the change of variables $(y,t) = \left(\varepsilon F^{-1}_{(t/\varepsilon^2)}\left(\frac{x}{\varepsilon}\right), t\right)$ in (17.1), we can straighten out the boundary of $Q_T^\varepsilon$ so that the problem (17.1) written in the

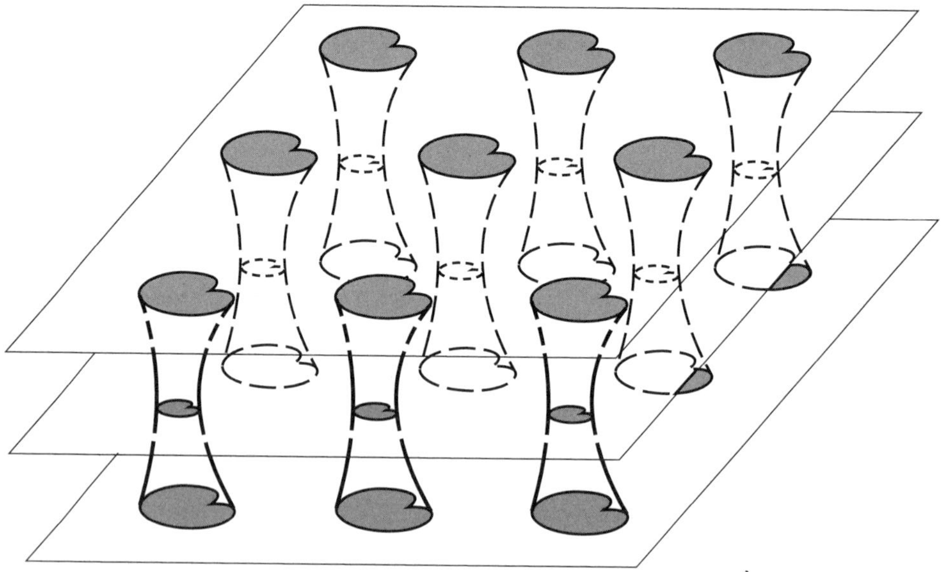

FIGURE 17.1

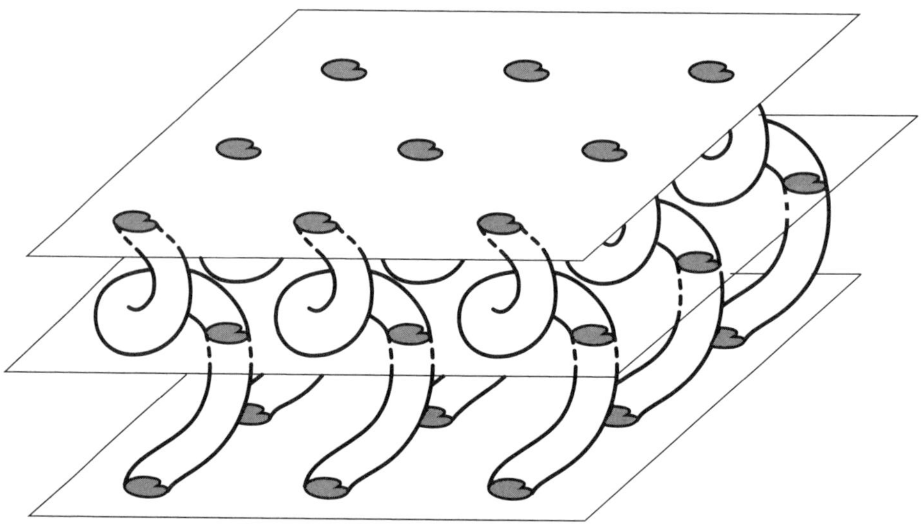

FIGURE 17.2

new variables $(y, t)$ will be given in the cylinder $\widetilde{Q}_T^\varepsilon = (\varepsilon G_{\text{ref}}) \times (0, T)$ and have the form

(17.3)

$$\begin{aligned} &\frac{\partial}{\partial t} u^\varepsilon(y,t) - a_{ij}^\varepsilon(y,t) \frac{\partial^2}{\partial y_i \partial y_j} u^\varepsilon(y,t) - b_i^\varepsilon(y,t) \frac{\partial}{\partial x_i} u^\varepsilon(y,t) = f(y,t) \quad \text{in} \quad \widetilde{Q}_T^\varepsilon, \\ &\frac{\partial u^\varepsilon}{\partial n_a^\varepsilon} = 0 \quad \text{on} \quad \varepsilon \partial G_{\text{ref}} \times (0,T), \quad u^\varepsilon \big|_{t=0} = g(y), \end{aligned}$$

where $f(y,t) = f(x(y),t)$, $g(y) = g(x(y))$, $n_a^\varepsilon(y,t)$ is the conormal to $\varepsilon\partial G_{\mathrm{ref}}$: $(n_a^\varepsilon)_i(y,t) = a_{ij}^\varepsilon(y,t)n_j^\varepsilon(y,t)$, and $n_j^\varepsilon(y,t)$ is the unit normal to $\varepsilon\partial G_{\mathrm{ref}}$. A direct computation shows that

$$a_{ij}^\varepsilon(y,t) = \sum_{k=1}^d \Big(\frac{\partial (F_{t/\varepsilon^2}^{-1})_i}{\partial x_k}\Big)\Big(\frac{x}{\varepsilon}\Big)\Big(\frac{\partial (F_{t/\varepsilon^2}^{-1})_j}{\partial x_k}\Big)\Big(\frac{x}{\varepsilon}\Big),$$

$$b_i^\varepsilon(y,t) = \frac{1}{\varepsilon}\sum_{k=1}^d \Big\{\frac{\partial}{\partial x_k}\Big(\frac{\partial (F_{t/\varepsilon^2}^{-1})_i}{\partial x_k}\Big)\Big\}\Big(\frac{x}{\varepsilon}\Big).$$

It is known (cf., for example, [**77**]) that the problem (17.3) is solvable. Under the above assumptions on the data of the problem (17.1), the solution belongs to the space $L_2(0,T;H^1(\varepsilon G_{\mathrm{ref}})) \cap L_\infty(0,T;L_2(\varepsilon G_{\mathrm{ref}}))$. Making the inverse change of variables, we arrive at the required assertion. □

REMARK 17.1. Since the coefficients in (17.3) increase in the absolute value as $\varepsilon \to 0$, the method of the proof of Proposition 17.1 does not yield estimates that are uniform with respect to $\varepsilon$ in the case of problem (17.1).

Denote by $G$ the perforated $(d+1)$-dimensional periodicity cell: $G = \{(z,\tau) \in [0,1]^{d+1} : z \in G(\tau)\}$. In the sequel, to derive *a priori* estimates, as well as to analyze the asymptotics for the problem (17.1), we essentially use the following periodic problem in $G$:

$$(17.4)\qquad \begin{aligned} &\frac{\partial p}{\partial \tau} + \Delta_z p = 0 \quad \text{in } G,\\ &\frac{\partial p}{\partial n_z(\tau)} + n_\tau p = 0 \quad \text{on } \partial G,\\ &p(z,\tau) \text{ is 1-periodic with respect to } z \text{ and } \tau, \end{aligned}$$

where $n_z$ is the spatial component and $n_\tau$ is the time component of the $(d+1)$-dimensional unit normal $n(z,\tau)$ to $\partial G$. The problem (17.4) is formally adjoint (i.e., is obtained by integrating by parts) to the following problem:

$$(17.5)\qquad \begin{aligned} &\frac{\partial v}{\partial \tau} - \Delta_z v = 0 \quad \text{in } G,\\ &\frac{\partial v}{\partial n_z(\tau)} = 0 \quad \text{on } \partial G,\\ &v(z,\tau) \text{ is 1-periodic with respect to } z \text{ and } \tau. \end{aligned}$$

We study the problem (17.4) in detail.

THEOREM 17.1. *The problem* (17.4) *has a nontrivial periodic solution* $p(z,\tau) \not\equiv 0$. *This solution is unique up to a multiplicative constant and preserves the sign so that* $p(z,\tau) > 0$ *for all* $z$ *and* $\tau$ *in a suitable normalization.*

PROOF. On the set $Q_- = \{(z,\tau) \in \mathbb{R}^{d+1} : \tau < 0,\ z \in G(\tau)\}$, we consider the initial-boundary value problem

$$(17.6)\qquad \begin{aligned} &\frac{\partial q}{\partial \tau} + \Delta_z q = 0 \quad \text{in } Q_-,\\ &\frac{\partial q}{\partial n_z(\tau)} + n_\tau q = 0 \quad \text{on } \partial Q_- \cap \{\tau < 0\},\\ &q(z,\tau)\big|_{\tau=0} = \psi(z) \quad \text{in } G(0), \end{aligned}$$

where $\psi(z)$ is 1-periodic with respect to $z$. The solution to this problem is also periodic with respect to $z$. As in the proof of Proposition 17.1, it is easy to show that the problem (17.6) has a unique solution in the space $C((-\infty,0);L_2(G(\tau)))$ if $\psi\in L_2(G(0))$, and in the space $C((-\infty,0);C(G(\tau)))$ if $\psi\in C(G(0))$. Consider a mapping over a period:

$$S:\psi(\cdot)\to q(\cdot,-1). \tag{17.7}$$

EXERCISE 17.1. Prove that the operator $S$ transforms the set of nonnegative functions to itself.

*Hint.* Use the maximum principle.

Further, by the smoothing properties of the parabolic equation, the operator $S$ continuously maps the space $C(G(0))$ to the space $C^1(G(-1))=C^1(G(0))$; the last equality holds in view of periodicity of $G(\tau)$ with respect to $\tau$. Since $C^1(G(0))$ is compactly embedded in $C(G(0))$, the operator $S$ is compact in $C(G(0))$. Thus, we have verified the assumptions of Theorems 2.9.2 and 2.11.1 in [**74**], asserting the existence of $\lambda_0>0$ and $\psi_0(\cdot)>0$ such that $S\psi_0(z)=\lambda_0\psi_0(z)$. Furthermore, the eigenvalue $\lambda_0$ is simple and the remaining eigenvalues $\lambda_j$, $j>0$, satisfy the estimate $|\lambda_k|<\lambda_0$.

We show that $\lambda_0=1$. For this purpose, we verify that the integral of $q(\cdot,\tau)$ over the section $G_{\mathrm{per}}(\tau)=G(\tau\cap[0,1]^d)$ is independent of the choice of the section, i.e., is constant. Indeed, by the Stokes formula and the problem (17.6), for arbitrary $\tau_1$ and $\tau_2$, $\tau_1<\tau_2$, we have

$$\begin{aligned}
&\int_{G_{\mathrm{per}}(\tau_2)} q(z,\tau_2)\,dz-\int_{G_{\mathrm{per}}(\tau_1)} q(z,\tau_1)\,dz\\
&\quad=\int_{\tau_1}^{\tau_2}\int_{G_{\mathrm{per}}(\tau)}\frac{\partial q(z,\tau)}{\partial\tau}\,dz\,d\tau-\int_{\tau_1}^{\tau_2}\int_{\partial G_{\mathrm{per}}(\tau)} q(z,\tau)n_\tau(z,\tau)\,dS_{z,\tau}\\
&\quad=-\int_{\tau_1}^{\tau_2}\int_{G_{\mathrm{per}}(\tau)}\Delta q(z,\tau)\,dz\,d\tau-\int_{\tau_1}^{\tau_2}\int_{\partial G_{\mathrm{per}}(\tau)} q(z,\tau)n_\tau(z,\tau)\,dS_{z,\tau}\\
&\quad=-\int_{\tau_1}^{\tau_2}\int_{\partial G_{\mathrm{per}}(\tau)}\left(\frac{\partial q(z,\tau)}{\partial n_z}+q(z,\tau)n_\tau(z,\tau)\right)dS_{z,\tau}=0,
\end{aligned}$$

where $dS_{z,\tau}$ is the volume element on $\partial G$. Let $q_0(z,\tau)$ be a solution to the problem (17.6) with the initial data $\psi_0(z)$. Without loss of generality, we assume that

$$\int_{G_{\mathrm{per}}(0)}\psi_0(z)\,dz=1. \tag{17.8}$$

Then

$$\begin{aligned}
\lambda_0&=\int_{G_{\mathrm{per}}(-1)}\lambda_0\psi_0(z)\,dz=\int_{G_{\mathrm{per}}(-1)} q_0(z,-1)\,dz\\
&=\int_{G_{\mathrm{per}}(0)} q_0(z,0)\,dz=\int_{G_{\mathrm{per}}(0)}\psi_0(z)\,dz=1.
\end{aligned}$$

Thus, setting $p(z,\tau)=q_0(z,\tau)$, we obtain the required positive solution to the problem (17.4). The uniqueness of a periodic solution follows because $\lambda_0$ is simple (check it!). □

EXERCISE 17.2. Prove that a periodic solution is unique.

We note that the normalization condition (17.8) implies

$$\int_{G_{\text{per}}(\tau)} p(z,\tau)\,dz\,d\tau = 1.$$

Hereinafter, we assume that this equality holds.

In addition to the problem (17.6), we consider the nonhomogeneous problem

$$\begin{aligned}
&\frac{\partial}{\partial\tau}v - \Delta_z v = f(z,\tau) \quad \text{in } G\\
&\frac{\partial v}{\partial n(\tau)}v = g(z,\tau) \quad \text{on } \partial G, \qquad (17.9)\\
&v(z,\tau) \text{ is 1-periodic with respect to } z \text{ and } \tau,
\end{aligned}$$

where $f$ and $g$ are $[0,1]^{d+1}$-periodic functions, $f \in L_2(G)$, and $g \in L_2(\partial G)$.

THEOREM 17.2. *The problem* (17.9) *is solvable in the space of periodic functions of $z$ and $\tau$ if and only if*

$$\int_{G_{\text{per}}} f(z,\tau)p(z,\tau)\,dz\,d\tau + \int_{\partial G_{\text{per}}} g(z,\tau)p(z,\tau)\,dS_z d\tau = 0. \qquad (17.10)$$

*If the equality* (17.10) *holds, the solution is unique up to an additive constant and belongs to the space* $L_2^{\text{per}}(0,1;H^1_{\text{per}}(G(\tau))) \cap L_\infty^{\text{per}}(0,1;L_2^{\text{per}}(G(\tau)))$.

REMARK 17.2. The boundary condition in the problem (17.9) can be written in the form

$$\frac{\partial v}{\partial n_z(\tau)}v = g(z,\tau) \quad \text{on } \partial G. \qquad (17.11)$$

In this case, the compatibility condition takes the form

$$\int_{G_{\text{per}}} f(z,\tau)p(z,\tau)dz\,d\tau + \int_{\partial G_{\text{per}}} g(z,\tau)p(z,\tau)\,dS_{z,\tau} = 0. \qquad (17.12)$$

PROOF OF THEOREM 17.2. We establish the necessity of the condition (17.10). For an arbitrary solution $v(z,\tau)$ we have

$$\begin{aligned}
0 &= \int_{G_{\text{per}}(1)} v(z,1)p(z,1)\,dz - \int_{G_{\text{per}}(0)} v(z,0)p(z,0)\,dz\\
&= \int_{G_{\text{per}}} \frac{\partial(v(z,\tau)p(z,\tau))}{\partial\tau}dz\,d\tau - \int_{\partial G_{\text{per}}} v(z,\tau)p(z,\tau)n_\tau(z,\tau)\,dS_{z,\tau}\\
&= \int_{G_{\text{per}}} p(z,\tau)\Delta_z v(z,\tau)\,dz\,d\tau - \int_{G_{\text{per}}} v(z,\tau)\Delta_z p(z,\tau)\,dz\,d\tau\\
&\quad + \int_{G_{\text{per}}} p(z,\tau)f(z,\tau)\,dz\,d\tau - \int_{\partial G_{\text{per}}} v(z,\tau)p(z,\tau)n_\tau(z,\tau)\,dS_{z,\tau}.
\end{aligned}$$

By the Stokes formula,

$$\begin{aligned}
&\int_{G_{\text{per}}} p(z,\tau)\Delta_z v(z,\tau)\,dz\,d\tau - \int_{G_{\text{per}}} v(z,\tau)\Delta_z p(z,\tau)\,dz\,d\tau\\
&\quad = \int_{\partial G_{\text{per}}} \frac{\partial v(z,\tau)}{\partial n(\tau)}p(z,\tau)\,dS_z d\tau - \int_{\partial G_{\text{per}}} \frac{\partial p(z,\tau)}{\partial n(\tau)}v(z,\tau)\,dS_z d\tau\\
&\quad = \int_{\partial G_{\text{per}}} g(z,\tau)p(z,\tau)\,dS_z d\tau - \int_{\partial G_{\text{per}}} \frac{\partial p(z,\tau)}{\partial n_z(\tau)}v(z,\tau)\,dS_{z,\tau}.
\end{aligned}$$

Substituting this equality in the above formula, we find that

$$\begin{aligned}0 &= \int_{G_{\mathrm{per}}} p(z,\tau)f(z,\tau)\,dz\,d\tau + \int_{\partial G_{\mathrm{per}}} g(z,\tau)p(z,\tau)\,dS_z d\tau \\ &\quad - \int_{\partial G_{\mathrm{per}}} v(z,\tau)\Big(\frac{\partial p(z,\tau)}{\partial n_z(\tau)} + p(z,\tau)n_\tau(z,\tau)\Big)dS_{z,\tau} \\ &= \int_{G_{\mathrm{per}}} p(z,\tau)f(z,\tau)\,dz\,d\tau + \int_{\partial G_{\mathrm{per}}} g(z,\tau)p(z,\tau)\,dS_z d\tau.\end{aligned}$$

Now we show that the condition (17.10) is sufficient. Assume that $f$ and $g$ satisfy (17.10). On the set $Q_+ = \{(z,\tau) \in \mathbb{R}^{d+1} : \tau > 0,\ z \in G(\tau)\}$, consider the following auxiliary initial-boundary value problem:

$$\begin{aligned}&\frac{\partial v_+}{\partial \tau} + \Delta_z v_+ = f \quad \text{in } Q_+, \\ &\frac{\partial v_+}{\partial n(\tau)} = g \quad \text{on } \partial Q_+ \cap \{\tau > 0\}, \\ &v_+(z,0) = 0 \quad \text{in } G(0).\end{aligned} \tag{17.13}$$

Using the Stokes formula, we obtain

$$\int_{G_{\mathrm{per}}(0)} v_+(z,0)p(z,0)\,dz = \int_{G_{\mathrm{per}}(1)} v_+(z,1)p(z,1)\,dz = 0.$$

By construction, $p(z,1) = p(z,0)$ belongs to the one-dimensional kernel of the operator $(\mathbf{I} - S)$, where $\mathbf{I}$ is the identity operator and $S$ is defined in (17.7). Thus, $v_+(z,1)$ is orthogonal to the one-dimensional kernel of the operator $(\mathbf{I} - S)$. Since $S$ is compact, the equation

$$(\mathbf{I} - S^*)\varphi = v_+(\cdot,1) \tag{17.14}$$

for the adjoint operator $S^*$ has a solution. We show that $S^*$ is a mapping over a period for the problem

$$\begin{aligned}&\frac{\partial w}{\partial \tau} - \Delta_z w = 0 \quad \text{in } Q_+, \\ &\frac{\partial w}{\partial n(\tau)} = 0 \quad \text{on } \partial Q_+ \cap \{\tau > 0\}, \\ &w(z,0) = \varphi \quad \text{in } G(0),\end{aligned} \tag{17.15}$$

i.e., $S^*\varphi = w(\cdot,1)$. We need to show that

$$J \stackrel{\text{def}}{=} \int_{G_{\mathrm{per}}(1)} \psi(z)w(z,1)\,dz - \int_{G_{\mathrm{per}}(0)} q(z,-1)\varphi(z)\,dz = 0 \quad \forall \varphi,\psi \in C_{\mathrm{per}}(G(0)),$$

where, by the definition of $S$, the functions $\psi$ and $q(z,\tau)$ are the initial data and solution to the problem (17.6) respectively. Using the Stokes formula repeatedly, we transform $J$ as follows:

$$J = \int_G \frac{\partial(q(z,\tau-1)w(z,\tau))}{\partial\tau} dz\,d\tau - \int_{\partial G} q(z,\tau-1)w(z,\tau)n_\tau(z,\tau)\,dS_{z,\tau}$$
$$= \int_G q(z,\tau-1)\Delta_z w(z,\tau)\,dz\,d\tau - \int_G w(z,\tau)\Delta_z q(z,\tau-1)\,dz\,d\tau$$
$$- \int_{\partial G} q(z,\tau-1)w(z,\tau)n_\tau(z,\tau)\,dS_{z,\tau} = \int_{\partial G} q(z,\tau-1)\frac{\partial w(z,\tau)}{\partial n_z(\tau)} dS_{z,\tau}$$
$$- \int_{\partial G} w(z,\tau)\Big(\frac{\partial q(z,\tau)}{\partial n_z(\tau)} + q(z,\tau)n_\tau(z,\tau)\Big)dS_{z,\tau} = 0.$$

Taking the solution to (17.14) for the initial data $\varphi$ in (17.15) and setting $v(z,\tau) = w(z,\tau) + v_+(z,\tau)$, we obtain the required periodic solution to the problem (17.9). Indeed, by (17.14) and (17.15), we have

$$v(z,1) = w(z,1) + v_+(z,1) = S^*\varphi + (\varphi - S^*\varphi) = \varphi,$$

whereas, by the choice of the initial conditions in (17.13) and (17.15), we have

$$v(z,0) = \varphi + 0 = \varphi.$$

The uniqueness of the solution follows from the maximum principle (check it!). □

**17.2. Energy estimates.** To prove energy estimates, we regard the solution to the problem (17.4) as a weight function. In addition to the problem (17.1), we also consider the problem with inhomogeneous Neumann condition on the boundary of perforation:

$$\begin{aligned} &\frac{\partial v^\varepsilon}{\partial t} - \Delta v^\varepsilon = F(x,t) \quad \text{in } Q_T^\varepsilon, \\ &\frac{\partial v^\varepsilon}{\partial n^\varepsilon(t)} = h(x,t) \quad \text{on } \partial Q_T^\varepsilon \cap \{t>0\}, \\ &v^\varepsilon(x,0) = H(x). \end{aligned} \tag{17.16}$$

LEMMA 17.1. *The solution to the problem* (17.1) *satisfies the estimate*

$$\begin{aligned} &\int_{Q_T^\varepsilon} |\nabla u^\varepsilon(x,t)|^2 dx\,dt + \max_{0\le t\le T}\int_{G^\varepsilon(t)} (u^\varepsilon(x,t))^2 dx \\ &\qquad \le C(\|f\|^2_{L_2(\mathbb{R}^d\times(0,T))} + \|q\|^2_{L_2(R^d)}). \end{aligned} \tag{17.17}$$

*Let $F \in L_2(\mathbb{R}^d \times (0,T))$, $h \in L_2(\partial Q_T^\varepsilon)$, $H \in L_2(\mathbb{R}^d)$. Then the solution $v^\varepsilon$ to the problem* (17.16) *satisfies the estimate*

$$\begin{aligned} &\int_{Q_T^\varepsilon} |\nabla v^\varepsilon(x,t)|^2 dx\,dt + \max_{0\le t\le T}\int_{G^\varepsilon(t)} (v^\varepsilon(x,t))^2 dx \\ &\qquad \le C(\|F\|^2_{L_2(\mathbb{R}^d\times(0,T))} + \|H\|^2_{L_2(R^d)} + \varepsilon^{-1}\|h\|^2_{L_2(\partial Q_T^\varepsilon)}). \end{aligned} \tag{17.18}$$

*If $h$ in* (17.16) *is the trace on $\partial Q_T^\varepsilon$ of some function $\widetilde{h} \in L_2(0,T;H^1(G^\varepsilon(t)))$, then*

$$\begin{aligned} &\int_{Q_T^\varepsilon} |\nabla v^\varepsilon(x,t)|^2 dx\,dt + \max_{0\le t\le T}\int_{G^\varepsilon(t)} (v^\varepsilon(x,t))^2 dx \\ &\qquad \le C(\|F\|^2_{L_2(\mathbb{R}^d\times(0,T))} + \|H\|^2_{L_2(R^d)} + \varepsilon^{-2}\|\widetilde{h}\|^2_{L_2(0,T;H^1(G^\varepsilon(t)))}). \end{aligned} \tag{17.19}$$

*In the estimates* (17.17)–(17.19), *the constant $C$ is independent of $\varepsilon$.*

PROOF. We multiply (17.16) by $v^\varepsilon(x,t)p(\frac{x}{\varepsilon},\frac{t}{\varepsilon^2})$ and integrate over $Q_t^\varepsilon = \{(x,s)\in Q_T^\varepsilon : s\le t\}$. Recalling (17.4) and using the Stokes formula, after simple, but awkward computations we find that

$$
\begin{aligned}
0 &= \frac{1}{2}\int_{Q_t^\varepsilon} p\Big(\frac{x}{\varepsilon},\frac{s}{\varepsilon^2}\Big)\frac{\partial}{\partial s}(v^\varepsilon)^2\,dx\,ds - \int_{Q_t^\varepsilon} p\Big(\frac{x}{\varepsilon},\frac{s}{\varepsilon^2}\Big)v^\varepsilon\Delta v^\varepsilon\,dx\,ds\\
&\quad - \int_{Q_t^\varepsilon} f(x,t)p\Big(\frac{x}{\varepsilon},\frac{s}{\varepsilon^2}\Big)v^\varepsilon\,dx\,ds\\
&= \frac{1}{2}\int_{G^\varepsilon(t)}(v^\varepsilon(x,t))^2 p\Big(\frac{x}{\varepsilon},\frac{t}{\varepsilon^2}\Big)\,dx\\
&\quad - \frac{1}{2}\int_{G^\varepsilon(0)} H^2(x)p\Big(\frac{x}{\varepsilon},0\Big)\,dx + \frac{1}{2}\int_{\partial Q_t^\varepsilon}(v^\varepsilon)^2 p\Big(\frac{x}{\varepsilon},\frac{s}{\varepsilon^2}\Big)n_s^\varepsilon(x,s)\,dS_{x,s}^\varepsilon\\
&\quad - \frac{1}{2\varepsilon^2}\int_{Q_t^\varepsilon}(v^\varepsilon)^2\Big(\frac{\partial}{\partial s}p\Big)\Big(\frac{x}{\varepsilon},\frac{s}{\varepsilon^2}\Big)\,dx\,ds + \int_{Q_t^\varepsilon} p\Big(\frac{x}{\varepsilon},\frac{s}{\varepsilon^2}\Big)|\nabla v^\varepsilon|^2\,dx\,ds\\
&\quad + \frac{1}{2\varepsilon}\int_{Q_t^\varepsilon}(\nabla_x p)\Big(\frac{x}{\varepsilon},\frac{s}{\varepsilon^2}\Big)\nabla((v^\varepsilon)^2)\,dx\,ds\\
&\quad - \int_{\partial Q_t^\varepsilon} p\Big(\frac{x}{\varepsilon},\frac{s}{\varepsilon^2}\Big)v^\varepsilon h(x,t)\,dS_{x,s}^\varepsilon - \int_{Q_t^\varepsilon} F(x,s)v^\varepsilon p\Big(\frac{x}{\varepsilon},\frac{s}{\varepsilon^2}\Big)\,dx\,ds\\
&= \frac{1}{2}\int_{G^\varepsilon(t)}(v^\varepsilon(x,t))^2 p\Big(\frac{x}{\varepsilon},\frac{t}{\varepsilon^2}\Big)\,dx\\
&\quad - \frac{1}{2}\int_{G^\varepsilon(0)} H^2(x)p\Big(\frac{x}{\varepsilon},0\Big)\,dx + \frac{1}{2}\int_{\partial Q_t^\varepsilon}(v^\varepsilon)^2 p\Big(\frac{x}{\varepsilon},\frac{s}{\varepsilon^2}\Big)n_s^\varepsilon(x,s)\,dS_{x,s}^\varepsilon\\
&\quad - \frac{1}{2\varepsilon^2}\int_{Q_t^\varepsilon}(v^\varepsilon)^2\Big(\frac{\partial}{\partial s}p\Big)\Big(\frac{x}{\varepsilon},\frac{s}{\varepsilon^2}\Big)\,dx\,ds + \int_{Q_t^\varepsilon} p\Big(\frac{x}{\varepsilon},\frac{s}{\varepsilon^2}\Big)|\nabla v^\varepsilon|^2\,dx\,ds\\
&\quad + \frac{1}{2}\int_{\partial Q_t^\varepsilon}(v^\varepsilon)^2\frac{\partial}{\partial n_s^\varepsilon}p\Big(\frac{x}{\varepsilon},\frac{s}{\varepsilon^2}\Big)dS_{x,s}^\varepsilon - \frac{1}{2\varepsilon^2}\int_{Q_t^\varepsilon}(v^\varepsilon)^2(\Delta p)\Big(\frac{x}{\varepsilon},\frac{s}{\varepsilon^2}\Big)\,dx\,ds\\
&\quad - \int_{\partial Q_t^\varepsilon} v^\varepsilon h(x,s)p\Big(\frac{x}{\varepsilon},\frac{s}{\varepsilon^2}\Big)\,dS_{x,s}^\varepsilon - \int_{Q_t^\varepsilon} F(x,s)v^\varepsilon p\Big(\frac{x}{\varepsilon},\frac{s}{\varepsilon^2}\Big)\,dx\,ds\\
&= \frac{1}{2}\int_{G^\varepsilon(t)}(v^\varepsilon(x,t))^2 p\Big(\frac{x}{\varepsilon},\frac{t}{\varepsilon^2}\Big)\,dx + \int_{Q_t^\varepsilon} p\Big(\frac{x}{\varepsilon},\frac{s}{\varepsilon^2}\Big)|\nabla v^\varepsilon|^2\,dx\,ds\\
&\quad - \frac{1}{2}\int_{G^\varepsilon(0)} H^2(x)p\Big(\frac{x}{\varepsilon},0\Big)\,dx - \int_{\partial Q_t^\varepsilon} v^\varepsilon h(x,s)p\Big(\frac{x}{\varepsilon},\frac{s}{\varepsilon^2}\Big)\,dS_{x,s}^\varepsilon\\
&\quad - \int_{Q_t^\varepsilon} F(x,s)v^\varepsilon p\Big(\frac{x}{\varepsilon},\frac{s}{\varepsilon^2}\Big)\,dx\,ds.
\end{aligned}
$$

From this equality, the inequalities $0 < c \le p(z,\tau) \le c^{-1}$, and the Cauchy–Schwarz–Bunyakovskii inequality, we find that

$$
\begin{aligned}
&\int_{G^\varepsilon(t)}(v^\varepsilon(x,t))^2dx + \int_{Q_t^\varepsilon}|\nabla v^\varepsilon|^2\,dx\,ds\\
&\qquad \le C(\|H\|^2_{L_2(\mathbb{R}^d)} + \|v^\varepsilon\|_{L_2(\partial Q_t^\varepsilon)}\|h\|_{L_2(\partial Q_t^\varepsilon)} + \|v^\varepsilon\|_{L_2(Q_t^\varepsilon)}\|F\|_{L_2(Q_t^\varepsilon)}).
\end{aligned}
\tag{17.20}
$$

By the trace theorem [**80**], in the fast coordinates $z = x/\varepsilon$, for every $t\in[0,T]$ we have $\|v^\varepsilon(\varepsilon z)\|^2_{L_2(\partial G(t))} \le C\|v^\varepsilon(\varepsilon z)\|^2_{H^1(G(t))}$, which, in the original coordinates

$x$, yields the estimate $\|v^\varepsilon\|^2_{L_2(\partial G^\varepsilon(t))} \le C\varepsilon^{-1}\|v^\varepsilon\|^2_{H^1(G^\varepsilon(t))}$. Consequently,

$$\|v^\varepsilon\|^2_{L_2(\partial Q_t^\varepsilon)} = \int_{\partial Q_t^\varepsilon} (v^\varepsilon)^2 dS^\varepsilon_{x,s} = \int_{\partial Q_t^\varepsilon} (v^\varepsilon)^2 \Big|\frac{dS^\varepsilon_{x,s}}{dS^\varepsilon_x ds}\Big| dS^\varepsilon_x ds \le C_1\varepsilon^{-1} \int_{\partial Q_t^\varepsilon} (v^\varepsilon)^2 dS^\varepsilon_x ds$$
$$\le C_2\varepsilon^{-2} \int_0^t \|v^\varepsilon\|^2_{L_2(G^\varepsilon(s))} ds = c_2\varepsilon^{-2}\|v^\varepsilon\|^2_{L_2(0,t;H^1(G^\varepsilon(s)))},$$

i.e.,

$$\|v^\varepsilon\|_{L_2(\partial Q_t^\varepsilon)} \le C\varepsilon^{-1}\|v^\varepsilon\|_{L_2(0,t;H^1(G^\varepsilon(s)))} \tag{17.21}$$

for all $t \in [0, T]$. From (17.20) and (17.21) we obtain the estimate (17.18) in a standard way, which implies (17.17). The estimate (17.19) is obtained from (17.18) in view of the inequality (17.21). □

**17.3. Formal asymptotic expansion.** We now turn to constructing a formal asymptotic expansion of the solution $u^\varepsilon$ to the problem (17.1). Following the general scheme presented in Section 7, we expand $u^\varepsilon$ into the asymptotic series

$$\begin{aligned} u^\varepsilon \sim u_0\Big(x - \frac{\alpha}{\varepsilon}t, t\Big) &+ \varepsilon\Big\{\nabla_x u_0\Big(x - \frac{\alpha}{\varepsilon}t, t\Big)\chi\Big(\frac{x}{\varepsilon}, \frac{t}{\varepsilon^2}\Big) + u_1\Big(x - \frac{\alpha}{\varepsilon}t, t\Big)\Big\} \\ &+ \varepsilon^2\Big\{\nabla_x\nabla_x u_0\Big(x - \frac{\alpha}{\varepsilon}t, t\Big)\xi\Big(\frac{x}{\varepsilon}, \frac{t}{\varepsilon^2}\Big) \\ &+ \nabla_x u_1\Big(x - \frac{\alpha}{\varepsilon}t, t\Big)\chi\Big(\frac{x}{\varepsilon}, \frac{t}{\varepsilon^2}\Big) + u_2\Big(x - \frac{\alpha}{\varepsilon}t, t\Big)\Big\} + \cdots, \end{aligned} \tag{17.22}$$

where $\chi(z, \tau)$ and $\xi(z, \tau)$ are vector-valued and matrix-valued $[0, 1]^{d+1}$-periodic functions respectively. A distinctive feature of this series is the presence of moving coordinates $(x - (\alpha/\varepsilon)t)$ with parameter $\alpha \in \mathbb{R}^d$, the purpose of which will be explained below.

Substituting (17.22) in (17.1), using the formulas

$$\frac{\partial}{\partial x}\gamma\Big(x, \frac{x}{\varepsilon}\Big) = \Big(\frac{\partial}{\partial x}\gamma(x, z) + \frac{1}{\varepsilon}\frac{\partial}{\partial z}\gamma(x, z)\Big)\Big|_{z=x/\varepsilon},$$
$$\frac{\partial}{\partial t}\gamma\Big(t, \frac{t}{\varepsilon^2}\Big) = \Big(\frac{\partial}{\partial t}\gamma(t, \tau) + \frac{1}{\varepsilon^2}\frac{\partial}{\partial \tau}\gamma(t, \tau)\Big)\Big|_{t=t/\varepsilon^2},$$

and collecting the terms with equal power of $\varepsilon$, we obtain a chain of problems, the first of which has the form

$$\begin{gathered} \Big(\frac{\partial}{\partial\tau}\chi - \Delta_z\chi\Big)\nabla_x u_0\Big(x - \frac{\alpha}{\varepsilon}t, t\Big) = -\alpha\nabla_x u_0\Big(x - \frac{\alpha}{\varepsilon}t, t\Big), \\ (x, t) \in \mathbb{R}^d \times (0, T),\ (z, \tau) \in G, \\ \frac{\partial}{\partial n(\tau)}\chi\nabla_x u_0\Big(x - \frac{\alpha}{\varepsilon}t, t\Big) + n(z, \tau)\nabla_x u_0\Big(x - \frac{\alpha}{\varepsilon}t, t\Big) = 0, \\ (x, t) \in \mathbb{R}^d \times (0, T),\ (z, \tau) \in \partial G. \end{gathered} \tag{17.23}$$

EXERCISE 17.3. Perform the substitution and derive the equation and boundary condition in (17.23).

Since $u_0$ is independent of $z$ and $\tau$, we have

$$\begin{gathered} \frac{\partial}{\partial\tau}\chi - \Delta_z\chi = -\alpha \quad \text{in } G, \\ \frac{\partial}{\partial n(\tau)}\chi = -n(z, \tau) \quad \text{on } \partial G. \end{gathered} \tag{17.24}$$

We look for a solution to the problem (17.24) in the space of functions periodic with respect to $z$ and $\tau$. By Theorem 17.2, the problem (17.24) is solvable if and only if

$$\alpha = -\int_{\partial G} p(z,\tau)n(z,\tau)\,dS_z d\tau, \tag{17.25}$$

which dictates the choice of $\alpha$ in a unique way.

The next problem (with variables $\tau$ and $z$ and parameters $x$ and $t$) in the chain of problems has the form

(17.26)

$$\begin{aligned}
&\Big(\frac{\partial}{\partial\tau}\xi_{ij} - \Delta_z\xi_{ij}\Big)\frac{\partial^2}{\partial x_i\partial x_j}u_0\Big(x-\frac{\alpha}{\varepsilon}t,t\Big) + \frac{\partial}{\partial s}u_0\Big(x-\frac{\alpha}{\varepsilon}t,s\Big)\Big|_{s=t}\\
&\quad - \Delta u_0\Big(x-\frac{\alpha}{\varepsilon}t,t\Big) + 2\frac{\partial}{\partial z_i}\chi_j\frac{\partial^2}{\partial x_i\partial x_j}u_0\Big(x-\frac{\alpha}{\varepsilon}t,t\Big)\\
&\quad - \alpha_i\chi_j\frac{\partial^2}{\partial x_i\partial x_j}u_0\Big(x-\frac{\alpha}{\varepsilon}t,t\Big) + \Big(\frac{\partial}{\partial\tau}\chi - \Delta_z\chi\Big)\nabla_x u_1\Big(x-\frac{\alpha}{\varepsilon}t,t\Big) = f(x,t),\\
&\qquad (x,t)\in\mathbb{R}^d\times(0,T),\ (z,\tau)\in G,\\
&\frac{\partial}{\partial n(\tau)}\xi_{ij}\frac{\partial^2}{\partial x_i\partial x_j}u_0\Big(x-\frac{\alpha}{\varepsilon}t,t\Big) + n_i(z,\tau)\chi_j\frac{\partial^2}{\partial x_i\partial x_j}u_0\Big(x-\frac{\alpha}{\varepsilon}t,t\Big)\\
&\quad + \frac{\partial}{\partial n(\tau)}\chi_i\frac{\partial}{\partial x_j}u_1\Big(x-\frac{\alpha}{\varepsilon}t,t\Big) + n_i(z,\tau)\frac{\partial}{\partial x_j}u_1\Big(x-\frac{\alpha}{\varepsilon}t,t\Big) = 0,\\
&\qquad (x,t)\in\mathbb{R}^d\times(0,T),\ (z,\tau)\in\partial G.
\end{aligned}$$

Note that the sum of terms containing the gradient of $u_1$ vanishes because $\chi(z,\tau)$ satisfies (17.24). We regard $\xi_{ij}(z,\tau)\frac{\partial^2}{\partial x_i\partial x_j}u_0\big(x-\frac{\alpha}{\varepsilon}t,t\big)$ as an unknown function of variables $z$ and $\tau$. Then by Theorem 17.2, the solvability condition for (17.26) takes the form

$$\frac{\partial}{\partial s}u_0\Big(x-\frac{\alpha}{\varepsilon}t,s\Big)\Big|_{s=t} - \widehat{a}_{ij}\frac{\partial^2}{\partial x_i\partial x_j}u_0\Big(x-\frac{\alpha}{\varepsilon}t,t\Big) = f(x,t), \tag{17.27}$$

where the homogenized matrix $\widehat{a}_{ij}$ is defined by the formula

$$\begin{aligned}
\widehat{a}_{ij} = \delta_{ij} + \alpha_i\int_G \chi_j(z,\tau)p(z,\tau)\,dzd\tau + 2\int_G\frac{\partial}{\partial z_j}\chi_j(z,\tau)p(z,\tau)\,dz\,d\tau\\
+ \int_{\partial G}\chi_j(z,\tau)n_i(z,\tau)\,dS_z d\tau.
\end{aligned}$$

Taking into account the definition of $\chi$ and $p$, we can transform the right-hand side of the latter equality as follows:

$$\widehat{a}_{ij} = \int_G (\mathbf{I}+\nabla_z\chi(z,\tau))p(z,\tau)(\mathbf{I}+\nabla_z\chi(z,\tau))^* dz\,d\tau,$$

which implies the positive definiteness of the matrix $\hat{a}_{ij}$.

EXERCISE 17.4. Prove that the matrix $\widehat{a}_{ij}$ is positive definite.

### 17.4. Homogenization theorem.

THEOREM 17.3 (on homogenization). (i) *If $g \in L_2(\mathbb{R}^d)$ and $f \equiv 0$ in (17.1), then*

$$\lim_{\varepsilon \downarrow 0} \left\| u^\varepsilon - u_0\Big(x - \frac{\alpha}{\varepsilon}t, t\Big) \right\|_{L_2(Q_T^\varepsilon)} = 0, \tag{17.28}$$

*where $u_0 = u_0(x,t)$ is the solution to the homogenized Cauchy problem,*

$$\frac{\partial}{\partial t}u_0 = \widehat{a}_{ij}\frac{\partial^2}{\partial x_i \partial x_j}u_0 \quad in\ \mathbb{R}^d \times (0,T), \qquad u_0(x,0) = g(x). \tag{17.29}$$

(ii) *If $g \in H^1(\mathbb{R}^d)$ and $f \equiv 0$, then*

$$\lim_{\varepsilon \downarrow 0} \left\| u^\varepsilon - u_0\Big(x - \frac{\alpha}{\varepsilon}t, t\Big) - \nabla u_0\Big(x - \frac{\alpha}{\varepsilon}t, t\Big)\chi\Big(\frac{x}{\varepsilon}, \frac{t}{\varepsilon^2}\Big) \right\|_{L_2(0,T;H^1(G^\varepsilon(t)))} = 0. \tag{17.30}$$

(iii) *If $f(x,t) = F\big(x - \frac{\alpha}{\varepsilon}t, t\big)$, $F \in L_2(\mathbb{R}^d \times (0,T))$, and $g \in L_2(\mathbb{R}^d)$ in (17.1), then (17.28) holds with a function $u^0$ such that*

$$\frac{\partial}{\partial t}u_0 = \widehat{a}_{ij}\frac{\partial^2}{\partial x_i \partial x_j}u_0 + F(x,t), \quad u_0(x,0) = g(x). \tag{17.31}$$

(iv) *If $f \in L_2(\mathbb{R}^d \times (0,T))$ is independent of $\varepsilon$ and $\alpha \neq 0$, then (17.28) and (17.29) remain valid.*

REMARK 17.3. Assertion (i) of Theorem 17.3 can be reformulated as follows: in the moving coordinates $y = x + \frac{\alpha}{\varepsilon}t$, the solutions $u^\varepsilon\big(x + \frac{\alpha}{\varepsilon}t, t\big)$ converge to the solution $u^0(x,t)$ of the problem (17.29) in the $L_2(0,T; L_2(G^\varepsilon(t) + \frac{\alpha}{\varepsilon}t))$-norm. Assertion (iii) also admits a similar reformulation.

REMARK 17.4. The function $U_\varepsilon^0 = u_0\big(x - \frac{\alpha}{\varepsilon}t, t\big)$ satisfies the equation

$$\frac{\partial}{\partial t}U_\varepsilon^0 = \widehat{a}_{ij}\frac{\partial^2}{\partial x_i \partial x_j}U_\varepsilon^0 + \frac{\alpha}{\varepsilon}\nabla U_\varepsilon^0 = f\Big(x + \frac{\alpha}{\varepsilon}t, t\Big),$$

i.e., for small $\varepsilon$ the effective dynamics includes large convections of order $\varepsilon^{-1}$ in the direction of the vector $\alpha$.

REMARK 17.5. If inclusions are symmetric, then $\alpha = 0$. Consider two examples.
1) For every $t$ the set $G(t)$ is invariant under rotations preserving the periodicity cube $[-1/2, 1/2]^d$.
2) For every motion $S$ preserving the periodicity cube there is $t_0 > 0$ such that $G(t + t_0) = SG(t)$ for all $t$.

EXERCISE 17.5. Prove that in both examples, the vector $\overline{\alpha}$ is invariant under motions preserving the periodicity cube and, consequently, vanishes.

PROOF OF THEOREM 17.3. Let $f$ and $g$ be smooth functions. Then $u_0$ is also a smooth function. We set

$$\begin{aligned} w^\varepsilon = u^\varepsilon - u_0\Big(x - \frac{\alpha}{\varepsilon}t, t\Big) - \varepsilon\Big\{\nabla_x u_0\Big(x - \frac{\alpha}{\varepsilon}t, t\Big)\chi\Big(\frac{x}{\varepsilon}, \frac{t}{\varepsilon^2}\Big) + u_1\Big(x - \frac{\alpha}{\varepsilon}t, t\Big)\Big\} \\ - \varepsilon^2\Big\{\nabla_x\nabla_x u_0\Big(x - \frac{\alpha}{\varepsilon}t, t\Big)\xi\Big(\frac{x}{\varepsilon}, \frac{t}{\varepsilon^2}\Big) \\ + \nabla_x u_1\Big(x - \frac{\alpha}{\varepsilon}t, t\Big)\chi\Big(\frac{x}{\varepsilon}, \frac{t}{\varepsilon^2}\Big) + u_2\Big(x - \frac{\alpha}{\varepsilon}t, t\Big)\Big\}. \end{aligned}$$

Substituting $w^\varepsilon$ in (17.1), we find that

$$\begin{aligned}
&\frac{\partial}{\partial t}w^\varepsilon - \Delta w^\varepsilon = f + \varepsilon^2 v_1^\varepsilon(x,t) \quad \text{in } Q_T^\varepsilon,\\
&\frac{\partial}{\partial n^\varepsilon(t)} w^\varepsilon(x,t) = \varepsilon^3 v_2^\varepsilon(x,t) \quad \text{on } \partial Q_T^\varepsilon \cap \{0<t<T\},\\
&w^\varepsilon\big|_{t=0} = g(x) + \varepsilon v_3^\varepsilon(x) \quad \text{in } \varepsilon G(0),
\end{aligned}$$

where $v_1^\varepsilon(x,t)$, $v_2^\varepsilon(x,t)$, and $v_3^\varepsilon(x)$ satisfy the estimates

$$\|v_1^\varepsilon\|_{L_2(Q_T^\varepsilon)} \le C, \quad \|v_2^\varepsilon\|_{L_2(\partial Q_T^\varepsilon)} \le C\varepsilon^{-1}, \quad \|v_3^\varepsilon\|_{L_2(G^\varepsilon(0))} \le C,$$

which can be verified by using (17.4), (17.24), and (17.27) (check it!). From (17.18) it follows that

$$\|w^\varepsilon\|_{L_2(0,T;H^1(G^\varepsilon(t)))} \le C\varepsilon.$$

Approximating $g \in L_2(\mathbb{R}^d)$ and $f \in L_2(\mathbb{R}^d \times (0,T))$ by smooth functions $g_n \to g$ in $L_2(\mathbb{R}^d)$ $(H^1(\mathbb{R}^d))$ and $f_n \to f$ in $L_2(\mathbb{R}^d \times (0,T))$ and using Lemma 17.1, we obtain all the required assertions. □

REMARK 17.6. Suppose that the data in (17.1) and the boundary of holes are smooth, as in Example 17.1. Then, successively solving problems in the chain, we construct a complete asymptotic expansion for $u^\varepsilon$.

REMARK 17.7. There are examples of structures with $\alpha \neq 0$ (see [**46**]).

## 18. Homogenization of nonautonomous parabolic equations with random coefficients

This section is devoted to homogenization of nonautonomous parabolic equations with random coefficients. Such equations arise in describing media with periodic microstructure whose characteristics alternate in time and are rapidly oscillating stationary random processes.

We consider model problems for operators in the divergence form, as well as operators with lower order terms.

Operators with asymptotically increasing lower order terms are of particular interest. Such operators naturally appear under automodel contractions of variables, for example, if we study the large time behavior of a process. We explain this by the example of the convection–diffusion equation

$$\frac{\partial}{\partial \tau} v(y,\tau) = \operatorname{div}(a(y,\tau)\nabla v(y,\tau)) + b(y,\tau)\nabla v(y,\tau),$$

where the entries of the matrix-valued function $a(y,\tau)$ and the components of the vector-valued function $b(y,\tau)$ are periodic with respect to the spatial variable $y$ and are random stationary functions of time $\tau$. We study the behavior of solutions at large times of order $\varepsilon^{-2}$, where $\varepsilon$ is a small (positive) parameter. Introducing a scale factor $\varepsilon$ and making the automodel change of variables $x = \varepsilon y$, $t = \varepsilon^2\tau$, we obtain the equation

$$\frac{\partial}{\partial t} v^\varepsilon(x,t) = \operatorname{div}\Big(a\Big(\frac{x}{\varepsilon},\frac{t}{\varepsilon^2}\Big)\nabla v^\varepsilon(x,t)\Big) + \frac{1}{\varepsilon} b\Big(\frac{x}{\varepsilon},\frac{t}{\varepsilon^2}\Big)\nabla v^\varepsilon(x,t). \tag{18.1}$$

Then the problem is reduced to the study of the limit behavior of the solution to equation (18.1) on a finite time interval as $\varepsilon \to 0$.

We consider partial differential evolution equations of parabolic type under the action of random exterior forces. Our goal is to study systematic alternations of the phase state of the system caused by the exterior perturbations. In many cases, such alternations can be described by stochastic partial differential equations and the corresponding measures on the function spaces of solutions. Necessary definitions are given in Section 2.

In the case of elliptic and parabolic operators in the divergence form, the homogenization results for equations with random, stochastically homogeneous coefficients look like those for equations with periodic coefficients. Indeed, in both cases, the existence of a limit operator with constant coefficients is asserted.

However, the situation sharply changes in the case of the automodel homogenization of operators with lower order terms. The results for periodic operators with lower order terms are not always generalized to analogous random operators. In particular, the super- and subdiffusion phenomena can be observed when studying the equation

$$\frac{\partial}{\partial t}v = \Delta v + \frac{1}{\varepsilon}b\Big(\frac{x}{\varepsilon}\Big)\nabla v$$

with statistically homogeneous random coefficients. The corresponding examples can be found in [**7, 119**].

Equation (18.1), as well as other equations considered in this section, corresponds to the intermediate case: the convergence almost surely (a.s.) of solutions to a deterministic limit is not always valid. We show that after homogenization, the limit dynamics may remain random; moreover, the solutions converge in law in the energy function space, whereas the convergence almost surely or in probability does not take place in the general case.

### 18.1. Equations with large potential.

18.1.1. *Statement of the problem.* We begin by considering the equation

$$\frac{\partial}{\partial t}u^\varepsilon(x,t) = \operatorname{div}\Big(a\Big(\frac{x}{\varepsilon},\frac{t}{\varepsilon^\alpha}\Big)\nabla u^\varepsilon(x,t)\Big) + \frac{1}{\varepsilon^{1\wedge\frac{\alpha}{2}}}g\Big(\frac{x}{\varepsilon},\frac{t}{\varepsilon^\alpha}\Big)u^\varepsilon(x,t) + f(x,t), \tag{18.2}$$

where $\alpha > 0$ is a positive parameter, $1\wedge\frac{\alpha}{2} = \min\{1, \alpha/2\}$. For the sake of definiteness, we consider the Cauchy problem in $\mathbb{R}^d\times(0,T)$ with the initial condition

$$u^\varepsilon(x,0) = u_0(x) \tag{18.3}$$

and assume that $u_0 \in L_2(\mathbb{R}^d)$ and $f \in L_2((0,T)\times\mathbb{R}^d)$. Note that the Dirichlet and Neumann initial-boundary value problems in a bounded domain can be treated in the same way. The corresponding results will be formulated later.

In an important special case, the time-dependence of the coefficients in (18.2) is governed by a finite-dimensional stationary diffusion process $\xi_s$ and the problem (18.2), (18.3) takes the form

(18.4)

$$\begin{gathered}\frac{\partial}{\partial t}u^\varepsilon(x,t) = \operatorname{div}\Big(a\Big(\frac{x}{\varepsilon},\xi_{\frac{t}{\varepsilon^\alpha}}\Big)\nabla u^\varepsilon(x,t)\Big) + \frac{1}{\varepsilon^{1\wedge\frac{\alpha}{2}}}g\Big(\frac{x}{\varepsilon},\xi_{\frac{t}{\varepsilon^\alpha}}\Big)u^\varepsilon(x,t) + f(x,t),\\ u^\varepsilon(x,0) = u_0(x).\end{gathered}$$

The main assumptions on the data of the problem (18.2), (18.3) are as follows.

**H1.** The coefficients $a_{ij}(y,s)$ and $g(y,s)$ are periodic with respect to $y$ with period 1 along each coordinate direction.

**H2.** The functions $a_{ij}(y,s)$ and $g(y,s)$ are random stationary processes with respect to $s$ and are defined on a probability space $(\Omega,\mathcal{F},\mathbf{P})$ with values in the space of periodic functions. The coefficients $a_{ij}(y,s)=a_{ij}(y,s,\omega)$, $g(y,s)=g(y,s,\omega)$ are measurable with respect to the $\sigma$-algebra $\mathcal{B}(\mathbb{T}^d)\times\mathcal{B}(-\infty,+\infty)\times\mathcal{F}$, where $\mathcal{B}$ is the Borel $\sigma$-algebra.

For the sake of simplicity, we introduce a random ergodic dynamical system $T_\tau$ in $\Omega$ and assume that the processes $a_{ij}(y,s)$ and $g(y,s)$ are the action of $T_\tau$ on the random variables $\widetilde{a}_{ij}(y)$ and $\widetilde{g}(y)$ with values in $L_\infty(\mathbb{T}^d)$, i.e.,

$$a_{ij}(y,s,\omega)=\widetilde{a}_{ij}(y,T_s\omega),\quad g(y,s,\omega)=\widetilde{g}(y,T_s\omega). \tag{18.5}$$

Recall that a random dynamical system and ergodicity were defined in Section 2.

**H3.** *Uniform parabolicity*:

$$\lambda^{-1}|\zeta|^2\le a_{ij}(y,\tau)\zeta_i\zeta_j\le\lambda|\zeta|^2,\quad \lambda>0,\qquad |g(y,s)|\le\lambda$$

for all $y,\tau,\zeta\in\mathbb{R}^d$.

**H4.** *Centering condition.* The average of $g(z,s)$ vanishes, i.e.,

$$\langle\overline{g}\rangle\stackrel{\text{def}}{=}\mathbf{E}\int_{[0,1]^d}g(z,s)\,dz=0\quad\forall\, s.$$

One more condition will be expressed in terms of mixing coefficients. First, we recall some definitions from Section 2.

Suppose $\xi_s$ is an arbitrary stationary random process on a probability space $(\Omega,\mathcal{F},\mathbf{P})$. Denote by $\mathcal{F}_{\le t}$ the $\sigma$-algebra generated by $\xi_s$ for all $s\le t$, and by $\mathcal{F}_{\ge t}$ the $\sigma$-algebra generated by $\xi_s$ for all $s\ge t$. In other words, $\mathcal{F}_{\le t}$ ($\mathcal{F}_{\ge t}$) is the minimal $\sigma$-algebra relative to which the process $\xi_s$ is measurable for all $s\le t$ ($s\ge t$). The existence of such a $\sigma$-algebra is well known in measure theory.

The *strong mixing coefficient* of a process $\xi_s$ is the function

$$\kappa(\gamma)=\sup_{E_1\in\mathcal{F}_{\le t},\,E_2\in\mathcal{F}_{\ge(t+\gamma)}}|\mathbf{P}(E_1)\mathbf{P}(E_2)-\mathbf{P}(E_1\cap E_2)|,$$

where the supremum is taken over all events $E_1$ and $E_2$ such that $E_1\in\mathcal{F}_{\le t}$ and $E_2\in\mathcal{F}_{\ge(t+\gamma)}$. Since the process is stationary, the right-hand side of the last formula is independent of $t$. In particular, for $t$ we can take 0.

The *uniform mixing coefficient* $\varphi(\gamma)$ is defined by the formula

$$\varphi(\gamma)=\sup_{\substack{E_1\in\mathcal{F}_{\le t},\,E_2\in\mathcal{F}_{\ge(t+\gamma)}\\ \mathbf{P}(E_2)\ne0}}\left|\mathbf{P}(E_1)-\frac{\mathbf{P}(E_1\cap E_2)}{\mathbf{P}(E_2)}\right|$$

and the *maximal correlation coefficient* $\rho(\gamma)$ is defined as follows:

$$\rho(\gamma)=\sup_{\eta_1,\eta_2}\left|\frac{\mathbf{E}((\eta_1-\mathbf{E}\eta_1)(\eta_2-\mathbf{E}\eta_2))}{\sqrt{\mathbf{E}\eta_1^2\,\mathbf{E}\eta_2^2}}\right|,$$

where $\eta_1\in L_2(\Omega,\mathcal{F}_{\le t},\mathbf{P})$ and $\eta_2\in L_2(\Omega,\mathcal{F}_{\ge(t+\gamma)},\mathbf{P})$.

We construct the $\sigma$-algebras $\mathcal{F}_{\le t}$ and $\mathcal{F}_{\ge t}$ generated by the coefficients of the operator in (18.2) and calculate the corresponding mixing coefficients. Suppose that the following condition is satisfied.

**H5.** *Mixing.* At least one of the following relations holds:

$$\int_0^\infty\kappa(\gamma)\,d\gamma<\infty,\quad\int_0^\infty\sqrt{\varphi(\gamma)}d\gamma<\infty\quad\int_0^\infty\rho(\gamma)\,d\gamma<\infty.$$

REMARK 18.1. For a more general stationary process $\{a_{ij}(s), g(s)\}$ with values in $L_\infty(\mathbb{T}^d)$ (thereby in $L_2(\mathbb{T}^d)$) we show how to construct an equivalent process over a new probability space $(\widetilde{\Omega}, \widetilde{\mathcal{F}}, \widetilde{\mathbf{P}})$ equipped with the action of a dynamical system $\widetilde{T}_\tau$ such that $\{\widetilde{a}_{ij}(s), \widetilde{g}(s)\} = \widetilde{T}_s(\{a_{ij}(0), g(0)\})$. For the sake of simplicity, we assume that the process $\{a_{ij}(s), g(s)\}$, regarded as a function with values in $L_2(\mathbb{T}^d)$, is continuous with respect to $s$.

Introduce a probability space $(\widetilde{\Omega}, \widetilde{\mathcal{F}}, \widetilde{\mathbf{P}})$ consisting of $\widetilde{\Omega} = (C(\mathbb{R}; L_2(\mathbb{T}^d)^{(n^2+1)}))$ with the topology of uniform convergence on compact sets with respect to the first argument, the Borel $\sigma$-algebra, and the probabilistic measure $\widetilde{\mathbf{P}}$ defined by the equality $\widetilde{\mathbf{P}}(\mathcal{A}) = \mathbf{P}\{(\{a_{ij}(\cdot)\}, g(\cdot)) \in \mathcal{A}\}$. Then the mapping $(\{a_{ij}(\cdot)\}, g(\cdot)) :$ $\Omega \to \widetilde{\Omega}$ is measurable. Therefore, the measure $\widetilde{\mathbf{P}}$ is well defined. On $\widetilde{\Omega}$, we set $a_{ij}(s, \widetilde{\omega}) = \widetilde{\omega}_{ij}(s)$ and $g(s, \widetilde{\omega}) = \widetilde{\omega}_{(n^2+1)}(s)$. By the definition of $\widetilde{\mathbf{P}}$, these random functions have the same finite-dimensional distributions as the initial coefficients (check it!). In the new probability space $\widetilde{\Omega}$, we introduce the group $\widetilde{T}_\tau$ of shifts with respect to the second argument: $\widetilde{T}_\tau(\widetilde{\omega}(y, s) = \widetilde{\omega}(y, s + \tau)$.

EXERCISE 18.1. Under the above assumptions, show that the group $\widetilde{T}_\tau$ preserves the measure $\widetilde{\mathbf{P}}$; moreover, it is continuous and ergodic.

REMARK 18.2. In fact, condition **H4** is not too restrictive: it can always be satisfied by the change of the unknown function $\widetilde{u}^\varepsilon(x, t) = \exp(\langle \overline{g} \rangle / \varepsilon) u(x, t)$, where $\langle \overline{g} \rangle$ is the full average of $g$. If $\langle \overline{g} \rangle \neq 0$, then the homogenization proceeds against the background of exponential increase or decrease.

Conditions **H1**–**H3** guarantee the well-posedness of the Cauchy problem for every $\varepsilon > 0$.

LEMMA 18.1. *Let conditions* **H1**–**H3** *hold. For every* $\varepsilon > 0$ *the problem* (18.2), (18.3) *is uniquely solvable for all* $\omega \in \Omega$ *and the solution* $u^\varepsilon$ *is a measurable mapping* $u^\varepsilon : (\Omega, \mathcal{F}) \to (L_2(0, T; H^1(\mathbb{R}^d)) \cap C((0, T); L_2(\mathbb{R}^d)), \mathcal{B})$. *Furthermore,*

$$\|u^\varepsilon\|_{C((0,T);L_2(\mathbb{R}^d))} + \|u^\varepsilon\|_{L_2((0,T);H^1(\mathbb{R}^d))} \leq C(\varepsilon). \tag{18.6}$$

PROOF. The unique solvability and the *a priori* estimate (18.6) can be obtained by standard methods of the theory of parabolic equations. Using the continuous dependence on the data, it is easy to prove that the solution is measurable. □

18.1.2. *Factorization.* The study of the asymptotic behavior of the solution to the problem (18.2), (18.3) can be considerably simplified by means of factorization. We represent $u^\varepsilon$ as the product

$$u^\varepsilon(x, t) = v^\varepsilon(x, t) \exp\biggl(\frac{1}{\varepsilon^{1\wedge\frac{\alpha}{2}}} \int_0^t \langle g \rangle\Bigl(\frac{s}{\varepsilon^\alpha}\Bigr)\, ds\biggr), \tag{18.7}$$

where $\langle g \rangle(s) \stackrel{\text{def}}{=} \int_{[0,1]^d} g(z, s)\, dz$. A direct computation shows that $v^\varepsilon$ satisfies the problem

$$\begin{aligned} \frac{\partial}{\partial t} v^\varepsilon(x, t) &= \operatorname{div}\Bigl(a\Bigl(\frac{x}{\varepsilon}, \frac{t}{\varepsilon^\alpha}\Bigr) \nabla v^\varepsilon(x, t)\Bigr) + \frac{1}{\varepsilon^{1\wedge\frac{\alpha}{2}}} \widetilde{g}\Bigl(\frac{x}{\varepsilon}, \frac{t}{\varepsilon^\alpha}\Bigr) v^\varepsilon(x, t) \\ &\quad + f(x, t) \exp\biggl(-\frac{1}{\varepsilon^{1\wedge\frac{\alpha}{2}}} \int_0^t \langle g \rangle\Bigl(\frac{s}{\varepsilon^\alpha}\Bigr)\, ds\biggr), \\ v^\varepsilon(x, 0) &= u_0(x), \end{aligned} \tag{18.8}$$

where $\widetilde{g}(z,s) = g(z,s) - \langle g\rangle(s)$. The problem (18.8) will be studied later. Now, we consider the exponential factor in (18.7).

LEMMA 18.2. *Suppose that at least one of the inequalities in condition* **H5** *holds. Then for* $\alpha \le 2$ *the process*

$$\zeta_t^\varepsilon = \frac{1}{\varepsilon^{1\wedge\frac{\alpha}{2}}}\int_0^t \langle g\rangle\Big(\frac{s}{\varepsilon^\alpha}\Big)\,ds \tag{18.9}$$

*satisfies the functional central limit theorem (invariance principle) with zero mean and the diffusion defined by the formula*

$$\sigma^2 = 2\int_0^\infty \mathbf{E}(\langle g\rangle(0)\langle g\rangle(s))\,ds, \tag{18.10}$$

*i.e., the sequence* $\zeta_t^\varepsilon$ *converges in law to the process* $\{\sigma W_t\}$ *in the space of continuous functions* $C[0,\infty)$*, where* $W_t$ *is a standard Wiener process.*

*If* $\alpha > 2$*, then the* $\zeta_\cdot^\varepsilon$ *converge almost surely to zero uniformly on compact sets.*

PROOF. The first assertion (the functional central limit theorem for stationary processes with good mixing properties) is a classical result of the theory of stationary processes (cf. Section 2). The proof can be found, for example, in [**81**, Chapter 9]. The second assertion easily follows from the first. □

In the case $\alpha \le 2$, a simple consequence of Lemma 18.2 is the convergence in law,

$$\exp\Big(\frac{1}{\varepsilon^{\alpha/2}}\int_0^t \langle g\rangle\Big(\frac{s}{\varepsilon^\alpha}\Big)\,ds\Big) \xrightarrow{\mathcal{L}} \exp(\sigma W_t), \tag{18.11}$$

in the space $C[0,\infty)$. For $\alpha > 2$ the uniform convergence on compact sets almost surely holds,

$$\exp\Big(\frac{1}{\varepsilon}\langle g\rangle\Big(\frac{t}{\varepsilon^\alpha}\Big)\Big) \to 1. \tag{18.12}$$

18.1.3. *A priori estimates for the factorized equation.* We need *a priori* estimates for the solution to the problem

$$\begin{aligned}&\frac{\partial}{\partial t}z^\varepsilon(x,t) = \operatorname{div}\Big(a\Big(\frac{x}{\varepsilon},\frac{t}{\varepsilon^\alpha}\Big)\nabla z^\varepsilon(x,t)\Big) + \frac{1}{\varepsilon^{1\wedge\frac{\alpha}{2}}}\widetilde{g}\Big(\frac{x}{\varepsilon},\frac{t}{\varepsilon^\alpha}\Big)z^\varepsilon(x,t) + h(x,t),\\ &z^\varepsilon(x,0) = z_0(x).\end{aligned} \tag{18.13}$$

Note that (18.13) differs from (18.8) by a more general right-hand side.

PROPOSITION 18.1. *The solution to the problem* (18.13) *satisfies the a priori estimate*

$$\|z^\varepsilon\|_{L_\infty((0,T);L_2(\mathbb{R}^d))} + \|z^\varepsilon\|_{L_2(0,T;H^1(\mathbb{R}^d))} \le C(\|z_0\|_{L_2(\mathbb{R}^d)} + \|h\|_{L_2(0,T;H^{-1}(\mathbb{R}^d))}), \tag{18.14}$$

*where the constant* $C$ *is independent of* $\varepsilon$.

PROOF. By construction, for every $s$ and $\omega$ the function $\widetilde{g}(y,s)$ has zero mean over a period. Therefore, the equation $\Delta\widetilde{Q} = \widetilde{g}$ is solvable in the space of periodic functions. We set $\widetilde{q} = \nabla_y\widetilde{Q}$. By the uniform boundedness of $\widetilde{g}$, we have $|\widetilde{q}(y,s)| \le C$ (check it!). Thus, we have constructed a uniformly bounded solution to the equation

$\operatorname{div}_y \widetilde{q}(y,s) = \widetilde{g}(y,s)$, which can be written in the coordinates $x = \varepsilon y$, $t = \varepsilon^2 s$ as follows:

$$\varepsilon \operatorname{div}_x \widetilde{q}^{\varepsilon}(x,t) = \widetilde{g}^{\varepsilon}(x,t). \tag{18.15}$$

Hereinafter, for an arbitrary function $F(y,s)$ we adopt the notation

$$\begin{aligned} F^{\varepsilon}(x,t) &= F\Big(\frac{x}{\varepsilon}, \frac{t}{\varepsilon^2}\Big), \\ \frac{\partial}{\partial y_i} F^{\varepsilon}(x,t) &= \frac{\partial}{\partial y_i} F\Big(y, \frac{t}{\varepsilon^2}\Big)\Big|_{y=x/\varepsilon}, \\ \frac{\partial}{\partial s} F^{\varepsilon}(x,t) &= \frac{\partial}{\partial s} F\Big(\frac{x}{\varepsilon}, s\Big)\Big|_{s=t/\varepsilon^2}. \end{aligned} \tag{18.16}$$

Multiplying the equation in (18.13) by $z^{\varepsilon}$ and integrating over the set $\mathbb{R}^d \times (0,T)$, we find that

$$\begin{aligned} \int_{\mathbb{R}^d} (z^{\varepsilon}(x,t))^2 dx - \int_{\mathbb{R}^d} (z_0(x))^2 dx &= -\int_0^t \int_{\mathbb{R}^d} a_{ij,\varepsilon}(x,\tau) \nabla z^{\varepsilon}(x,\tau) \cdot \nabla z^{\varepsilon}(x,\tau)\, dx\, d\tau \\ &+ \varepsilon^{-1} \int_0^t \int_{\mathbb{R}^d} \widetilde{g}_{\varepsilon}(x,\tau)(z^{\varepsilon}(x,\tau))^2 dx\, d\tau + \int_0^t \int_{\mathbb{R}^d} z^{\varepsilon}(x,\tau) h(x,\tau)\, dx\, d\tau. \end{aligned}$$

Further, using (18.15) and integrating by parts, we see that

$$\begin{aligned} \int_{\mathbb{R}^d} (z^{\varepsilon}(x,t))^2 dx &+ \int_0^t \int_{\mathbb{R}^d} a_{ij,\varepsilon}(x,\tau) \nabla z^{\varepsilon}(x,\tau) \cdot \nabla z^{\varepsilon}(x,\tau)\, dx\, d\tau \\ &= \int_{\mathbb{R}^d} (z_0(x))^2 dx + \int_0^t \int_{\mathbb{R}^d} z^{\varepsilon}(x,\tau) \widetilde{q}_{\varepsilon}(x,\tau) \cdot \nabla z^{\varepsilon}(x,\tau)\, dx\, d\tau \\ &+ \int_0^t \int_{\mathbb{R}^d} z^{\varepsilon}(x,\tau) h(x,\tau)\, dx\, d\tau. \end{aligned} \tag{18.17}$$

Denote by $R^{\varepsilon}(t)$ the right-hand side of (18.17). For every $\gamma > 0$ we have

$$\begin{aligned} |R^{\varepsilon}(t)| &\le \int_{\mathbb{R}^d} (z_0(x))^2 dx + \gamma^{-1} \int_0^t \int_{\mathbb{R}^d} |z^{\varepsilon}(x,\tau) \widetilde{q}_{\varepsilon}(x,\tau)|^2 dx\, d\tau \\ &+ \gamma \int_0^t \int_{\mathbb{R}^d} |\nabla z^{\varepsilon}(x,\tau)|^2 dx\, d\tau + \int_0^t \|z^{\varepsilon}(\cdot,\tau)\|_{H^1(\mathbb{R}^d)} \|h(\cdot,\tau)\|_{H^{-1}(\mathbb{R}^d)} d\tau \\ &\le \int_{\mathbb{R}^d} (z_0(x))^2 dx + C\gamma^{-1} \int_0^t \int_{\mathbb{R}^d} |z^{\varepsilon}(x,\tau)|^2 dx\, d\tau \\ &+ \gamma \int_0^t \int_{\mathbb{R}^d} |\nabla z^{\varepsilon}(x,\tau)|^2 dx d\tau + \gamma^{-1} \int_0^t \|h(\cdot,\tau)\|^2_{H^{-1}(\mathbb{R}^d)} d\tau \\ &+ \gamma \int_0^t \int_{\mathbb{R}^d} (|z^{\varepsilon}(\cdot,\tau)|^2 + |\nabla z^{\varepsilon}(\cdot,\tau)|^2)\, dx\, d\tau. \end{aligned}$$

It remains to choose sufficiently small $\gamma$ and use this inequality in (18.17). Then (18.14) follows from the Gronwall lemma. □

18.1.4. *Auxiliary problems.* To pass to the limit in (18.8), we need a number of auxiliary problems that are, in fact, cell problems for (18.8). We introduce auxiliary equations and study their properties.

We set $A = \frac{\partial}{\partial y_i} a_{ij}(y,s) \frac{\partial}{\partial y_j}$ and consider two equations in the cylinder $\mathbb{T}^d \times (-\infty, +\infty)$:

$$\frac{\partial}{\partial s}\chi^j(y,s) - A\chi^j(y,s) = \frac{\partial}{\partial y_i} a_{ij}(y,s), \tag{18.18}$$

$$\frac{\partial}{\partial s}G(y,s) - AG(y,s) = \widetilde{g}(y,s). \tag{18.19}$$

PROPOSITION 18.2. *Equations* (18.18) *and* (18.19) *have stationary solutions in the space* $L_\infty((-\infty,+\infty); C(\mathbb{T}^d)) \cap L_2^{\mathrm{loc}}((-\infty,+\infty); H^1(\mathbb{T}^d))$. *For each of these equations the stationary solution is unique up to an additive constant. The following estimates hold:*

$$\begin{aligned} &\|\chi\|_{L_2((N,N+1);H^1(\mathbb{T}^d))} \le C, \quad \|\chi\|_{L_\infty((-\infty,\infty);L_2(\mathbb{T}^d))} \le C, \\ &\|G\|_{L_2((N,N+1);H^1(\mathbb{T}^d))} \le C, \quad \|G\|_{L_\infty((-\infty,\infty);L_2(\mathbb{T}^d))} \le C \end{aligned} \tag{18.20}$$

*uniformly with respect to* $N \in \mathbb{R}$; *moreover, the constant* $C$ *is not random and depends only on* $\lambda$ *in condition* **H3**.

PROOF. For the sake of definiteness, we consider equation (18.18) and establish the existence of a stationary solution. To this end, we consider the sequence of Cauchy problems

$$\begin{aligned} &\frac{\partial}{\partial s}\chi_N^j(y,s) - A\chi_N^j(y,s) = \frac{\partial}{\partial y_i} a_{ij}(y,s)\mathbf{1}_{[N,N+1)}(s), \ (y,s) \in \mathbb{T}^d \times (N,+\infty), \\ &\chi_N^j(y,N) = 0. \end{aligned} \tag{18.21}$$

For $s < N$ we extend $\chi_N^j(y,s)$ by zero. In view of the well-known Nash estimates (cf. [**77**]), the function $\chi_N^j$ is continuous and satisfies the estimate

$$\|\chi_N^j\|_{L_\infty((-\infty,N+1]\times\mathbb{T}^d)} \le c_1(\lambda).$$

Since the right-hand side of (18.21) vanishes for $s > N+1$, the same estimate holds for all $s$ in view of the maximum principle:

$$\|\chi_N^j\|_{L_\infty((-\infty,+\infty)\times\mathbb{T}^d)} \le c_1(\lambda).$$

We need to show that $\chi_N^j(y,s)$ exponentially decreases as $(s-N) \to \infty$.

LEMMA 18.3. *There are deterministic constants* $c_2(\lambda) > 0$ *and* $c_3(\lambda) > 0$ *that are independent of* $N$ *and such that the following estimate holds*:

$$|\chi_N^j(y,s)| \le c_2 \exp(-c_3(s-N)). \tag{18.22}$$

PROOF. First of all, we note that

$$\int_{\mathbb{T}^d} \chi_N^j(y,s)\,dy = 0 \tag{18.23}$$

for every $s \in \mathbb{R}$. Indeed, integrating (18.18) over the cylinder $(s_1,s_2) \times \mathbb{T}^d$, we find

$$\int_{\mathbb{T}^d} \chi_N^j(y,s_1)\,dy = \int_{\mathbb{T}^d} \chi_N^j(y,s_2)\,dy.$$

This implies (18.23) since $\chi_N^j(y,N) = 0$.

In view of (18.23), we can apply the Poincaré inequality to $\chi_N^j(\cdot,s)$ for every $s$. Together with **H3**, this yields

$$\begin{aligned}\int_{\mathbb{T}^d} a_{ij}(y,s)\frac{\partial}{\partial y_i}\chi_N^k(y,s)\frac{\partial}{\partial y_j}\chi_N^k(y,s)\,dy \\ \geq \lambda^{-1}\int_{\mathbb{T}^d}|\nabla_y\chi_N^k(y,s)|^2\,dy \geq c_4\lambda^{-1}\int_{\mathbb{T}^d}|\chi_N^k(y,s)|^2dy.\end{aligned} \tag{18.24}$$

Multiplying (18.18) by $\chi_N^j(y,s)$ and integrating over the cylinder $\mathbb{T}^d\times(s_1,s_2)$, we obtain

$$\|\chi_N^k(\cdot,s_2)\|^2_{L_2(\mathbb{T}^d)} - \|\chi_N^k(\cdot,s_1)\|^2_{L_2(\mathbb{T}^d)} \leq -c_4\lambda^{-1}\int_{s_1}^{s_2}\|\chi_N^k(\cdot,s)\|^2_{L_2(\mathbb{T}^d)}ds \tag{18.25}$$

for all $s_1$ and $s_2$ such that $N+1\leq s_1\leq s_2$.

EXERCISE 18.2. Under the assumption (18.25), prove that $\|\chi_N^k(\cdot,s)\|^2_{L_2(\mathbb{T}^d)}$ monotonically decreases for $s\geq N+1$ and satisfies the estimate

$$\|\chi_N^k(\cdot,s)\|^2_{L_2(\mathbb{T}^d)} \leq c_2\exp(-c_3(s-(N+1)).$$

To complete the proof, it remains to use the Nash estimate. □

Define $\chi(y,s)$ by the formula

$$\chi^k(y,s) = \sum_{N=-\infty}^{+\infty}\chi_N^k(y,s). \tag{18.26}$$

By construction, Lemma 18.3 implies that $\chi^j(y,s)$ is a solution to equation (18.18) and satisfies the estimate (18.20). Let us verify that this solution is stationary. By construction and the stationarity of $a_{ij}(\cdot,s)$, any finite-dimensional distribution is invariant under shifts by integers. Now, we repeat the above procedure, taking the partition with a rational step $q=1/l$ instead of the integer partition of the time axis. Then the constructed solution coincides with $\chi(y,s)$. Indeed, it suffices to collect auxiliary solutions in blocks of length $l$ and use the uniqueness of a solution to the Cauchy problem. This implies the invariance of finite-dimensional distributions under rational shifts, which, in turn, implies that the solution is stationary.

The uniqueness up to a constant (which is, in general, random) of a stationary solution can be proved in the same way as in Lemma 18.3. Indeed, if there are two different stationary solutions with zero means over a torus, then the difference exponentially decreases as $s\to\infty$, which contradicts the fact that the solution is stationary. □

The choice of an additive constant for $\chi^k(y,s)$ and $G(y,s)$ can be fixed by the normalization conditions

$$\int_{\mathbb{T}^d}\chi^j(y,s)\,dy = 0, \qquad \int_{\mathbb{T}^d}G(y,s)\,dy = 0. \tag{18.27}$$

Hereinafter, the conditions (18.27) are assumed to be satisfied. Note that conditions (18.27) hold because, as we can see in the proof of Proposition 18.2, the corresponding integrals are independent of $s$.

We set $\widetilde{\chi}^k(y,\omega)=\chi^k(y,0,\omega)$ and $\widetilde{G}(y,\omega)=G(y,0,\omega)$. These functions are well defined because $\chi^k(y,s)$ and $G(y,s)$ are continuous with respect to $s$.

Since $a_{ij}(y,s+\tau,\omega)=a_{ij}(y,s,T_\tau\omega)$ by definition, from the uniqueness of the solution to the problem (18.18) it follows that $\chi^k(y,s+\tau,\omega)=\chi^k(y,s,T_\tau\omega)$.

In particular, $\chi^k(y,s,\omega) = \chi^k(y,0,T_s\omega) = \widetilde{\chi}^k(y,T_s\omega)$. Similarly, $G(y,s,\omega) = G(y,0,T_s\omega) = \widetilde{G}(y,T_s\omega)$.

In the further consideration, we need two families of auxiliary functions. The first family is obtained for every $s \in \mathbb{R}$ from the equation

$$\frac{\partial}{\partial y_i} a_{ij}(y,s) \frac{\partial}{\partial y_j} \psi^k(y,s) = -\frac{\partial}{\partial y_i} a_{ik}(y,s). \tag{18.28}$$

The existence and uniqueness of periodic solutions were discussed in Section 7. We only recall that these problems are solvable for any $s \in \mathbb{R}$ and $\omega \in \Omega$ because the right-hand side has zero mean. The solution $\psi^k(y,s)$ is defined up to an additive random function of the variable $s$, which serves as a parameter in (18.28).

To construct the second family, we introduce the functions

$$\overline{a}_{ij}(y) = \mathbf{E} a_{ij}(y,s), \quad \overline{g}(y) = \mathbf{E}\widetilde{g}(y,s) = \mathbf{E} g(y,s), \tag{18.29}$$

which are independent of $s$ since the random functions $a_{ij}$ and $g$ are stationary. After that, we find $\overline{\psi}^k(y)$ and $\overline{G}(y)$ from the equations

$$\frac{\partial}{\partial y_i} \overline{a}_{ij}(y) \frac{\partial}{\partial y_j} \overline{\psi}^k(y) = -\frac{\partial}{\partial y_i} \overline{a}_{ik}(y), \tag{18.30}$$

$$\frac{\partial}{\partial y_i} \overline{a}_{ij}(y) \frac{\partial}{\partial y_j} \overline{G}(y) = -\overline{g}(y). \tag{18.31}$$

18.1.5. *Homogenization of the factorized equation. I: Case $f = 0$.* Consider a special case of the homogeneous equation (18.2). We set $f = 0$ in (18.2) and show that the homogenization result is valid almost surely for the factorized problem (18.8) for every $\alpha > 0$. However, the procedure for constructing the coefficients of the limit operator, as well as the form of the operator, depends on the choice of $\alpha$. Therefore, we consider independently three possibilities: $0 < \alpha < 2$, $\alpha > 2$, and the critical case $\alpha = 2$.

In the case $\alpha = 2$, the following assertion holds.

THEOREM 18.1. *Suppose that $\alpha = 2$, $f = 0$, and $v^\varepsilon$ is the solution to problem (18.8). Let $\varepsilon \to 0$. Then the sequence of solutions $v^\varepsilon$ converges almost surely in the space $L_\infty(0,T;L_2(\mathbb{R}^d))$ to the solution of the Cauchy problem*

$$\begin{aligned} &\frac{\partial}{\partial t} v^0(x,t) = \operatorname{div}(\widehat{a}\nabla v^0(x,t)) + \widehat{b}\cdot\nabla v^0(x,t) + \widehat{G} v^0(x,t), \\ &v^0(x,0) = u_0(x) \end{aligned} \tag{18.32}$$

*with the constant coefficients*

$$\widehat{a}_{ij} = \mathbf{E}\int_{\mathbb{T}^d} a_{ik}(y,s)\Big(\delta_{kj} + \frac{\partial}{\partial y_k}\chi^j(y,s)\Big)\,dy, \tag{18.33}$$

$$\widehat{b}_i = \mathbf{E}\int_{\mathbb{T}^d} \Big(\widetilde{g}(y,s)\chi^i(y,s) + a_{ij}(y,s)\frac{\partial}{\partial y_j} G(y,s)\Big)dy, \tag{18.34}$$

$$\widehat{G} = \mathbf{E}\int_{\mathbb{T}^d} \widetilde{g}(y,s) G(y,s)\,dy. \tag{18.35}$$

Similar results hold for the remaining values of $\alpha$.

THEOREM 18.2. *Suppose that $\alpha < 2$, $f = 0$, and $v^\varepsilon$ is the solution to the problem (18.8). Let $\varepsilon \to 0$. Then the sequence $v^\varepsilon$ converges in probability in the space*

$L_\infty(0,T;L_2(\mathbb{R}^d))$ *to the solution of the Cauchy problem* (18.32) *with the constant coefficients*

$$\widehat{a}_{ij} = \mathbf{E}\int_{\mathbb{T}^d} a_{ik}(y,s)\Big(\delta_{kj} + \frac{\partial}{\partial y_k}\psi^j(y,s)\Big)\,dy, \tag{18.36}$$

$$\widehat{b} = 0, \quad \widehat{G} = 0. \tag{18.37}$$

THEOREM 18.3. *Suppose that $\alpha > 2$, $f = 0$, and $v^\varepsilon$ is the solution to the problem* (18.8). *Let $\varepsilon \to 0$. Then the sequence $v^\varepsilon$ converges in probability in the space $L_\infty(0,T;L_2(\mathbb{R}^d))$ to the solution of the Cauchy problem* (18.32) *with the constant coefficients*

$$\widehat{a}_{ij} = \int_{\mathbb{T}^d} \overline{a}_{ik}(y)\Big(\delta_{kj} + \frac{\partial}{\partial y_k}\overline{\psi}^j(y)\Big)\,dy, \tag{18.38}$$

$$\widehat{b} = 0, \quad \widehat{G} = \mathbf{E}\int_{\mathbb{T}^d} \overline{g}(y)\overline{G}(y)\,dy. \tag{18.39}$$

PROOF OF THEOREM 18.1. Using (18.15), we write (18.8) in the form (recall that $f = 0$)

$$\frac{\partial}{\partial t}v^\varepsilon(x,t) = \operatorname{div}(a^\varepsilon(x,t)\nabla v^\varepsilon(x,t)) + \operatorname{div}(\widetilde{q}^\varepsilon(x,t)v^\varepsilon(x,t)) - \widetilde{q}^\varepsilon(x,t)\nabla v^\varepsilon(x,t).$$

Taking into account the estimate (18.14), we obtain the inequality

$$\|\partial_t v^\varepsilon\|_{L_2(0,T;H^{-1}(\mathbb{R}^n))} \le C. \tag{18.40}$$

By [**78**, Theorem 5.1], the family $\{v^\varepsilon\}$ is compact in the space $L_2(0,T;L_2^{\mathrm{loc}}(\mathbb{R}^n))$.

Assume that $u_0 \in C_0^\infty$, so that the solution $v^0$ to the problem (18.32) is infinitely differentiable and, together with all its derivatives, decreases exponentially as $|x| \to \infty$. Introduce the notation

$$\begin{aligned}
\breve{a}_{ij}(s) &= \int_{\mathbb{T}^d} a_{ik}(y,s)\Big(\delta_{kj} + \frac{\partial}{\partial y_k}\chi^j(y,s)\Big)\,dy,\\
\breve{b}_i(s) &= \mathbf{E}\int_{\mathbb{T}^d} \Big(\widetilde{g}(y,s)\chi^i(y,s) + a_{ij}(y,s)\frac{\partial}{\partial y_j}G(y,s)\Big)dy,\\
\breve{G}(s) &= \mathbf{E}\int_{\mathbb{T}^d} \widetilde{g}(y,s)G(y,s)\,dy.
\end{aligned}$$

Substitute the expression

$$\breve{v}^\varepsilon(x,t) = v^0(x,t) + \varepsilon\chi\Big(\frac{x}{\varepsilon},\frac{t}{\varepsilon^2}\Big)\cdot\nabla v^0 + \varepsilon G\Big(\frac{x}{\varepsilon},\frac{t}{\varepsilon^2}\Big)v^0$$

in (18.8). Taking into account (18.18) and (18.19), we find after simple transformations that

$$\begin{aligned}
&\frac{\partial}{\partial t}(\breve{v}^\varepsilon(x,t) - v^\varepsilon(x,t)) - \operatorname{div}\Big(a\Big(\frac{x}{\varepsilon},\frac{t}{\varepsilon^2}\Big)\nabla(\breve{v}^\varepsilon(x,t) - v^\varepsilon(x,t))\Big)\\
&\quad - \frac{1}{\varepsilon}\widetilde{g}\Big(\frac{x}{\varepsilon},\frac{t}{\varepsilon^2}\Big)(\breve{v}^\varepsilon(x,t) - v^\varepsilon(x,t))\\
&= \frac{\partial}{\partial t}v^0 + \frac{1}{\varepsilon}\frac{\partial}{\partial s}\chi^k\Big(\frac{x}{\varepsilon},\frac{t}{\varepsilon^2}\Big)\frac{\partial}{\partial x_k}v^0 + \frac{1}{\varepsilon}\frac{\partial}{\partial s}G\Big(\frac{x}{\varepsilon},\frac{t}{\varepsilon^2}\Big)v^0 + \varepsilon\chi^k\Big(\frac{x}{\varepsilon},\frac{t}{\varepsilon^2}\Big)\frac{\partial}{\partial t}\frac{\partial}{\partial x_k}v^0\\
&\quad + \varepsilon G\Big(\frac{x}{\varepsilon},\frac{t}{\varepsilon^2}\Big)\frac{\partial}{\partial t}v^0 - a_{ij}\Big(\frac{x}{\varepsilon},\frac{t}{\varepsilon^2}\Big)\frac{\partial^2}{\partial x_i\partial x_j}v^0
\end{aligned}$$

$$
\begin{aligned}
&-\frac{1}{\varepsilon}\frac{\partial}{\partial y_i}a_{ij}\Big(\frac{x}{\varepsilon},\frac{t}{\varepsilon^2}\Big)\frac{\partial}{\partial x_j}v^0-\frac{1}{\varepsilon}\widetilde{g}\Big(\frac{x}{\varepsilon},\frac{t}{\varepsilon^2}\Big)v^0\\
&-\frac{1}{\varepsilon}\frac{\partial}{\partial y_i}a_{ij}\Big(\frac{x}{\varepsilon},\frac{t}{\varepsilon^2}\Big)\frac{\partial}{\partial y_j}\chi^k\Big(\frac{x}{\varepsilon},\frac{t}{\varepsilon^2}\Big)\frac{\partial}{\partial x_k}v^0\\
&-a_{ij}\Big(\frac{x}{\varepsilon},\frac{t}{\varepsilon^2}\Big)\frac{\partial}{\partial y_i}\chi^k\Big(\frac{x}{\varepsilon},\frac{t}{\varepsilon^2}\Big)\frac{\partial^2}{\partial x_k\partial x_j}v^0\\
&-\frac{\partial}{\partial y_i}\Big(a_{ij}\Big(\frac{x}{\varepsilon},\frac{t}{\varepsilon^2}\Big)\chi^k\Big(\frac{x}{\varepsilon},\frac{t}{\varepsilon^2}\Big)\Big)\frac{\partial^2}{\partial x_k\partial x_j}v^0\\
&-\varepsilon a_{ij}\Big(\frac{x}{\varepsilon},\frac{t}{\varepsilon^2}\Big)\chi^k\Big(\frac{x}{\varepsilon},\frac{t}{\varepsilon^2}\Big)\frac{\partial^3}{\partial x_k\partial x_i\partial x_j}v^0\\
&-\widetilde{g}\Big(\frac{x}{\varepsilon},\frac{t}{\varepsilon^2}\Big)\chi^k\Big(\frac{x}{\varepsilon},\frac{t}{\varepsilon^2}\Big)\frac{\partial}{\partial x_k}v^0-\frac{1}{\varepsilon}\frac{\partial}{\partial y_i}a_{ij}\Big(\frac{x}{\varepsilon},\frac{t}{\varepsilon^2}\Big)\frac{\partial}{\partial y_j}G\Big(\frac{x}{\varepsilon},\frac{t}{\varepsilon^2}\Big)v^0\\
&-a_{ij}\Big(\frac{x}{\varepsilon},\frac{t}{\varepsilon^2}\Big)\frac{\partial}{\partial y_i}G\Big(\frac{x}{\varepsilon},\frac{t}{\varepsilon^2}\Big)\frac{\partial}{\partial x_j}v^0-\frac{\partial}{\partial y_i}\Big(\Big(\frac{x}{\varepsilon},\frac{t}{\varepsilon^2}\Big)G\Big(\frac{x}{\varepsilon},\frac{t}{\varepsilon^2}\Big)\Big)\frac{\partial}{\partial x_j}v^0\\
&-\varepsilon a_{ij}\Big(\frac{x}{\varepsilon},\frac{t}{\varepsilon^2}\Big)G\Big(\frac{x}{\varepsilon},\frac{t}{\varepsilon^2}\Big)\frac{\partial^2}{\partial x_i\partial x_j}v^0-\widetilde{g}\Big(\frac{x}{\varepsilon},\frac{t}{\varepsilon^2}\Big)G\Big(\frac{x}{\varepsilon},\frac{t}{\varepsilon^2}\Big)v^0\\
=&\ \frac{\partial}{\partial t}v^0-a_{ij}\Big(\frac{x}{\varepsilon},\frac{t}{\varepsilon^2}\Big)\frac{\partial^2}{\partial x_i\partial x_j}v^0-a_{ij}\Big(\frac{x}{\varepsilon},\frac{t}{\varepsilon^2}\Big)\frac{\partial}{\partial y_i}\chi^k\Big(\frac{x}{\varepsilon},\frac{t}{\varepsilon^2}\Big)\frac{\partial^2}{\partial x_k\partial x_j}v^0\\
&-\frac{\partial}{\partial y_i}\Big(a_{ij}\Big(\frac{x}{\varepsilon},\frac{t}{\varepsilon^2}\Big)\chi^k\Big(\frac{x}{\varepsilon},\frac{t}{\varepsilon^2}\Big)\Big)\frac{\partial^2}{\partial x_k\partial x_j}v^0-\widetilde{g}\Big(\frac{x}{\varepsilon},\frac{t}{\varepsilon^2}\Big)\chi^k\Big(\frac{x}{\varepsilon},\frac{t}{\varepsilon^2}\Big)\frac{\partial}{\partial x_k}v^0\\
&-a_{ij}\Big(\frac{x}{\varepsilon},\frac{t}{\varepsilon^2}\Big)\frac{\partial}{\partial y_i}G\Big(\frac{x}{\varepsilon},\frac{t}{\varepsilon^2}\Big)\frac{\partial}{\partial x_j}v^0-\frac{\partial}{\partial y_i}\Big(a_{ij}\Big(\frac{x}{\varepsilon},\frac{t}{\varepsilon^2}\Big)G\Big(\frac{x}{\varepsilon},\frac{t}{\varepsilon^2}\Big)\Big)\frac{\partial}{\partial x_j}v^0\\
&-\widetilde{g}\Big(\frac{x}{\varepsilon},\frac{t}{\varepsilon^2}\Big)G\Big(\frac{x}{\varepsilon},\frac{t}{\varepsilon^2}\Big)v^0+\varepsilon\chi^k\Big(\frac{x}{\varepsilon},\frac{t}{\varepsilon^2}\Big)\frac{\partial}{\partial t}\frac{\partial}{\partial x_k}v^0+\varepsilon G\Big(\frac{x}{\varepsilon},\frac{t}{\varepsilon^2}\Big)\frac{\partial}{\partial t}v^0\\
&-\varepsilon a_{ij}\Big(\frac{x}{\varepsilon},\frac{t}{\varepsilon^2}\Big)\chi^k\Big(\frac{x}{\varepsilon},\frac{t}{\varepsilon^2}\Big)\frac{\partial^3}{\partial x_k\partial x_i\partial x_j}v^0-\varepsilon a_{ij}\Big(\frac{x}{\varepsilon},\frac{t}{\varepsilon^2}\Big)G\Big(\frac{x}{\varepsilon},\frac{t}{\varepsilon^2}\Big)\frac{\partial^2}{\partial x_i\partial x_j}v^0.
\end{aligned}
$$

We substitute the right-hand side of (18.32) for $\frac{\partial}{\partial t}v^0$ and regroup the terms:

$$
\begin{aligned}
&\frac{\partial}{\partial t}(\breve{v}^\varepsilon(x,t)-v^\varepsilon(x,t))-\operatorname{div}\Big(a\Big(\frac{x}{\varepsilon},\frac{t}{\varepsilon^2}\Big)\nabla(\breve{v}^\varepsilon(x,t)-v^\varepsilon(x,t))\Big)\\
&\qquad-\frac{1}{\varepsilon}\widetilde{g}\Big(\frac{x}{\varepsilon},\frac{t}{\varepsilon^2}\Big)(\breve{v}^\varepsilon(x,t)-v^\varepsilon(x,t))\\
&\quad=\Big[\Big(\widehat{a}_{ij}-\breve{a}_{ij}\Big(\frac{t}{\varepsilon^2}\Big)\Big)\frac{\partial^2}{\partial x_i\partial x_j}v^0+\Big(\widehat{b}_i-\breve{b}_i\Big(\frac{t}{\varepsilon^2}\Big)\Big)\frac{\partial}{\partial x_i}v^0+\Big(\widehat{G}-\breve{G}\Big(\frac{t}{\varepsilon^2}\Big)\Big)v^0\Big]\\
&\qquad+\Big\{\breve{a}_{ij}\Big(\frac{t}{\varepsilon^2}\Big)-a_{ij}\Big(\frac{x}{\varepsilon},\frac{t}{\varepsilon^2}\Big)-a_{ik}\Big(\frac{x}{\varepsilon},\frac{t}{\varepsilon^2}\Big)\frac{\partial}{\partial y_k}\chi^j\Big(\frac{x}{\varepsilon},\frac{t}{\varepsilon^2}\Big)\\
&\qquad\qquad\qquad-\frac{\partial}{\partial y_k}\Big(a_{kj}\Big(\frac{x}{\varepsilon},\frac{t}{\varepsilon^2}\Big)\chi^i\Big(\frac{x}{\varepsilon},\frac{t}{\varepsilon^2}\Big)\Big)\Big\}\frac{\partial^2}{\partial x_i\partial x_j}v^0\\
&\qquad+\Big\{\breve{b}_i\Big(\frac{t}{\varepsilon^2}\Big)-\widetilde{g}\Big(\frac{x}{\varepsilon},\frac{t}{\varepsilon^2}\Big)\chi^j\Big(\frac{x}{\varepsilon},\frac{t}{\varepsilon^2}\Big)-a_{ji}\Big(\frac{x}{\varepsilon},\frac{t}{\varepsilon^2}\Big)\frac{\partial}{\partial y_i}G\Big(\frac{x}{\varepsilon},\frac{t}{\varepsilon^2}\Big)\\
&\qquad\qquad\qquad-\frac{\partial}{\partial y_i}\Big(a_{ij}\Big(\frac{x}{\varepsilon},\frac{t}{\varepsilon^2}\Big)G\Big(\frac{x}{\varepsilon},\frac{t}{\varepsilon^2}\Big)\Big)\Big\}\frac{\partial}{\partial x_j}v^0
\end{aligned}
$$

$$+\Big\{\breve{G}\Big(\frac{t}{\varepsilon^2}\Big)-\widetilde{g}\Big(\frac{x}{\varepsilon},\frac{t}{\varepsilon^2}\Big)G\Big(\frac{x}{\varepsilon},\frac{t}{\varepsilon^2}\Big)\Big\}v^0$$
$$+\varepsilon\chi^k\Big(\frac{x}{\varepsilon},\frac{t}{\varepsilon^2}\Big)\frac{\partial}{\partial t}\frac{\partial}{\partial x_k}v^0+\varepsilon G\Big(\frac{x}{\varepsilon},\frac{t}{\varepsilon^2}\Big)\frac{\partial}{\partial t}v^0$$
$$-\varepsilon a_{ij}\Big(\frac{x}{\varepsilon},\frac{t}{\varepsilon^2}\Big)\chi^k\Big(\frac{x}{\varepsilon},\frac{t}{\varepsilon^2}\Big)\frac{\partial^3}{\partial x_k\partial x_i\partial x_j}v^0-\varepsilon a_{ij}\Big(\frac{x}{\varepsilon},\frac{t}{\varepsilon^2}\Big)G\Big(\frac{x}{\varepsilon},\frac{t}{\varepsilon^2}\Big)\frac{\partial^2}{\partial x_i\partial x_j}v^0.$$

Denote by $R_1^\varepsilon$ the expression in the square brackets on the right-hand side of this equality and by $R_2^\varepsilon$ the sum of the remaining terms. Since the expressions in the curly braces are periodic with respect to the spatial variables and have zero average for all $t$, each of these expressions admits a representation in the form (18.15) and $R_2^\varepsilon$ becomes

$$\begin{aligned}R_2^\varepsilon&=\varepsilon\,\mathrm{div}_x\,\kappa_{1,ij}\Big(\frac{x}{\varepsilon},\frac{t}{\varepsilon^2}\Big)\frac{\partial^2}{\partial x_i\partial x_j}v^0+\varepsilon\,\mathrm{div}_x\,\kappa_{2,i}\Big(\frac{x}{\varepsilon},\frac{t}{\varepsilon^2}\Big)\frac{\partial}{\partial x_i}v^0\\&+\varepsilon\,\mathrm{div}_x\,\kappa_3\Big(\frac{x}{\varepsilon},\frac{t}{\varepsilon^2}\Big)v^0+\varepsilon\chi^k\Big(\frac{x}{\varepsilon},\frac{t}{\varepsilon^2}\Big)\frac{\partial}{\partial t}\frac{\partial}{\partial x_k}v^0+\varepsilon G\Big(\frac{x}{\varepsilon},\frac{t}{\varepsilon^2}\Big)\frac{\partial}{\partial t}v^0\\&-\varepsilon a_{ij}\Big(\frac{x}{\varepsilon},\frac{t}{\varepsilon^2}\Big)\chi^k\Big(\frac{x}{\varepsilon},\frac{t}{\varepsilon^2}\Big)\frac{\partial^3}{\partial x_k\partial x_i\partial x_j}v^0\\&-\varepsilon a_{ij}\Big(\frac{x}{\varepsilon},\frac{t}{\varepsilon^2}\Big)G\Big(\frac{x}{\varepsilon},\frac{t}{\varepsilon^2}\Big)\frac{\partial^2}{\partial x_i\partial x_j}v^0,\end{aligned}\tag{18.41}$$

where $\kappa_{1,ij}(y,s)$, $\kappa_{2,i}(y,s)$, and $\kappa_3(y,s)$ are periodic with respect to $y$ and satisfy the estimates

$$\|\kappa_{1,ij}\|_{L_2((s,s+1)\times\mathbb{T}^d)}\le C,\quad \|\kappa_{2,i}\|_{L_2((s,s+1)\times\mathbb{T}^d)}\le C,\quad \|\kappa_3\|_{L_2((s,s+1)\times\mathbb{T}^d)}\le C$$

uniformly with respect to $s\in\mathbb{R}$. Since $v^0$ is smooth and decreases at infinity, we obtain the estimate

$$\|R_2^\varepsilon\|_{L_2((0,T);H^{-1}(\mathbb{R}^d))}\le C\varepsilon,$$

which allows us to derive the estimate

$$\sup_{0\le t\le T}\Big|\int_0^t\int_{\mathbb{R}^d}R_2^\varepsilon(\breve{v}^\varepsilon-v^\varepsilon)\,dx\,dt\Big|\le C\varepsilon.\tag{18.42}$$

Let us estimate the contribution of $R_1^\varepsilon$. By the Birkhoff theorem, the functions $(\widehat{a}_{ij}-\breve{a}_{ij}(t/\varepsilon^2))$, $(\widehat{b}_i-\breve{b}_i(t/\varepsilon^2))$, and $(\widehat{G}-\breve{G}(t/\varepsilon^2))$ weakly converge a.s. to zero in the space $L_2(0,T)$ as $\varepsilon\to 0$. Consequently, the sequence $R_1^\varepsilon$ weakly converges to zero in $L_2(0,T;L_2(\mathbb{R}^d))$.

EXERCISE 18.3. Prove that

$$\lim_{\varepsilon\to 0}\sup_{0\le t\le T}\Big|\int_0^t\int_{\mathbb{R}^d}R_1^\varepsilon(\breve{v}^\varepsilon-v^\varepsilon)\,dx\,dt\Big|=0\quad\text{a.s.}\tag{18.43}$$

*Hint.* Use the following facts: the family $\{(\breve{v}^\varepsilon-v^\varepsilon)\}$ is compact in the space $C(0,T;L_2^{\mathrm{loc}}(\mathbb{R}^d))$, the function $v^0$, together with its derivatives, rapidly decreases at infinity, and the sequence $R_1^\varepsilon$ weakly converges to zero.

Multiplying the equation

$$\frac{\partial}{\partial t}(\check{v}^\varepsilon(x,t) - v^\varepsilon(x,t)) - \operatorname{div}\Big(a\Big(\frac{x}{\varepsilon}, \frac{t}{\varepsilon^2}\Big)\nabla(\check{v}^\varepsilon(x,t) - v^\varepsilon(x,t))\Big)$$
$$- \frac{1}{\varepsilon}\widetilde{g}\Big(\frac{x}{\varepsilon}, \frac{t}{\varepsilon^2}\Big)(\check{v}^\varepsilon(x,t) - v^\varepsilon(x,t)) = R_1^\varepsilon + R_2^\varepsilon$$

by $(\check{v}^\varepsilon - v^\varepsilon)$, using (18.42)–(18.43) with the obvious inequality

$$\|\check{v}^\varepsilon(\cdot,0) - v^0(\cdot,0)\|_{L_\infty(\mathbb{R}^d)} \le C\varepsilon,$$

and arguing in the same way as in the proof of Proposition 18.1, we find that

$$\lim_{\varepsilon\to 0}\big(\|\check{v}^\varepsilon - v^\varepsilon\|_{L_2((0,T);H^1(\mathbb{R}^d))} + \|\check{v}^\varepsilon - v^\varepsilon\|_{L_\infty((0,T);L_2(\mathbb{R}^d))}\big) = 0 \quad \text{a.s.,} \tag{18.44}$$

which completes the proof in the case of smooth compactly supported initial data.

In a more general case $u_0 \in L_2(\mathbb{R}^d)$, the convergence can be established by using approximations. Indeed, let $u_0^\delta \in C_0^\infty(\mathbb{R}^d)$ satisfy $\|u_0^\delta - u_0\|_{L_2(\mathbb{R}^d)} \le \delta$. A solution to the problem (18.8) with the initial data $u_0^\delta$ is denoted by $v^{\delta,\varepsilon}$. By Proposition 18.1, we have

$$\|v^{\delta,\varepsilon} - v^\varepsilon\|_{L_\infty(0,T;L_2(\mathbb{R}^d))} \le C\delta.$$

It is clear that

$$\|v^{\delta,0} - v^0\|_{L_\infty(0,T;L_2(\mathbb{R}^d))} \le C\delta.$$

By the above arguments, the sequence $v^{\delta,\varepsilon}$ converges to $v^{\delta,0}$ in $L_\infty(0,T;L_2(\mathbb{R}^d))$. Hence

$$\limsup_{\varepsilon\to 0}\|v^\varepsilon - v^0\|_{L_\infty(0,T;L_2(\mathbb{R}^d))} \le C\delta.$$

Since $\delta$ is arbitrary, we arrive at the required assertion. □

To prove similar results in the case $\alpha > 2$ or $\alpha < 2$, we need additional arguments, which will be given below.

PROOF OF THEOREM 18.2. As in the proof of Theorem 18.1, we begin with additional smoothness assumptions. Namely, suppose that the initial data are presented by a smooth compactly supported function.

Let $\Psi^\varepsilon(z,s)$ be a stationary solution to the problem

$$\frac{\partial}{\partial s}\Psi^\varepsilon(z,s) = \frac{\partial}{\partial z_i}a_{ij}(z,\varepsilon^{2-\alpha}s)\frac{\partial}{\partial z_j}(\Psi^\varepsilon(z,s) + z), \qquad \int_{\mathbb{T}^d}\Psi^\varepsilon(z,s)\,dz = 0, \tag{18.45}$$

periodic with respect to $z$.

The existence of such a solution is guaranteed by Proposition 18.2.

LEMMA 18.4. *We have convergence in probability,*

$$a_{ij}\Big(\frac{x}{\varepsilon}, \frac{t}{\varepsilon^\alpha}\Big)\Big(\nabla_z\Psi^\varepsilon\Big(\frac{x}{\varepsilon}, \frac{t}{\varepsilon^2}\Big) + I\Big) \rightharpoonup \widehat{a} \text{ weakly in } (L_2^{\mathrm{loc}}((0,\infty)\times\mathbb{R}^d))^{n^2}, \tag{18.46}$$

*where the matrix* $\widehat{a}$ *is defined by formula* (18.36).

PROOF. Consider the function

$$\nu^\varepsilon(z,s) = \varepsilon^{\frac{2-\alpha}{2}}\Psi^\varepsilon\Big(\frac{z}{\varepsilon^{(2-\alpha)/2}}, \frac{s}{\varepsilon^{2-\alpha}}\Big).$$

It is easy to check that $\nu^\varepsilon(z,s)$ satisfies the equation

$$\frac{\partial}{\partial s}\nu^\varepsilon(z,s) = \frac{\partial}{\partial z_i}a_{ij}\Big(\frac{z}{\varepsilon^{(2-\alpha)/2}}, s\Big)\frac{\partial}{\partial z_j}(\nu^\varepsilon(z,s) + z), \tag{18.47}$$

which is a special case of the equation with locally periodic coefficients; the coefficients are periodic and rapidly oscillate with respect to $z$, whereas there is no scaling with respect to the variable $s$. For $\nu^\varepsilon(z,s)$ we have

$$(18.48)\quad |\nu^\varepsilon(z,s)|\le C,\qquad \int_N^{N+1}\int_{[0,1]^d}\Big|a\Big(\frac{z}{\varepsilon^{(2-\alpha)/2}},s\Big)\nabla_z(\nu^\varepsilon(z,s)+z)\Big|^2 dz\,ds\le C,$$

where $C$ is a nonrandom constant independent of $\varepsilon$. These estimates were obtained in Proposition 18.2.

We introduce the fast variables $y=\frac{z}{\varepsilon^{(2-\alpha)/2}}$. The homogenized coefficients of equation (18.47) are defined by the formula

$$(18.49)\qquad a^{\mathrm{sp}}(s)=\int_{\mathbb{T}^d} a(z,s)(\nabla_z\psi(z,s)+I)\,dz,$$

where $\psi(z,s)$ is a solution to equation (18.28).

EXERCISE 18.4. Applying the two-scale convergence method and using the estimates (18.48), prove that equation (18.47) admits almost surely a homogenization such that the coefficients of the homogenized equation are defined by formula (18.49) and the following weak two-scale convergence almost surely holds:

$$(18.50)\qquad a\Big(\frac{z}{\varepsilon^{(2-\alpha)/2}},s\Big)(\nabla\nu^\varepsilon(z,s)+I)\overset{2}{\rightharpoonup} a(y,s)(\nabla_y\psi(y,s)+I).$$

We introduce the notation

$$S^\varepsilon(s)=\int_{\mathbb{T}^d} a(y,s)\nu^\varepsilon(\varepsilon^{(2-\alpha)/2}y,s)\,dy.$$

From (18.50) it follows that $S^\varepsilon(s)\rightharpoonup a^{\mathrm{sp}}(s)$ a.s. weakly in $L_2^{\mathrm{loc}}(\mathbb{R})$ as $\varepsilon\to 0$.

EXERCISE 18.5. Suppose that a random process $\eta^\varepsilon(s)$ is stationary, $\|\eta^\varepsilon\|_{L_2(0,1)}\le C$, and the sequence $\eta^\varepsilon(s)$ weakly converges a.s. to zero in $L_2^{\mathrm{loc}}(\mathbb{R})$ as $\varepsilon\to 0$. Then the sequence $\eta^\varepsilon(s/\varepsilon^\gamma)$ weakly converges in probability to zero in $L_2^{\mathrm{loc}}(\mathbb{R})$ for every $\gamma>0$.

Thus, the sequence $S^\varepsilon(s/\varepsilon^\alpha)-a^{\mathrm{sp}}(s/\varepsilon^\alpha)$ weakly converges in probability to zero in $L_2^{\mathrm{loc}}(\mathbb{R})$. By the Birkhoff theorem,

$$a^{\mathrm{sp}}(s/\varepsilon^\alpha)\rightharpoonup\widehat{a}\quad\text{weakly in } L_2^{\mathrm{loc}}(\mathbb{R})$$

almost surely as $\varepsilon\to 0$, which, together with the above convergence, yields

$$S^\varepsilon(s/\varepsilon^\alpha)\rightharpoonup\widehat{a}\quad\text{weakly in } L_2^{\mathrm{loc}}(\mathbb{R})$$

in probability as $\varepsilon\to 0$. Finally, by the definition of $S^\varepsilon$, $\nu^\varepsilon$ and periodicity of $\Psi^\varepsilon(y,s)$ with respect to the first argument, we have

$$a_{ij}\Big(\frac{x}{\varepsilon},\frac{t}{\varepsilon^\alpha}\Big)\Big(\nabla_z\Psi^\varepsilon\Big(\frac{x}{\varepsilon},\frac{t}{\varepsilon^2}\Big)+I\Big)-S^\varepsilon\Big(\frac{s}{\varepsilon^\alpha}\Big)\rightharpoonup 0\quad\text{weakly in } L_2^{\mathrm{loc}}(\mathbb{R}^{d+1}).$$

The last two relations imply the required assertion. □

We substitute the difference $(\breve{v}^\varepsilon-v^\varepsilon)$ in the problem (18.8), where

$$\breve{v}^\varepsilon=v^0(x,t)+\varepsilon\Psi^{\varepsilon,j}\Big(\frac{x}{\varepsilon},\frac{t}{\varepsilon^2}\Big)\frac{\partial}{\partial x_j}v^0.$$

Taking into account (18.45), (18.32), (18.36), and (18.37), after simple transformations we find that

(18.51)

$$\begin{aligned}
&\frac{\partial}{\partial t}(\breve{v}^\varepsilon(x,t)-v^\varepsilon(x,t))-\operatorname{div}\Big(a\Big(\frac{x}{\varepsilon},\frac{t}{\varepsilon^\alpha}\Big)\nabla(\breve{v}^\varepsilon(x,t)-v^\varepsilon(x,t))\Big)\\
&\qquad-\frac{1}{\varepsilon^{\alpha/2}}\widetilde{g}\Big(\frac{x}{\varepsilon},\frac{t}{\varepsilon^\alpha}\Big)(\breve{v}^\varepsilon(x,t)-v^\varepsilon(x,t))\\
&\qquad=\Big\{\widehat{a}_{ij}-a_{ij}\Big(\frac{x}{\varepsilon},\frac{t}{\varepsilon^\alpha}\Big)-a_{ik}\Big(\frac{x}{\varepsilon},\frac{t}{\varepsilon^\alpha}\Big)\frac{\partial}{\partial y_k}\Psi^{\varepsilon,j}\Big(\frac{x}{\varepsilon},\frac{t}{\varepsilon^2}\Big)\\
&\qquad\qquad-\frac{\partial}{\partial y_k}\Big(a_{kj}\Big(\frac{x}{\varepsilon},\frac{t}{\varepsilon^\alpha}\Big)\Psi^{\varepsilon,i}\Big(\frac{x}{\varepsilon},\frac{t}{\varepsilon^2}\Big)\Big)\Big\}\frac{\partial^2}{\partial x_i\partial x_j}v^0\\
&\qquad-\varepsilon^{-\alpha/2}\widetilde{g}\Big(\frac{x}{\varepsilon},\frac{t}{\varepsilon^\alpha}\Big)v^0-\varepsilon^{1-\alpha/2}\widetilde{g}\Big(\frac{x}{\varepsilon},\frac{t}{\varepsilon^\alpha}\Big)\Psi^{\varepsilon,j}\Big(\frac{x}{\varepsilon},\frac{t}{\varepsilon^2}\Big)\frac{\partial}{\partial x_j}v^0\\
&\qquad-\varepsilon\Psi^{\varepsilon,k}\Big(\frac{x}{\varepsilon},\frac{t}{\varepsilon^2}\Big)\frac{\partial}{\partial t}\frac{\partial}{\partial x_k}v^0-\varepsilon a_{ij}\Big(\frac{x}{\varepsilon},\frac{t}{\varepsilon^\alpha}\Big)\Psi^{\varepsilon,k}\Big(\frac{x}{\varepsilon},\frac{t}{\varepsilon^\alpha}\Big)\frac{\partial^3}{\partial x_k\partial x_i\partial x_j}v^0
\end{aligned}$$

By Lemma 18.4, the sequence

$$\Big\{a_{ij}-a_{ij}\Big(\frac{x}{\varepsilon},\frac{t}{\varepsilon^\alpha}\Big)-a_{ik}\Big(\frac{x}{\varepsilon},\frac{t}{\varepsilon^\alpha}\Big)\frac{\partial}{\partial y_k}\Psi^{\varepsilon,j}\Big(\frac{x}{\varepsilon},\frac{t}{\varepsilon^2}\Big)\Big\}\frac{\partial^2}{\partial x_i\partial x_j}v^0$$

weakly converges in probability to zero in $L_2(0,T;L_2(\mathbb{R}^d))$. The remaining terms on the right-hand side of (18.51) converge almost surely to zero in probability or in the $L_2(0,T;H^{-1}(\mathbb{R}^d))$-norm. Therefore, as in the case $\alpha=2$, $(\breve{v}^\varepsilon-v^\varepsilon)$ converges to zero. The case of the general initial condition is treated in the same way as in the proof of Theorem 18.1. □

Consider the case $\alpha>2$.

PROOF OF THEOREM 18.3. As above, we begin with smooth compactly supported initial data $u_0$.

Consider auxiliary equations

$$\frac{\partial}{\partial s}\zeta^\varepsilon(z,s)=\frac{\partial}{\partial z_i}a_{ij}\left(z,\frac{s}{\varepsilon^{\alpha-2}}\right)\frac{\partial}{\partial z_j}(\zeta^\varepsilon(z,s)+z),\tag{18.52}$$

$$\frac{\partial}{\partial s}\mathcal{G}^\varepsilon(z,s)=\frac{\partial}{\partial z_i}a_{ij}\left(z,\frac{s}{\varepsilon^{\alpha-2}}\right)\frac{\partial}{\partial z_j}\mathcal{G}^\varepsilon(z,s)-\widetilde{g}\left(z,\frac{s}{\varepsilon^{\alpha-2}}\right),\tag{18.53}$$

each having a unique stationary solution with zero mean in view of Lemma 18.3.

LEMMA 18.5. *We have the following convergence in probability:*

$$a\Big(\frac{x}{\varepsilon},\frac{t}{\varepsilon^\alpha}\Big)\Big(\nabla_z\zeta^\varepsilon\Big(\frac{x}{\varepsilon},\frac{t}{\varepsilon^2}\Big)+I\Big)\rightharpoonup\widehat{a}\ \textit{weakly in}\ (L_2^{\mathrm{loc}}((0,\infty)\times\mathbb{R}^d))^{n^2},\tag{18.54}$$

$$\begin{gathered}a\Big(\frac{x}{\varepsilon},\frac{t}{\varepsilon^\alpha}\Big)\nabla_z\mathcal{G}^\varepsilon\Big(\frac{x}{\varepsilon},\frac{t}{\varepsilon^2}\Big)+\widetilde{g}(\frac{x}{\varepsilon},\frac{t}{\varepsilon^\alpha})\nabla_z\zeta^\varepsilon\Big(\frac{x}{\varepsilon},\frac{t}{\varepsilon^2}\Big)\rightharpoonup0\\ \textit{weakly in}\ (L_2^{\mathrm{loc}}((0,\infty)\times\mathbb{R}^d))^d,\end{gathered}\tag{18.55}$$

$$\mathcal{G}^\varepsilon\Big(\frac{x}{\varepsilon},\frac{t}{\varepsilon^2}\Big)\widetilde{g}(\frac{x}{\varepsilon},\frac{t}{\varepsilon^\alpha})\rightharpoonup\widehat{G}\ \textit{weakly in}\ L_2^{\mathrm{loc}}((0,\infty)\times\mathbb{R}^d),\tag{18.56}$$

*where $\widehat{a}$ and $\widehat{G}$ are defined by formulas* (18.38) *and* (18.39) *respectively.*

PROOF. We show that

$$a_{ij}\Big(z,\frac{s}{\varepsilon^{(2-\alpha)}}\Big)\Big(\nabla\zeta^{\varepsilon}(z,s)+I\Big)\rightharpoonup\overline{a}(z)(\nabla\overline{\psi}(z)+I)\quad\text{weakly in } L_2^{\mathrm{loc}}(\mathbb{R}^1\times\mathbb{T}^d)\text{ a.s.,}\tag{18.57}$$

where $\overline{a}(z)=\mathbf{E}\,a(z,s)$ and $\overline{\psi}(z)$ is a solution to equation (18.30). To do this, consider the Cauchy problem

$$\begin{aligned}&\frac{\partial}{\partial s}\zeta_{s_0}^{\varepsilon}(z,s)=\frac{\partial}{\partial z_i}a_{ij}\Big(z,\frac{s}{\varepsilon^{(2-\alpha)}}\Big)\Big(\frac{\partial}{\partial z_j}\zeta_{s_0}^{\varepsilon}(z,s)+I\Big),\\&\zeta_{s_0}^{\varepsilon}(z,s_0)=0,\end{aligned}\tag{18.58}$$

in the cylinder $\mathbb{T}^d\times(s_0,\infty)$. Denote by $\zeta_{s_0}^{0}(z,s)$ the solution to the similar Cauchy problem with deterministic coefficients $\overline{a}(z)$ defined by formula (18.29):

$$\begin{aligned}&\frac{\partial}{\partial s}\zeta_{s_0}^{0}(z,s)=\frac{\partial}{\partial z_i}\overline{a}_{ij}(z)\Big(\frac{\partial}{\partial z_j}\zeta_{s_0}^{0}(z,s)+I\Big),\\&\zeta_{s_0}^{\varepsilon}(z,s_0)=0.\end{aligned}\tag{18.59}$$

By Lemma 18.3,

$$\|\zeta^{\varepsilon}-\zeta_{s_0}^{\varepsilon}\|_{L_2(s,s+1;H^1(\mathbb{T}^d))}+\|\zeta^{\varepsilon}(\cdot,s)-\zeta_{s_0}^{\varepsilon}(\cdot,s)\|_{L_\infty(\mathbb{T}^d)}\le c_1\exp(-c_2(s-s_0))$$

for $s>s_0$, where the constants $c_1>0$ and $c_2>0$ are nonrandom and are independent of $\varepsilon$. It is easy to verify that the sequence $\zeta_{s_0}^{0}(z,s)$ converges exponentially to $\overline{\psi}(z)$ as $(s-s_0)\to\infty$ (check it!). Therefore, to obtain (18.57), it suffices to prove that

$$a_{ij}\Big(z,\frac{s}{\varepsilon^{(2-\alpha)}}\Big)(\nabla\zeta_{s_0}^{\varepsilon}(z,s)+I)\rightharpoonup\overline{a}(z)(\nabla\zeta_{s_0}^{0}(z,s)+I)\quad\text{weakly in } L_{2,\mathrm{loc}}(\mathbb{R}^1\times\mathbb{T}^d)\text{ a.s.}\tag{18.60}$$

for all $s_0$. It is easy to establish this convergence if the coefficients $a(z,s)$ are continuously differentiable with respect to $z$ and $|\nabla a(z,s)|\le C$. In this case, differentiating (18.58) with respect to $z$, we obtain the estimates (check it!)

$$\|\nabla\zeta_{s_0}^{\varepsilon}\|_{L_2(s,s+1;H^1(\mathbb{T}^d))}\le C,\quad\|\partial_t\nabla\zeta_{s_0}^{\varepsilon}\|_{L_2(s,s+1;H^{-1}(\mathbb{T}^d))}\le C.$$

By [**78**, Theorem 5.1], these estimates imply that the family of functions $\nabla\zeta_{s_0}^{\varepsilon}(z,s)$ is compact in the space $L_2((s_0,\infty);L_2(\mathbb{T}^d))$.

By the Birkhoff theorem, $a\big(z,\frac{s}{\varepsilon^{(2-\alpha)}}\big)$ weakly converges a.s. to $\overline{a}(z)$ in the space $L_2^{\mathrm{loc}}(\mathbb{R}^1\times\mathbb{T}^d)$.

Passing to the limit in the integral identity corresponding to the problem (18.58), it is easy to see that any limit point of the family $\{\zeta_{s_0}^{\varepsilon}\}$ almost surely satisfies (18.59). Then the convergence (18.60) follows from the compactness of the family $\{\nabla\zeta_{s_0}^{\varepsilon}(z,s)\}$ and the Birkhoff theorem.

For general coefficients the convergence (18.60) can be established by using approximations. Let $\delta_n$ be a decreasing sequence of positive numbers such that $\delta_n\to0$ as $n\to\infty$. Denote by $a^{\delta}(z,s)$ and $\overline{a}^{\delta}(z)$ the regularizations of $a(z,s)$ and $\overline{a}(z)$ with respect to $z$ (cf. the definition of regularization in (2.6)). Then, on a set of full measure in $\Omega$, we have

$$\Big(a^{\delta_n}\Big(z,\frac{s}{\varepsilon^{\alpha-2}}\Big)-a^{\delta_n}(z)\Big)\rightharpoonup0\quad\text{weakly in } L_{2,\mathrm{loc}}(\mathbb{R}^1;H^1(\mathbb{T}^d))\text{ as }\varepsilon\to0,\tag{18.61}$$

$$\lim_{\varepsilon\to 0}\Big\|\Big(a^{\delta_n}\Big(z,\frac{s}{\varepsilon^{\alpha-2}}\Big)-a^{\delta_n}\Big(z,\frac{s}{\varepsilon^{\alpha-2}}\Big)\Big)\Big\|^2_{L_2((s_1,s_2)\times\mathbb{T}^d)}$$
$$= (s_2-s_1)\,\mathbf{E}\,\|a^{\delta_n}(z,s)-a(z,s)\|^2_{L_2(\mathbb{T}^d)} \stackrel{\text{def}}{=} \kappa(\delta_n). \tag{18.62}$$

Considering the fact that for almost all $\omega$ the norms $\|a^{\delta_n}(z,s)-a(z,s)\|^2_{L_2(\mathbb{T}^d)}$ converge to zero as $n\to\infty$ and using the Lebesgue theorem, we find

$$\lim_{n\to\infty}\kappa(\delta_n)=0. \tag{18.63}$$

Denote by $\zeta_{s_0}(z,s)$ an arbitrary limit point of the family $\{\zeta^\varepsilon_{s_0}(z,s)\}$. Let

$$\Phi_\varphi(u)\stackrel{\text{def}}{=}\Big|\int_0^\infty\int_{\mathbb{T}^d}(u(z,s)\partial_s\varphi(z,s)-\overline{a}(z)(\nabla u(z,s)+I)\cdot\nabla\varphi(z,s))\,dz\,ds\Big|.$$

Then for an arbitrary test function $\varphi\in C_0^\infty((s_0,\infty)\times\mathbb{T}^d)$ we have

$$\Phi_\varphi(\zeta_{s_0})=\lim_{\varepsilon\to 0}\Phi_\varphi(\zeta^\varepsilon_{s_0}).$$

We estimate $\Phi_\varphi(\zeta^\varepsilon_{s_0})$ as follows:

$$\begin{aligned}\Phi_\varphi(\zeta^\varepsilon_{s_0})\le{}&\Big|\int_0^\infty\int_{\mathbb{T}^d}\Big(\zeta^\varepsilon_{s_0}(z,s)\partial_s\varphi(z,s)-a\Big(z,\frac{s}{\varepsilon^{\alpha-2}}\Big)(\nabla\zeta^\varepsilon_{s_0}(z,s)+I)\nabla\varphi(z,s)\Big)dz\,ds\Big|\\&+\Big|\int_0^\infty\int_{\mathbb{T}^d}\Big(a\Big(z,\frac{s}{\varepsilon^{\alpha-2}}\Big)-a^\delta\Big(z,\frac{s}{\varepsilon^{\alpha-2}}\Big)\Big)(\nabla\zeta^\varepsilon_{s_0}(z,s)+I)\nabla\varphi(z,s)\,dz\,ds\Big|\\&+\Big|\int_0^\infty\int_{\mathbb{T}^d}\Big(a^\delta\Big(z,\frac{s}{\varepsilon^{\alpha-2}}\Big)-\overline{a}^\delta(z)\Big)(\nabla\zeta^\varepsilon_{s_0}(z,s)+I)\nabla\varphi(z,s)\,dz\,ds\Big|\\&+\Big|\int_0^\infty\int_{\mathbb{T}^d}\Big(\overline{a}^\delta(z)-\overline{a}(z)\Big)(\nabla\zeta^\varepsilon_{s_0}(z,s)+I)\nabla\varphi(z,s)\,dz\,ds\Big|.\end{aligned}$$

The first term on the right-hand side of the last inequality vanishes in view of the equation for $\zeta^\varepsilon_{s_0}$. We choose $s_1$ and $s_2$ so that $\operatorname{supp}(\varphi)\subset(s_1,s_2)\times\mathbb{T}^d$. Taking into account the estimates in Proposition 18.2, we have

$$\begin{aligned}\Phi_\varphi(\zeta^\varepsilon_{s_0})\le{}&C\int_{s_1}^{s_2}\int_{\mathbb{T}^d}\Big(a\Big(z,\frac{s}{\varepsilon^{\alpha-2}}\Big)-a^{\delta_n}\Big(z,\frac{s}{\varepsilon^{\alpha-2}}\Big)\Big)^2dz\,ds\\&+\Big|\int_0^\infty\int_{\mathbb{T}^d}\Big(a^{\delta_n}\Big(z,\frac{s}{\varepsilon^{\alpha-2}}\Big)-\overline{a}^{\delta_n}(z)\Big)(\nabla\zeta^\varepsilon_{s_0}(z,s)+I)\nabla\varphi(z,s)\,dz\,ds\Big|\\&+C\int_{\mathbb{T}^d}(a^{\delta_n}(z)-a(z))^2dz.\end{aligned}$$

Integrating by parts and taking into account (18.61) and the compactness of $\zeta^\varepsilon_{s_0}$ in $C(s_1,s_2;L_2(\mathbb{T}^d))$, we conclude that the second term on the right-hand side converges to zero as $\varepsilon\to 0$. Therefore,

$$\limsup_{\varepsilon\to 0}\Phi_\varphi(\zeta^\varepsilon_{s_0})\le C\kappa(\delta_n)+C\|a^{\delta_n}(z)-a(z)\|^2_{L_2(\mathbb{T}^d)}.$$

Since $n$ is arbitrary, we have $\lim_{\varepsilon\to 0}\Phi_\varphi(\zeta^\varepsilon_{s_0})=0$. Thus, $\zeta_{s_0}$ coincides with $\zeta^0_{s_0}$ and the entire sequence $\zeta^\varepsilon_{s_0}$ converges a.s. to $\zeta^0_{s_0}$.

Simultaneously, we proved (18.60) and (18.57). As in the proof of Lemma 18.4, we obtain (18.54). The convergences (18.55) and (18.56) can be established in a similar way provided that we have the equality

$$\int_{\mathbb{T}^d}(\overline{a}(z)\nabla_z\overline{G}(z)+\overline{g}(z)\overline{\psi}(z))\,dz=0.$$

By equations (18.30) and (18.31),

$$\begin{aligned}\int_{\mathbb{T}^d} \overline{g}(z)\overline{\psi}(z)\,dz &= -\int_{\mathbb{T}^d} \operatorname{div}(\overline{a}(z)\nabla_z\overline{G}(z))\overline{\psi}(z)\,dz\\ &= -\int_{\mathbb{T}^d} \operatorname{div}(\overline{a}(z)\nabla_z\overline{\psi}(z))\overline{G}(z)\,dz\\ &= \int_{\mathbb{T}^d} \operatorname{div}\overline{a}(z)\overline{G}(z)\,dz = -\int_{\mathbb{T}^d} \overline{a}(z)\nabla_z\overline{G}(z)\,dz,\end{aligned}$$

and the required relation follows.

The proof of the remaining part is practically the same as in the previous theorems, and it is left to the reader. □

REMARK 18.3. In the proof of Theorems 18.1–18.3, the mixing condition **H5** was not used at all. Hence Theorems 18.1–18.3 remain valid for general stationary fields of coefficients that are time-ergodic and periodic with respect to the spatial variables provided that condition **H3** holds and $\int_{\mathbb{T}^d} \widetilde{g}(y,s)\,dy = 0$ a.s. for all $s$.

REMARK 18.4. The methods used in the proof of Theorems 18.1–18.3 can be applied, without any essential modifications, to the problem (18.13) if $h(x,t)$ is a smooth function compactly supported in $x$. By the uniform *a priori* estimate (18.14), the corresponding assertions remain valid for arbitrary functions $h(x,t) \in L_2((0,T)\times\mathbb{R}^d)$. A precise formulation will be given at the beginning of Subsection 18.1.6 below.

18.1.6. *Homogenization of the factorized equation. II: General case.* Consider a homogenization of the factorized problem (18.8) with a nonzero right-hand side. First, consider the auxiliary problem (18.13) with a fixed $h(x,t) \in L_2((0,T)\times\mathbb{R}^d)$. As was mentioned in Remark 18.4, the above methods still work in this case.

THEOREM 18.4. *Suppose that $h(x,t) \in L_2((0,T)\times\mathbb{R}^d)$, $z_0 \in L_2(\mathbb{R}^d)$, and $z^\varepsilon$ is the solution to the problem* (18.13). *Then the sequence $z^\varepsilon$ converges almost surely if $\alpha = 2$ or in probability if $\alpha \neq 2$, to the solution of the problem*

$$\begin{aligned}&\frac{\partial}{\partial t} z^0(x,t) = \operatorname{div}(\widehat{a}\nabla z^0(x,t)) + \widehat{b}\cdot\nabla z^0(x,t) + \widehat{G}z^0(x,t) + h(x,t),\\ &z^0(x,0) = z_0(x),\end{aligned} \tag{18.64}$$

*where the deterministic constant coefficients are defined by formulas* (18.33)–(18.35) *if $\alpha = 2$,* (18.36)–(18.37) *if $0 < \alpha < 2$, or* (18.38)–(18.39) *if $\alpha > 2$.*

Denote

$$\zeta_t^\varepsilon = \frac{1}{\varepsilon^{1\wedge\frac{\alpha}{2}}}\int_0^t \langle g\rangle\Big(\frac{s}{\varepsilon^\alpha}\Big)\,ds.$$

Let $\alpha \le 2$. We represent the solution $v^\varepsilon$ to the problem (18.8) as the sum $v^\varepsilon = V^\varepsilon + (v^\varepsilon - V^\varepsilon)$, where $V^\varepsilon(x,t)$ is a solution to the Cauchy problem in $\mathbb{R}^d\times(0,T)$,

$$\begin{aligned}\frac{\partial}{\partial t} V^\varepsilon(x,t) &= \operatorname{div}(\widehat{a}\nabla V^\varepsilon(x,t)) + \widehat{b}\cdot\nabla V^\varepsilon(x,t) + \widehat{G}V^\varepsilon(x,t)\\ &\qquad + f(x,t)\exp(-\zeta_t^\varepsilon),\\ V^\varepsilon(x,0) &= u_0(x),\end{aligned} \tag{18.65}$$

with coefficients defined by formulas (18.33)–(18.35) or (18.36)–(18.37) depending on the value of $\alpha$. Let us check that the sequence $(v^\varepsilon - V^\varepsilon)$ converges in probability to zero in $L_2((0,T)\times\mathbb{R}^d)$ if $V^\varepsilon$ converges in law in $L_2((0,T)\times\mathbb{R}^d)$ to the solution of the following Cauchy problem with a random right-hand side:

$$\begin{aligned}\frac{\partial}{\partial t}v^0(x,t) &= \operatorname{div}(\widehat{a}\nabla v^0(x,t)) + \widehat{b}\cdot\nabla v^0(x,t) + \widehat{G}v^0(x,t)\\ &\qquad + f(x,t)\exp(-\sigma W_t),\\ v^0(x,0) &= u_0(x),\end{aligned}\tag{18.66}$$

where $\sigma$ is defined by formula (18.10).

PROPOSITION 18.3. *As $\varepsilon\to 0$, the sequence $(v^\varepsilon - V^\varepsilon)$ converges in probability to zero in the $L_2((0,T)\times\mathbb{R}^d)$-norm.*

PROOF. By Lemma 18.2, $\zeta_t^\varepsilon$ converges in law to $\sigma W_t$ in the space $C(0,T)$ as $\varepsilon\to 0$. Therefore, the random functions $f(x,t)\zeta_t^\varepsilon$ converge in law to $\sigma f(x,t)W_t$ in the space $L_2((0,T)\times\mathbb{R}^d)$. By the Prokhorov theorem, for any $\delta>0$ there is a compact set $K^\delta\subset L_2((0,T)\times\mathbb{R}^d)$ such that $\mathbf{P}\{f(x,t)\zeta_t^\varepsilon\notin K^\delta\}\le\delta$. We choose a finite $\delta$-net $\{h_j\}$, $j=1,2,\dots,N(\delta)$, in $K^\delta$ and denote by $z_j^\varepsilon(x,t)$ and $z_j^0(x,t)$ the solutions to the problem (18.13) and (18.64), respectively, with the right-hand side $h_j(x,t)$ and initial data $u_0(x)$. By Theorems 18.1 and 18.2, for every $\delta>0$ there is $\varepsilon_0(\delta)>0$ such that for all $\varepsilon<\varepsilon_0(\delta)$,

$$\max_j \mathbf{P}\{\|z_j^\varepsilon - z_j^0\|_{L_2(0,T;\mathbb{R}^d)}>\delta\}<\frac{\delta}{N(\delta)}.$$

We introduce

$$\mathcal{E}_j=\{\omega:\|f(x,t)\zeta_t^\varepsilon - h_j(x,t)\|_{L_2(0,T;\mathbb{R}^d)}<\delta\}.$$

By construction, $\mathbf{P}(\Omega\setminus\bigcup_{j=1}^{N(\delta)}\mathcal{E}_j)<\delta$. By the estimate (18.14) and a similar estimate for the homogenized problem, for all $\omega\in\mathcal{E}_j$ we have

$$\begin{aligned}\|V^\varepsilon - v^\varepsilon\|_{L_2} &\le \|V^\varepsilon - z_j^0\|_{L_2} + \|z_j^0 - z_j^\varepsilon\|_{L_2} + \|z_j^\varepsilon - v^\varepsilon\|_{L_2}\\ &\le 2C\delta + \|z_j^0 - z_j^\varepsilon\|_{L_2}\end{aligned}$$

with a universal constant $C$. Hence for all $\varepsilon<\varepsilon_0(\delta)$ we have

$$\begin{aligned}&\mathbf{P}\{\|V^\varepsilon - v^\varepsilon\|_{L_2}>(2C+1)\delta\}\\ &\quad\le \mathbf{P}\Big(\Omega\setminus\bigcup_{j=1}^{N(\delta)}\mathcal{E}_j\Big)+\sum_{j=1}^{N(\delta)}\mathbf{P}\{\mathcal{E}_j\cap(\|V^\varepsilon - v^\varepsilon\|_{L_2}>(2C+1)\delta)\}\\ &\quad\le \delta+\sum_{j=1}^{N(\delta)}\mathbf{P}\{\|z_j^0 - z_j^\varepsilon\|_{L_2}>\delta\}\le\delta+\sum_{j=1}^{N(\delta)}\frac{\delta}{N(\delta)}=2\delta,\end{aligned}$$

which shows the convergence of $(v^\varepsilon - V^\varepsilon)$ in probability. □

PROPOSITION 18.4. *The sequence $V^\varepsilon$ converges in law to the solution $v^0$ of the problem* (18.66) *in the space $L_2((0,T)\times\mathbb{R}^d)$ as $\varepsilon\to 0$.*

PROOF. This assertion follows from the continuous dependence of the solution to the problem (18.64) on the right-hand side in the $L_2((0,T)\times\mathbb{R}^d)$-norm. By continuity, the convergence in law of the right-hand sides implies the required convergence of the solutions. □

Now we can formulate and prove a convergence result for the inhomogeneous factorized problem (18.8).

THEOREM 18.5. *Let conditions* **H1**–**H5** *hold. Suppose that* $\alpha \le 2$ *and* $v^\varepsilon$ *is the solution to the factorized problem* (18.8). *Let* $\varepsilon \to 0$. *Then the sequence* $v^\varepsilon$ *converges in law to the solution of the problem* (18.66) *in the space* $L_2((0,T) \times L_2(\mathbb{R}^d))$; *moreover, the coefficients of the limit equation are found from formulas* (18.33)–(18.35) *if* $\alpha = 2$ *and from* (18.36)–(18.37) *if* $\alpha < 2$.

*If* $\alpha > 2$, *then the sequence* $v^\varepsilon$ *converges in probability in the space* $L_2((0,T) \times L_2(\mathbb{R}^d))$ *to the solution of the deterministic problem*

$$\begin{aligned}&\frac{\partial}{\partial t} v^0(x,t) = \operatorname{div}(\widehat{a}\nabla v^0(x,t)) + \widehat{b}\cdot\nabla v^0(x,t) + \widehat{G}v^0(x,t) + f(x,t),\\ &v^0(x,0) = u_0(x),\end{aligned} \tag{18.67}$$

*where the coefficients are defined by formulas* (18.38)–(18.39).

PROOF. In the case $\alpha \le 2$, it suffices to use the representation $v^\varepsilon = V^\varepsilon + (v^\varepsilon - V^\varepsilon)$ and Propositions 18.3 and 18.4.

In the case $\alpha > 2$, the required assertion follows from Theorem 18.4, Lemma 18.2, and the *a priori* estimate (18.14). □

18.1.7. *Homogenization of a nonfactorized equation.* Our goal is to obtain basic results concerning a homogenization of the problem (18.2), (18.3). The necessary technical assertions have already been proved.

First of all, we note that for $\alpha \le 2$ the random processes $(\exp(\zeta_t^\varepsilon), \exp(-\zeta_t^\varepsilon))$ converge in law to the process $(\exp(\sigma W_t), \exp(-\sigma W_t))$, where $\zeta_t^\varepsilon$ and $\sigma$ are defined by formulas (18.9), (18.10).

THEOREM 18.6. *Suppose conditions* **H1**–**H5** *hold. Let* $\alpha \le 2$, *and let* $u^\varepsilon$ *be the solution to the problem* (18.2), (18.3). *Let* $\varepsilon \to 0$. *Then the sequence* $u^\varepsilon$ *converges in law in the strong topology of the space* $L_2(\mathbb{R}^d \times (0,T))$ *to the solution of the problem for the homogenized stochastic partial differential equation*

$$\begin{aligned}&d\widehat{u} = \Big(\widehat{a}_{ij}\frac{\partial^2 \widehat{u}}{\partial x_i \partial x_j} + \widehat{b}_i \frac{\partial \widehat{u}}{\partial x_i} + \widehat{g}\widehat{u}\Big)\,dt + \sigma\widehat{u}\,dW_t + f(x,t),\\ &\widehat{u}(x,0) = u_0(x),\end{aligned} \tag{18.68}$$

*where* $\widehat{g} = \widehat{G} + \sigma^2$, *the coefficients* $\widehat{a}_{ij}$, $\widehat{b}_i$, *and* $\widehat{G}$ *are defined by formulas* (18.33)–(18.35) *if* $\alpha = 2$ *and by formulas* (18.36), (18.37) *if* $\alpha < 2$, *and* $\sigma$ *is defined by* (18.10).

By [**48**, Theorem 6.7], the problem (18.68) is well posed and has a unique solution. Therefore, the limit distribution is uniquely determined. □

REMARK 18.5. If, in the problem (18.2), (18.3), the mean value of $g(z,s)$ with respect to $z$ vanishes a.s. for every $s$, then we can use Theorem 18.4. Therefore, the limit problem is deterministic and the solutions $u^\varepsilon$ converge in probability (and even almost surely if $\alpha = 2$) to a nonrandom function. In this case, it is easy to verify that the coefficient $\sigma$ in (18.68) vanishes.

In the case $\alpha > 2$, the homogenized equation is always deterministic.

THEOREM 18.7. *Suppose conditions* **H1**–**H5** *hold. Let* $\alpha > 2$, *and let* $u^\varepsilon$ *be the solution to the problem* (18.2), (18.3). *Let* $\varepsilon \to 0$. *Then the sequence* $u^\varepsilon$ *converges* *in probability in the* $L_2(\mathbb{R}^d\times(0,T))$*-norm to the solution of the homogenized Cauchy problem*

$$
\begin{aligned}
d\widehat{u} &= \Big(\widehat{a}_{ij}\frac{\partial^2\widehat{u}}{\partial x_i\partial x_j}+\widehat{b}_i\frac{\partial\widehat{u}}{\partial x_i}+\widehat{G}\widehat{u}\Big)\,dt+f(x,t),\\
\widehat{u}(x,0) &= u_0(x),
\end{aligned}
\tag{18.69}
$$

*where the coefficients* $\widehat{a}_{ij}$, $\widehat{b}_i$, *and* $\widehat{G}$ *are defined by formulas* (18.38), (18.39).

We represent the solution $u^\varepsilon$ to the problem (18.2), (18.3) in the form

$$
u^\varepsilon(x,t)=(v^\varepsilon(x,t)-V^\varepsilon(x,t))\exp(\zeta^\varepsilon_t)+V^\varepsilon(x,t)\exp(\zeta^\varepsilon_t),
$$

where $v^\varepsilon$ and $V^\varepsilon$ are the solutions to the problems (18.8) and (18.65), respectively, and $\zeta^\varepsilon_t$ is defined by formula (18.9). As was proved in Proposition 18.3, the sequence $(v^\varepsilon(x,t)-V^\varepsilon(x,t))$ converges in probability to zero in the $L_2(\mathbb{R}^d\times(0,T))$-norm. Since the sequence $\zeta^\varepsilon_t$ converges in law in the space $C[0,T]$ by Lemma 18.2, the sequence $(v^\varepsilon(x,t)-V^\varepsilon(x,t))\exp(\zeta^\varepsilon_t)$ converges in probability to zero in the space $L_2(\mathbb{R}^d\times(0,T))$ (check it!).

The factor $V^\varepsilon$ in the second term continuously (in the $C[0,T]$-norm) maps the trajectories $\zeta^\varepsilon_\cdot$ to the space $L_2(\mathbb{R}^d\times(0,T))$. Therefore, the product $V^\varepsilon(x,t)\exp(\zeta^\varepsilon_t)$ is a continuous functional on the processes $\zeta^\varepsilon_\cdot$ in the topology of the space $C[0,T]$ with values in $L_2(\mathbb{R}^d\times(0,T))$. By continuity, the convergence in law of $\zeta^\varepsilon_\cdot$ to $\sigma W_\cdot$ in the space $C[0,T]$ implies the convergence in law of $V^\varepsilon\exp(\zeta^\varepsilon_\cdot)$ to the random function $v^0\exp(\sigma W_\cdot)$ in the space $L_2(\mathbb{R}^d\times(0,T))$, where $v^0$ is the solution to the problem (18.66).

We set $\widehat{u}=v^0\exp(\sigma W_\cdot)$ and

$$
\widehat{A}=\widehat{a}_{ij}\frac{\partial^2}{\partial x_i\partial x_j}+\widehat{b}_i\frac{\partial}{\partial x_i}+\widehat{G}
$$

and show that $\widehat{u}$ is a solution to the problem (18.68). To this end, for arbitrary $\varphi(x)\in C_0^\infty(\mathbb{R}^d)$ we regard the inner product $(\widehat{u}(t),\varphi)$ in $L_2(\mathbb{R}^d)$ as a function of time and apply the Itô formula:

$$
\begin{aligned}
d(\widehat{u},\varphi) &= \exp(\sigma W_t)\,d(v^0(t),\varphi)+\sigma(v^0(t),\varphi)\exp(\sigma W_t)\,dW_t\\
&\quad+(1/2)\sigma^2(v^0(t),\varphi)\exp(\sigma W_t)\,dt\\
&= \exp(\sigma W_t)(\widehat{A}v^0(t),\varphi)\,dt+(f(x,t),\varphi)\exp(\sigma W_t)\exp(-\sigma W_t)\,dt\\
&\quad+\sigma(\widehat{u}(t),\varphi)\,dW_t+(1/2)\sigma^2(\widehat{u}(t),\varphi)\,dt\\
&= (\widehat{A}\widehat{u}(t),\varphi)\,dt+(1/2)\sigma^2(\widehat{u}(t),\varphi)\,dt+(f(x,t),\varphi)\,dt+\sigma(\widehat{u}(t),\varphi)\,dW_t.
\end{aligned}
$$

This relation, combined with the obvious equality $\widehat{u}(0)=u_0$, means that $\widehat{u}$ is a solution to the problem (18.68). By [**48**, Theorem 6.7], the solution is unique and, consequently, it uniquely characterizes the limit distribution.

REMARK 18.6. Another proof (cf. [**48**, Theorem 2.4]) is based on the Skorokhod theorem. The interested reader can reproduce it.

Theorem 18.7 follows from Lemma 18.2 and Theorem 18.5.

Now we formulate similar results for solutions to initial-boundary value problems. We begin with the Dirichlet condition. In an arbitrary Lipschitz domain

$Q \subset \mathbb{R}^d$, we consider the following initial-boundary value problem on $Q \times (0, T)$:

$$
\begin{aligned}
&\frac{\partial}{\partial t} u^\varepsilon(x,t) = \operatorname{div}\Big(a\Big(\frac{x}{\varepsilon}, \frac{t}{\varepsilon^\alpha}\Big)\nabla u^\varepsilon(x,t)\Big) + \frac{1}{\varepsilon^{1\wedge\frac{\alpha}{2}}} g\Big(\frac{x}{\varepsilon}, \frac{t}{\varepsilon^\alpha}\Big) u^\varepsilon(x,t) + f(x,t), \\
&u^\varepsilon(x,t) = 0 \text{ for } x \in \partial Q, \quad u(x,0) = u_0(x).
\end{aligned} \tag{18.70}
$$

As above, we assume that $u_0 \in L_2(Q)$ and $f \in L_2(Q \times (0,T))$. It is well known that for every $\varepsilon > 0$ there exists a unique solution to the problem (18.70) in the space $L_2((0,T); \overset{\circ}{H}{}^1(Q)) \cap C((0,T); L_2(Q))$ if condition **H3** is satisfied (cf., for example, [**77**]).

The assertions below can be proved in the same way as in the case of Cauchy problem, and we omit the proof. In the case $\alpha \le 2$, the following assertion holds.

THEOREM 18.8. *Suppose that conditions* **H1**–**H5** *hold. Let* $\alpha \le 2$, *and let* $u^\varepsilon$ *be the solution to the problem* (18.70). *Let* $\varepsilon \to 0$. *Then the sequence* $u^\varepsilon$ *converges in law in the strong topology of the space* $L_2(Q \times (0,T))$ *to the solution of the following problem for the homogenized stochastic partial differential equation:*

$$
\begin{aligned}
&d\widehat{u} = \Big(\widehat{a}_{ij}\frac{\partial^2 \widehat{u}}{\partial x_i \partial x_j} + \widehat{b}_i \frac{\partial \widehat{u}}{\partial x_i} + \widehat{g}\widehat{u}\Big)\, dt + \sigma \widehat{u}\, dW_t + f(x,t), \\
&\widehat{u}(x,t) = 0 \textit{ for } x \in \partial Q, \quad \widehat{u}(x,0) = u_0(x),
\end{aligned} \tag{18.71}
$$

*where the coefficients are defined in Theorem* 18.6.

In the case $\alpha > 2$, the following analog of Theorem 18.7 holds.

THEOREM 18.9. *Suppose conditions* **H1**–**H5** *hold. Let* $\alpha > 2$, *and let* $u^\varepsilon$ *be the solution to the problem* (18.70). *Let* $\varepsilon \to 0$. *Then the sequence* $u^\varepsilon$ *converges in probability in the* $L_2(Q \times (0,T))$*-norm to the solution of the homogenized Dirichlet initial-boundary value problem*

$$
\begin{aligned}
&d\widehat{u} = \Big(\widehat{a}_{ij}\frac{\partial^2 \widehat{u}}{\partial x_i \partial x_j} + \widehat{b}_i \frac{\partial \widehat{u}}{\partial x_i} + \widehat{G}\widehat{u}\Big)\, dt + f(x,t), \\
&\widehat{u}(x,t) = 0 \quad \textit{for} \quad x \in \partial Q, \quad \widehat{u}(x,0) = u_0(x),
\end{aligned} \tag{18.72}
$$

*where the coefficients are defined in Theorem* 18.7.

The same results are true for the problems with the Neumann or Fourier boundary conditions.

### 18.2. Equations with large convection.

18.2.1. *Statement of the problem.* Consider the convection-diffusion equation

$$
\frac{\partial}{\partial t} u^\varepsilon(x,t) = \operatorname{div}\Big(a\Big(\frac{x}{\varepsilon}, \frac{t}{\varepsilon^2}\Big)\nabla u^\varepsilon(x,t)\Big) + \frac{1}{\varepsilon} b\Big(\frac{x}{\varepsilon}, \frac{t}{\varepsilon^2}\Big)\nabla u^\varepsilon(x,t) \tag{18.73}
$$

in $\mathbb{R}^d \times (0,T)$ with the initial condition

$$
u^\varepsilon(x,0) = u_0(x) \in L_2(\mathbb{R}^d). \tag{18.74}
$$

First, we note that we deal only with automodel equations. The study of the nonautomodel case encounters serious difficulties and is outside the framework of this book.

We list conditions on the coefficients of the operator (18.73). They are only slightly different from conditions **H1**–**H5** in Subsection 18.1.1, except for **H4**.

**A1.** The coefficients $a_{ij}(y,s)$ and $b_i(y,s)$ are periodic with respect to $y$ with period 1 along each coordinate direction.

**A2.** The functions $a_{ij}(y,s)$ and $b_i(y,s)$ are random stationary processes with respect to $s$ on a probability space $(\Omega, \mathcal{F}, \mathbf{P})$ with values in the space of periodic functions. The coefficients $a_{ij}(y,s) = a_{ij}(y,s,\omega)$ and $b_i(y,s) = b_i(y,s,\omega)$ are measurable with respect to the $\sigma$-algebra $\mathcal{B}(\mathbb{T}^d) \times \mathcal{B}(-\infty,+\infty) \times \mathcal{F}$, where $\mathcal{B}$ is the Borel $\sigma$-algebra.

Furthermore, the processes $a_{ij}(y,s)$ and $b_i(y,s)$ can be interpreted as the action of some dynamical system $T_s$ on the random variables $\widetilde{a}_{ij}(y)$ and $\widetilde{b}_i(y)$ with values in $L_\infty(\mathbb{T}^d)$, i.e.,

$$a_{ij}(y,s,\omega) = \widetilde{a}_{ij}(y, T_s\omega), \quad b_i(y,s,\omega) = \widetilde{b}_i(y, T_s\omega). \tag{18.75}$$

**A3.** *Uniform ellipticity*:

$$\lambda^{-1}|\zeta|^2 \le a_{ij}(y,\tau)\zeta_i\zeta_j \le \lambda|\zeta|^2, \quad \lambda > 0, \quad |b_i(y,s)| \le \lambda$$

for all $y$, $\tau$, and $\zeta \in \mathbb{R}^d$.

**A4.** *Mixing.* At least one of the following relations holds:

$$\int_0^\infty \kappa(\gamma)\,d\gamma < \infty, \quad \int_0^\infty \sqrt{\varphi(\gamma)}d\gamma < \infty, \quad \int_0^\infty \rho(\gamma)\,d\gamma < \infty.$$

The existence and uniqueness of a solution in the class $L_2((0,T); H^1(\mathbb{R}^d))$ for every $\varepsilon > 0$ is a classical result and can be found, for example, in [**77**]. However, standard methods for obtaining *a priori* estimates do not provide uniform estimates with respect to the parameter $\varepsilon$.

18.2.2. *Auxiliary equations and technical results.* First, we consider the auxiliary equation

$$\frac{\partial}{\partial s}p(y,s) + \operatorname{div}(a(y,s)\nabla_y p(y,s)) - \operatorname{div}(b(y,s)p(y,s)) = 0 \tag{18.76}$$

in $\mathbb{T}^d \times (-\infty,+\infty)$. We also consider the sequence of the Cauchy problems

$$\begin{gathered} \tfrac{\partial}{\partial s}p^N(y,s) + \operatorname{div}(a(y,s)\nabla_y p^N(y,s)) - \operatorname{div}(b(y,s)p^N(y,s)) = 0, \\ p^N(y,N) = 1 \end{gathered} \tag{18.77}$$

in $\mathbb{T}^d \times (-\infty, N)$.

For the sake of brevity, we write $A_s\varphi(y) = \operatorname{div}(a(y,s)\nabla_y\varphi(y)) + b(y,s)\nabla_y\varphi(y)$ and $A_s^*\varphi(y) = \operatorname{div}(a(y,s)\nabla_y\varphi(y)) - \operatorname{div}(b(y,s)\varphi(y))$.

PROPOSITION 18.5. *The functions $p^N$ converge uniformly in compact sets $\mathbb{T}^d \times (-\infty,+\infty)$ as $N \to \infty$ to the stationary positive solution $p(y,s)$ of equation* (18.76); *moreover,*

$$\int_{\mathbb{T}^d} p(y,s)\,dy = 1 \text{ a.s.}, \quad s \in (-\infty,+\infty). \tag{18.78}$$

*The solution $p(y,s)$ is continuous with respect to $y$, $s$ and satisfies the estimates*

$$\|p\|_{L_\infty(\mathbb{T}^d\times(-\infty,+\infty))} \le C, \quad \sup_{k\in\mathbb{R}} \|p\|_{L_2((k,k+1);H^1(\mathbb{T}^d))} \le C, \quad p(y,s) \ge c_1, \tag{18.79}$$

$$\max_{y\in\mathbb{T}^d,\ k\le s\le k+1} |p(y,s) - p^N(y,s)| \le C\exp(-c_2(N-k)), \tag{18.80}$$

*where $C$, $c_1$, and $c_2$ are nonrandom positive constants.*

PROOF. We apply the maximum principle (cf. [**77**]) to the problem (18.77). Since the data are positive, the solutions are positive for all $y$ and $s \le N$. Furthermore, taking into account the structure of equation (18.77) and integrating over the set $\mathbb{T}^d \times (s, N)$, $s < N$, we find that

$$\int_{\mathbb{T}^d} p^N(y,s)\,dy = 1$$

for all $s \le N$. Using the Harnack inequality (cf. [**127**]), we obtain the estimate

$$0 < c_1 \le p^N(y,s) \le c_3 < \infty, \tag{18.81}$$

where the constants $c_1$ and $c_3$ are independent of $N$ and $\omega$.

To justify the convergence of $p^N(y,s)$ and the estimate (18.80), we consider the auxiliary Cauchy problem

$$\begin{aligned} &\frac{\partial}{\partial s} q^N(y,s) + A_s^* q^N(y,s) = 0, \quad (y,s) \in \mathbb{T}^d \times (-\infty, N), \\ &q^N(y,N) = q_0, \end{aligned} \tag{18.82}$$

where $q_0 \in L_2(\mathbb{T}^d)$ is such that

$$\int_{\mathbb{T}^d} q_0(y)\,dy = 0. \tag{18.83}$$

LEMMA 18.6. *For $s - N \le -1$,*

$$|q^N(y,s)| \le c_1 \|q_0\|_{L_2(\mathbb{T}^d)} \exp(C(s-N)). \tag{18.84}$$

*If $q_0 \in L_\infty(\mathbb{T}^d)$, then*

$$|q^N(y,s)| \le c_1 \|q_0\|_{L_\infty(\mathbb{T}^d)} \exp(C(s-N)) \tag{18.85}$$

*for all $s \le N$.*

The proof of this lemma will be given later, but now we continue the proof of Proposition 18.5.

The difference $(p^{N+k} - p^N)$ is a solution to the equation

$$\frac{\partial}{\partial s}(p^{N+k} - p^N) + A_s^*(p^{N+k} - p^N) = 0, \quad s < N,$$

and

$$\int_{\mathbb{T}^d} (p^{N+k}(y,N) - p^N(y,N))\,dy = 0.$$

Consequently,

$$\|p^{N+k} - p^N\|_{C(\mathbb{T}^d \times (s,s+1))} \le c_1 \exp(C(N-s)).$$

By the Cauchy criterion, the $p^N$ uniformly converge on compact sets to a continuous function, denoted by $p(y,s)$, as $N \to \infty$. Passing to the limit in the last inequality as $k \to \infty$, we obtain (18.80), (18.78), and the first inequality in (18.79).

Further, using the standard technique for obtaining *a priori* elliptic estimates, it is easy to see that $\|\nabla p^N\|_{L_2(\mathbb{T}^d \times (s,s+1))} \le c$ uniformly with respect to $\omega \in \Omega$ and $s \le N - 1$. Hence we have the weak convergence of $\nabla p^N$ to $\nabla p$ in the space $L_2^{\mathrm{loc}}$ and, as a consequence, the second estimate in (18.79). Passing to the limit in the integral identity corresponding to the problem (18.77), we conclude that $p(y,s)$ is a solution to equation (18.76). It is clear that for any $N \in \mathbb{R}$ the function $p^{N+s}(y,s)$ is stationary with respect to $s$. As has already been proved, the $p^{N+s}(y,s)$ converge to $p(y,s)$. Hence $p(y,s)$ is also stationary, which completes the proof of Proposition 18.5. □

PROOF OF LEMMA 18.6. Consider the adjoint problem to (18.82):

$$\frac{\partial}{\partial s}\nu^{s_0}(y,s) - A_s\nu^{s_0}(y,s) = 0, \quad (y,s) \in \mathbb{T}^d \times (s_0,\infty), \qquad (18.86)$$
$$\nu^{s_0}(y,s_0) = \nu_0(y),$$

where $\nu_0$ is an arbitrary function in $L_2(\mathbb{T}^d)$. We multiply the equation in (18.86) by the function $q^N(y,s)$ defined by formula (18.82) and integrate over the set $\mathbb{T}^d \times (s_0,N)$. Integrating by parts and taking into account (18.82), after simple transformations we find that

$$\begin{aligned}
&\int_{\mathbb{T}^d} \nu^{s_0}(y,N)q_0(y)\,dy - \int_{\mathbb{T}^d} \nu_0(y)q^N(y,s_0)\,dy\\
&= \int_{s_0}^{d}\int_{\mathbb{T}^d}\Big(q^N(y,s)\frac{\partial}{\partial s}\nu^{s_0}(y,s) + \nu^{s_0}(y,s)\frac{\partial}{\partial s}q^N(y,s)\Big)\\
&= \int_{s_0}^{d}\int_{\mathbb{T}^d}\Big(q^N(y,s)A_s\nu^{s_0}(y,s) - \nu^{s_0}(y,s)A_s^*q^N(y,s)\Big) = 0.
\end{aligned} \qquad (18.87)$$

By the Nash inequality (cf., for example, [**77**]), the solution to the problem (18.86) satisfies the estimate

$$|\nu^{s_0}(y,s_0+1)| \le c\|\nu_0\|_{L_2(\mathbb{T}^d)}, \qquad (18.88)$$

where the constant $c$ depends only on the dimension and on $\lambda$ in condition **A3**. Let $\nu^+(y,s_1)$ and $\nu^-(y,s_1)$ be positive and negative parts of $\nu(y,s_1)$ respectively, i.e., $\nu^+(y,s_1) = \max(\nu^{s_0}(y,s_1),0)$ and $\nu^-(y,s_1) = \min(\nu^{s_0}(y,s_1),0)$. Adding, if necessary, a suitable constant, we can assume that

$$\max_{y\in\mathbb{T}^d}\nu^{s_0}(y,s_1) = -\min_{y\in\mathbb{T}^d}\nu^{s_0}(y,s_1)$$

and, as a consequence,

$$\|\nu^+(\cdot,s_1)\|_{L_\infty} = \|\nu^-(\cdot,s_1)\|_{L_\infty}.$$

By the choice of $\nu^+(y,s_1)$ and the maximum principle, the solution to the problem

$$\frac{\partial}{\partial s}\widetilde{\nu}^1(y,s) - A_s\widetilde{\nu}^1(y,s) = 0, \quad (y,s) \in \mathbb{T}^d \times (s_1,\infty),$$
$$\widetilde{\nu}^1(y,s_1) = \nu^+(y,s_1),$$

is of a fixed sign. By the Harnack inequality,

$$\max_{y\in\mathbb{T}^d}\widetilde{\nu}^1(y,s_1+1) \le c_2\min_{y\in\mathbb{T}^d}\widetilde{\nu}^1(y,s_1+1),$$

where $c_2$ depends only on the dimension and the constant $\lambda$ in condition **A3**. Thus,

$$c_2^{-1}\|\widetilde{\nu}^1(\cdot,s_1+1)\|_{L_\infty} \le \min_{y\in\mathbb{T}^d}\widetilde{\nu}^1(y,s_1+1).$$

Similarly, the solution to the problem

$$\frac{\partial}{\partial s}\widetilde{\nu}^2(y,s) - A_s\widetilde{\nu}^2(y,s) = 0, \quad (y,s) \in \mathbb{T}^d \times (s_1,\infty),$$
$$\widetilde{\nu}^2(y,s_1) = \nu^-(y,s_1),$$

satisfies the inequality

$$-c_2^{-1}\|\widetilde{\nu}^2(\cdot, s_1+1)\|_{L_\infty} \geq \max_{y\in\mathbb{T}^d} \widetilde{\nu}^2(y, s_1+1).$$

Since $\nu^{s_0}(y,s) = \nu^1(y,s) + \nu^2(y,s)$ for $s \geq s_0$, we have

$$\begin{aligned} - \|\widetilde{\nu}^2(\cdot, s_1+1)\|_{L_\infty} + c_2^{-1}\|\widetilde{\nu}^1(\cdot, s_1+1)\|_{L_\infty} &\leq \nu^{s_0}(y, s_1+1) \\ &\leq \|\widetilde{\nu}^1(\cdot, s_1+1)\|_{L_\infty} - c_2^{-1}\|\widetilde{\nu}^2(\cdot, s_1+1)\|_{L_\infty}, \quad c_2 > 0. \end{aligned}$$

Consequently,

$$\begin{aligned} \operatorname*{osc}_{y\in\mathbb{T}^d} \nu^{s_0}(y, s_1+1) &\leq (1-c_2^{-1})(\|\widetilde{\nu}^1(\cdot, s_1+1)\|_{L_\infty} + \|\widetilde{\nu}^2(\cdot, s_1+1)\|_{L_\infty}) \\ &\leq (1-c_2^{-1})(\|\widetilde{\nu}^1(\cdot, s_1)\|_{L_\infty} + \|\widetilde{\nu}^2(\cdot, s_1)\|_{L_\infty}) \\ &= (1-c_2^{-1}) \operatorname*{osc}_{y\in\mathbb{T}^d} \nu^{s_0}(y, s_1). \end{aligned}$$

Taking into account (18.88), we obtain the estimate

$$\operatorname*{osc}_{y\in\mathbb{T}^d} \nu^{s_0}(\cdot, s) \leq c_3 \exp(-c(s-s_0))\|\nu_0\|_{L_2(\mathbb{T}^d)}, \quad s \geq s_0 + 1. \tag{18.89}$$

This estimate and (18.87) imply

$$\begin{aligned} \left| \int_{\mathbb{T}^d} \nu_0(y) q^N(y, s_0)\, dy \right| &= \left| \int_{\mathbb{T}^d} \nu^{s_0}(y, N) q_0(y)\, dy \right| \\ &\leq c_3 \exp(-c(N-s_0))\|q_0\|_{L_2(\mathbb{T}^d)}\|\nu_0\|_{L_2(\mathbb{T}^d)}. \end{aligned}$$

Since $\nu_0$ is arbitrary, we have

$$\|q^N(\cdot, s_0)\|_{L_2(\mathbb{T}^d)} \leq c_3 \exp(-c(N-s_0))\|q_0\|_{L_2(\mathbb{T}^d)}.$$

Using the Nash estimate again, we find that

$$\|q^N(\cdot, s_0)\|_{L_\infty(\mathbb{T}^d)} \leq c_4 \exp(-c(N-s_0))\|q_0\|_{L_2(\mathbb{T}^d)}, \quad N - s_0 \geq 1.$$

The lemma is proved. □

We define a constant vector $\overline{b}$ by the equality

$$\begin{aligned} \overline{b}^i &= \mathbf{E} \int_{\mathbb{T}^d} \Big(\frac{\partial}{\partial y_j} a^{ij}(y,s) + b^i(y,s)\Big) p(y,s)\, dy \\ &= \mathbf{E} \int_{\mathbb{T}^d} \Big( - a^{ij}(y,s)\frac{\partial}{\partial y_j} p(y,s) + b^i(y,s)p(y,s)\Big)\, dy. \end{aligned} \tag{18.90}$$

Note that the integrand is stationary, and consequently the right-hand side of (18.90) is independent of $s$. Hence the vector $\overline{b}$ is well defined. We show that the vector $\varepsilon^{-1}\overline{b}$ plays the role of the effective convection in the problem (18.73). All the processes in the definition of $\overline{b}$ are stationary and ergodic. By the Birkhoff ergodic theorem, we have

$$\begin{aligned} \overline{b}^i &= \lim_{T\to\infty} \frac{1}{T} \int_0^T \int_{\mathbb{T}^d} \Big(\frac{\partial}{\partial y_j} a^{ij}(y,s) + b^i(y,s)\Big) p(y,s)\, dy\, ds \\ &= \lim_{T\to\infty} \frac{1}{T} \int_0^T \int_{\mathbb{T}^d} \Big( - a^{ij}(y,s)\frac{\partial}{\partial y_j} p(y,s) + b^i(y,s)p(y,s)\Big) dy\, ds \quad \text{a.s.} \end{aligned} \tag{18.91}$$

For further analysis, we need the random processes

$$\overline{\eta}^i(s) = \int_{\mathbb{T}^d} \Big(\frac{\partial}{\partial y_j} a^{ij}(y,s) + b^i(y,s)\Big) p(y,s)\, dy \tag{18.92}$$
$$= \int_{\mathbb{T}^d} \Big( - a^{ij}(y,s) \frac{\partial}{\partial y_j} p(y,s) + b^i(y,s) p(y,s)\Big)\, dy.$$

Using these processes, we introduce the matrix

$$(\sigma^2)^{ij} = \int_0^\infty \mathbf{E}(\eta^i(s)\eta^j(0) + \eta^j(s)\eta^i(0))\, ds. \tag{18.93}$$

The following assertion will play the key role in the passage to the limit and the description of the limit dynamics.

LEMMA 18.7. *The integral on the right-hand side of* (18.93) *converges. The process $\eta(s)$ satisfies the functional central limit theorem* (*invariance principle*) *with the covariance matrix $\sigma^2$, i.e.,*

$$\varepsilon \int_0^{t/\varepsilon^2} \eta(s)\, ds \xrightarrow{\mathcal{L}} \sigma W_t \quad \textit{as } \varepsilon \to 0$$

*in $(C[0,T])^d$, where $W_\cdot$ is a standard $n$-dimensional Wiener process.*

PROOF. We show that for all $T > 0$,

$$\|\mathbf{E}\{\eta(0)|\mathcal{F}_{\geq T}\}\|_{L_2(\Omega)} \leq c_1(\exp(-cT) + \rho(T/2)), \tag{18.94}$$

where $c > 0$. We represent $p(y,s)$ on $(0, T/2)$ as the sum $p(y,s) = p_1(y,s) + p_2(y,s)$, where $p_1$ and $p_2$ satisfy the equation

$$\frac{\partial}{\partial s} p_m(y,s) + A_s^* p_m(y,s) = 0, \quad s \leq \frac{T}{2},$$

and the initial conditions

$$p_1(y, T/2) = 1, \quad p_2(y, T/2) = p(y, T/2) - 1.$$

Then $\eta(0) = \eta_1(0) + \eta_2(0)$, where

$$\eta^i_m(0) = \int_{\mathbb{T}^d} \Big(\frac{\partial}{\partial y_j} a^{ij}(y,0) + b^i(y,0)\Big) p^m(y,0) dy, \quad m = 1, 2.$$

By definition, the function $p_1(0)$ and the random vector $\eta_1(0)$ are measurable with respect to the $\sigma$-algebra $\mathcal{F}_{\leq T/2}$. Therefore, if one of the mixing conditions in **A4** (say, the last one) is satisfied, then

$$\|\mathbf{E}\{\eta_1(0)|\mathcal{F}_{\geq T}\}\|_{L_2(\Omega)} \leq \rho(T/2)\|\eta_1(0)\|_{L_2(\Omega)} \leq C\rho(T/2). \tag{18.95}$$

By Lemma 18.6, we have

$$|p_2(y,0)| \leq c_1 \exp(-CT/2).$$

Consequently,

$$\|\mathbf{E}\{\eta_2(0)|\mathcal{F}_{\geq T}\}\|_{L_2(\Omega)} \leq \|\eta_2(0)\|_{L_2(\Omega)} \leq c_2 \exp(-CT/2),$$

which, together with the estimate (18.95), yields the required inequality (18.94).

As was shown in [**81**, Section 9.2], inequality (18.94) implies the functional central limit theorem for $\eta(-s)$, which, in turn, implies the functional central limit theorem for $\eta(s)$. Indeed, by the Prokhorov theorem (cf. Section 2 and [**18**]) and the fact that $\eta(s)$ is a stationary process, the weak compactness of the distributions $\{\varepsilon \int_0^{t/\varepsilon^2} \eta(-s)\, ds\}$ in $(C[0,T])^d$ is equivalent to the weak compactness of the distributions $\{\varepsilon \int_0^{t/\varepsilon^2} \eta(s)\, ds\}$. The convergence of finite-dimensional distributions of the processes $\{\varepsilon \int_0^{t/\varepsilon^2} \eta(-s)\, ds\}$ obviously follows from the convergence of the corresponding finite-dimensional distributions $\{\varepsilon \int_0^{t/\varepsilon^2} \eta(s)\, ds\}$.

In the case of the other mixing condition in **A4**, the required assertion is proved in a similar way. □

In addition to the above auxiliary equation, we also need the equation

$$\frac{\partial}{\partial s}\chi^i(y,s) - A_s\chi^i(y,s) = \frac{\partial}{\partial y_j}a^{ij}(y,s) + b^i(y,s) - \overline{b}^i - \eta^i(s). \tag{18.96}$$

It also has a stationary ergodic solution.

LEMMA 18.8. *Equation* (18.96) *has a unique stationary solution such that*

$$\int_{\mathbb{T}^d} \chi(y,s)p(y,s)\, dy = 0. \tag{18.97}$$

*This solution is ergodic.*

PROOF. Multiply equation (18.96) by $p(y,s)$ and integrate over the cylinder $\mathbb{T}^d \times (s_1, s_2)$. Integrating by parts and taking into account the definition of $p(y,s)$, $\eta(s)$, and $\overline{b}$, we find

$$\int_{\mathbb{T}^d} \chi(y,s_2)p(y,s_2)\, dy - \int_{\mathbb{T}^d} \chi(y,s_1)p(y,s_1)\, dy = 0.$$

Thus,

$$\int_{\mathbb{T}^d} \chi(y,s)p(y,s)\, dy = \text{const}.$$

Now the existence and uniqueness of a stationary solution, as well as its ergodicity, under the normalization condition (18.97) can be proved in the same way as in Proposition 18.2 by using the estimate (18.89) and the maximum principle. □

18.2.3. *A priori estimates.* To obtain *a priori* estimates for the problem (18.73), (18.74) and the problem

$$\frac{\partial}{\partial t}v^\varepsilon(x,t) = \operatorname{div}\Big(a\Big(\frac{x}{\varepsilon},\frac{t}{\varepsilon^2}\Big)\nabla v^\varepsilon(x,t)\Big) + \frac{1}{\varepsilon}b\Big(\frac{x}{\varepsilon},\frac{t}{\varepsilon^2}\Big)\nabla v^\varepsilon(x,t) + f(x,t), \tag{18.98}$$

$$v(x,0) = v_0(x), \quad v_0 \in L_2(\mathbb{R}^d), \quad f \in L_2(\mathbb{R}^d \times (0,T)), \tag{18.99}$$

we regard $p(y,s)$ as a weight function. Multiplying (18.98) by $v^\varepsilon(x,t)p(x/\varepsilon, t/\varepsilon^2)$, integrating over $\mathbb{R}^d \times (0,T)$, and repeatedly integrating by parts, we obtain the equality

$$\frac{1}{2}\int_{\mathbb{R}^d} p\Big(\frac{x}{\varepsilon},\frac{t}{\varepsilon^2}\Big)(v^\varepsilon(x,t))^2dx - \frac{1}{2}\int_{\mathbb{R}^d} p\Big(\frac{x}{\varepsilon},0\Big)(v_0(x))^2dx$$
$$+\int_0^t\int_{\mathbb{R}^d} a^{ij}\Big(\frac{x}{\varepsilon},\frac{s}{\varepsilon^2}\Big)p\Big(\frac{x}{\varepsilon},\frac{s}{\varepsilon^2}\Big)\frac{\partial}{\partial x_i}v^\varepsilon(x,s)\frac{\partial}{\partial x_j}v^\varepsilon(x,s)\,dx\,ds$$
$$-\frac{1}{2}\int_0^t\int_{\mathbb{R}^d}(v^\varepsilon(x,t))^2\Big[\frac{\partial}{\partial s}p\Big(\frac{x}{\varepsilon},\frac{s}{\varepsilon^2}\Big)+\frac{\partial}{\partial x_i}\Big(a^{ij}\Big(\frac{x}{\varepsilon},\frac{s}{\varepsilon^2}\Big)\frac{\partial}{\partial x_j}p\Big(\frac{x}{\varepsilon},\frac{s}{\varepsilon^2}\Big)\Big)$$
$$-\frac{1}{\varepsilon}\frac{\partial}{\partial x_i}\Big(b^i\Big(\frac{x}{\varepsilon},\frac{s}{\varepsilon^2}\Big)p\Big(\frac{x}{\varepsilon},\frac{s}{\varepsilon^2}\Big)\Big)\Big]\,dx\,ds$$
$$+\int_0^t\int_{\mathbb{R}^d} v^\varepsilon(x,t)f(x,t)\,dx\,dt = 0.$$

By (18.76), the expression in the square brackets (and the integral itself) vanishes. Therefore, by the Cauchy–Schwarz–Bunyakovskii inequality and (18.79), we have

$$\|v^\varepsilon\|^2_{L_\infty(0,T;L_2(\mathbb{R}^d))}+\|\nabla v^\varepsilon\|^2_{L_2(\mathbb{R}^d\times(0,T))}\le C\|v_0\|_{L_2(\mathbb{R}^d)}+C\|f\|^2_{L_2(\mathbb{R}^d\times(0,T))}. \tag{18.100}$$

In particular, for the solution to the problem (18.73), (18.74) we have

$$\|u^\varepsilon\|^2_{L_\infty(0,T;L_2(\mathbb{R}^d))}+\|\nabla u^\varepsilon\|^2_{L_2(\mathbb{R}^d\times(0,T))}\le C\|u_0\|_{L_2(\mathbb{R}^d)}. \tag{18.101}$$

18.2.4. *Passage to the limit. Construction of an effective problem.* Here, we find the first two terms of the asymptotic expansion of the solution $u^\varepsilon$ to the problem (18.73), (18.74) and characterize the limit problem.

Define $\overline{a}^{ij}$ by the formula

$$\overline{a}^{ij} = \mathbf{E}\int_{\mathbb{T}^d}\Big\{a^{ij}(y,s)+\frac{\partial}{\partial y_k}a^{ik}(y,s)\chi^j(y,s)+a^{ik}(y,s)\frac{\partial}{\partial y_k}\chi^j(y,s)$$
$$+b^i(y,s)\chi^j(y,s)\Big\}p(y,s)\,dy.$$

Introduce the notation

$$a^{ij}_{\mathrm{aux}}(y,s)=\Big(a^{y,s}+\frac{\partial}{\partial y_k}\{a^{ik}(y,s)\chi^j(y,s)\}+a^{ik}(y,s)\frac{\partial}{\partial y_k}\chi^j(y,s)$$
$$+b^i(y,s)\chi^j(y,s)-(\overline{b}^i+\eta^i(s))\chi^j(y,s)\Big)p(y,s).$$

Let $u^0(x,t)$ be a solution to the problem

$$\frac{\partial}{\partial t}u^0(x,t)-\operatorname{div}(\overline{a}\nabla u^0(x,t))=0,\quad u^0(x,0)=u_0(x),\quad (x,t)\in\mathbb{R}^d\times(0,T). \tag{18.102}$$

THEOREM 18.10. *Let $u_0\in L_2(\mathbb{R}^d)$. Then*

$$\lim_{\varepsilon\to0}\Big\|u^0\Big(x+\frac{t}{\varepsilon}\overline{b}+\frac{1}{\varepsilon}\int_0^t\eta\Big(\frac{s}{\varepsilon^2}\Big)\,ds,t\Big)-u^\varepsilon\Big\|_{L_2(\mathbb{R}^d\times(0,T))}=0\quad a.s.$$

PROOF. We show that for $u_0\in C_0^\infty(\mathbb{R}^d)$ the functions

$$V^\varepsilon(x,t)=u^\varepsilon(x,t)-u^0\Big(x+\frac{t}{\varepsilon}\overline{b}+\frac{1}{\varepsilon}\int_0^t\eta\Big(\frac{s}{\varepsilon^2}\Big)ds,t\Big)$$
$$-\varepsilon\nabla_x u^0\Big(x+\frac{t}{\varepsilon}\overline{b}+\frac{1}{\varepsilon}\int_0^t\eta\Big(\frac{s}{\varepsilon^2}\Big)\,ds,t\Big)\chi\Big(\frac{x}{\varepsilon},\frac{t}{\varepsilon^2}\Big)$$

converge a.s. to zero in the $L_\infty(\mathbb{R}^d\times[0,T])\cap L_2(0,T;H^1(\mathbb{R}^d))$-norm, i.e., the sum of the second and third terms on the right-hand side of the last equality approximates $u^\varepsilon$ in this space. For this purpose, we put $V^\varepsilon$ in equation (18.73). We set

$$\frac{\partial}{\partial y_i}g\Big(\frac{x}{\varepsilon},\frac{t}{\varepsilon^2}\Big)=\frac{\partial}{\partial y_i}g\Big(y,\frac{t}{\varepsilon^2}\Big)\Big|_{y=\frac{x}{\varepsilon}},$$
$$\frac{\partial}{\partial s}g\Big(\frac{x}{\varepsilon},\frac{t}{\varepsilon^2}\Big)=\frac{\partial}{\partial y_i}g\Big(\frac{x}{\varepsilon},s\Big)\Big|_{s=\frac{t}{\varepsilon^2}},$$

for an arbitrary function $g(y,s)$, and

$$\widetilde{u}^0_\varepsilon(x,t)=u^0\Big(x+\frac{t}{\varepsilon}\bar{b}+\frac{1}{\varepsilon}\int_0^t\eta\Big(\frac{s}{\varepsilon^2}\Big)\,ds,t\Big).$$

After simple transformations we find that

$$\begin{aligned}
&\frac{\partial}{\partial t}V^\varepsilon-\operatorname{div}\Big(a\Big(\frac{x}{\varepsilon},\frac{t}{\varepsilon^2}\Big)\nabla V^\varepsilon\Big)\\
&\quad=\frac{1}{\varepsilon}\Big\{\Big(-\bar{b}-\eta\Big(\frac{t}{\varepsilon^2}\Big)\Big)\cdot\nabla_x\widetilde{u}^0_\varepsilon(x,t)\\
&\qquad-\Big(\frac{\partial}{\partial y_i}a^{ij}\Big(\frac{x}{\varepsilon},\frac{t}{\varepsilon^2}\Big)+b^j\Big(\frac{x}{\varepsilon},\frac{t}{\varepsilon^2}\Big)\Big)\frac{\partial}{\partial x_j}\widetilde{u}^0_\varepsilon(x,t)\\
&\qquad+\Big(\frac{\partial}{\partial s}\chi\Big(\frac{x}{\varepsilon},\frac{t}{\varepsilon^2}\Big)-\frac{\partial}{\partial y_i}a^{ij}\Big(\frac{x}{\varepsilon},\frac{t}{\varepsilon^2}\Big)\frac{\partial}{\partial y_j}\chi\Big(\frac{x}{\varepsilon},\frac{t}{\varepsilon^2}\Big)\\
&\qquad\qquad-b^i\Big(\frac{x}{\varepsilon},\frac{t}{\varepsilon^2}\Big)\frac{\partial}{\partial y_i}\chi\Big(\frac{x}{\varepsilon},\frac{t}{\varepsilon^2}\Big)\Big)\nabla_x\widetilde{u}^0_\varepsilon(x,t)\Big\}\\
&\quad+\frac{\partial}{\partial t}u^0\Big(x+\frac{\tau}{\varepsilon}\bar{b}+\frac{1}{\varepsilon}\int_0^\tau\eta\Big(\frac{s}{\varepsilon^2}\Big)\,ds,t\Big)\Big|_{\tau=t}\\
&\quad-\Big\{a^{ij}\Big(\frac{x}{\varepsilon},\frac{t}{\varepsilon^2}\Big)+\bar{b}^i\chi^j\Big(\frac{x}{\varepsilon},\frac{t}{\varepsilon^2}\Big)+\eta^i\Big(\frac{t}{\varepsilon^2}\Big)\chi^j\Big(\frac{x}{\varepsilon},\frac{t}{\varepsilon^2}\Big)\\
&\qquad+\frac{\partial}{\partial y_k}\Big[a^{kj}\Big(\frac{x}{\varepsilon},\frac{t}{\varepsilon^2}\Big)\chi^i\Big(\frac{x}{\varepsilon},\frac{t}{\varepsilon^2}\Big)\Big]+a^{ik}\Big(\frac{x}{\varepsilon},\frac{t}{\varepsilon^2}\Big)\frac{\partial}{\partial y_k}\chi^j\Big(\frac{x}{\varepsilon},\frac{t}{\varepsilon^2}\Big)\\
&\qquad\qquad+b^i\Big(\frac{x}{\varepsilon},\frac{t}{\varepsilon^2}\Big)\chi^j\Big(\frac{x}{\varepsilon},\frac{t}{\varepsilon^2}\Big)\Big\}\frac{\partial}{\partial x_i}\frac{\partial}{\partial x_j}\widetilde{u}^0_\varepsilon(x,t)\\
&\quad+\varepsilon\nabla_x\Big(\frac{\partial}{\partial t}u^0\Big(x+\frac{\tau}{\varepsilon}\bar{b}+\frac{1}{\varepsilon}\int_0^\tau\eta\Big(\frac{s}{\varepsilon^2}\Big)\,ds,t\Big)\Big|_{\tau=t}\Big)\chi\Big(\frac{x}{\varepsilon},\frac{t}{\varepsilon^2}\Big)\\
&\quad+\varepsilon a^{ij}\Big(\frac{x}{\varepsilon},\frac{t}{\varepsilon^2}\Big)\chi^k\Big(\frac{x}{\varepsilon},\frac{t}{\varepsilon^2}\Big)\frac{\partial^3}{\partial x_i\partial x_j\partial x_k}\widetilde{u}^0_\varepsilon(x,t)\\
&\quad=-\Big\{a^{ij}\Big(\frac{x}{\varepsilon},\frac{t}{\varepsilon^2}\Big)+\bar{b}^i\chi^j\Big(\frac{x}{\varepsilon},\frac{t}{\varepsilon^2}\Big)+\eta^i\Big(\frac{t}{\varepsilon^2}\Big)\chi^j\Big(\frac{x}{\varepsilon},\frac{t}{\varepsilon^2}\Big)\\
&\qquad+\frac{\partial}{\partial y_k}\Big[a^{kj}\Big(\frac{x}{\varepsilon},\frac{t}{\varepsilon^2}\Big)\chi^i\Big(\frac{x}{\varepsilon},\frac{t}{\varepsilon^2}\Big)\Big]+a^{ik}\Big(\frac{x}{\varepsilon},\frac{t}{\varepsilon^2}\Big)\frac{\partial}{\partial y_k}\chi^j\Big(\frac{x}{\varepsilon},\frac{t}{\varepsilon^2}\Big)\\
&\qquad\qquad+b^i\Big(\frac{x}{\varepsilon},\frac{t}{\varepsilon^2}\Big)\chi^j\Big(\frac{x}{\varepsilon},\frac{t}{\varepsilon^2}\Big)-\bar{a}^{ij}\Big\}\frac{\partial}{\partial x_i}\frac{\partial}{\partial x_j}\widetilde{u}^0_\varepsilon(x,t)\\
&\quad+\varepsilon\nabla_x\Big(\frac{\partial}{\partial t}u^0\Big(x+\frac{\tau}{\varepsilon}\bar{b}+\frac{1}{\varepsilon}\int_0^\tau\eta\Big(\frac{s}{\varepsilon^2}\Big)\,ds,t\Big)\Big|_{\tau=t}\Big)\chi\Big(\frac{x}{\varepsilon},\frac{t}{\varepsilon^2}\Big)\\
&\quad+\varepsilon a^{ij}\Big(\frac{x}{\varepsilon},\frac{t}{\varepsilon^2}\Big)\chi^k\Big(\frac{x}{\varepsilon},\frac{t}{\varepsilon^2}\Big)\frac{\partial^3}{\partial x_i\partial x_j\partial x_k}\widetilde{u}^0_\varepsilon(x,t)
\end{aligned}$$

$$= \Big(p^{-1}\Big(\frac{x}{\varepsilon}, \frac{t}{\varepsilon^2}\Big) a^{ij}_{\rm aux}\Big(\frac{x}{\varepsilon}, \frac{t}{\varepsilon^2}\Big) - \overline{a}^{ij}\Big) \frac{\partial}{\partial x_i}\frac{\partial}{\partial x_j}\widetilde{u}^0_\varepsilon(x,t) + R_\varepsilon(x,t),$$

where $\|R^\varepsilon\|_{L_2(\mathbb{R}^d\times(0,T))} \le C\varepsilon$ and $\|V^\varepsilon(\cdot,0)\|_{L_\infty(\mathbb{R}^d)} \le C\varepsilon$. By (18.100), the contributions of $R^\varepsilon$ and $V^\varepsilon(\cdot,0)$ to the solution are negligible as $\varepsilon \to 0$. Therefore, it suffices to estimate the solution to the problem

$$\begin{aligned}\frac{\partial}{\partial t}\widetilde{V}^\varepsilon(x,t) = \operatorname{div}\Big(a\Big(\frac{x}{\varepsilon}, \frac{t}{\varepsilon^2}\Big)\nabla\widetilde{V}^\varepsilon(x,t)\Big) + \frac{1}{\varepsilon}b\Big(\frac{x}{\varepsilon}, \frac{t}{\varepsilon^2}\Big)\nabla\widetilde{V}^\varepsilon(x,t)\\ + \Big(p^{-1}\Big(\frac{x}{\varepsilon}, \frac{t}{\varepsilon^2}\Big)a^{ij}_{\rm aux}\Big(\frac{x}{\varepsilon}, \frac{t}{\varepsilon^2}\Big) - \overline{a}^{ij}\Big)\frac{\partial}{\partial x_i}\frac{\partial}{\partial x_j}\widetilde{u}^0_\varepsilon(x,t),\end{aligned} \tag{18.103}$$

$$\widetilde{V}^\varepsilon(x,0) = 0. \tag{18.104}$$

As in Subsection 18.2.3, we multiply this equation by $p(x/\varepsilon, t/\varepsilon^2)\widetilde{V}^\varepsilon(x,t)$ and integrate by parts. Then

$$\begin{aligned}&\int_{\mathbb{R}^d}(\widetilde{V}^\varepsilon(x,t))^2dx + \int_0^t\int_{\mathbb{R}^d}|\nabla\widetilde{V}^\varepsilon(x,s)|^2\,dx\,ds\\ &\quad\le C\Big|\int_0^t\int_{\mathbb{R}^d}\widetilde{V}^\varepsilon(x,t)\Big(a^{ij}_{\rm aux}\Big(\frac{x}{\varepsilon}, \frac{s}{\varepsilon^2}\Big) - p\Big(\frac{x}{\varepsilon}, \frac{s}{\varepsilon^2}\Big)\overline{a}^{ij}\Big)\frac{\partial}{\partial x_i}\frac{\partial}{\partial x_j}\widetilde{u}^0_\varepsilon(x,s)\,dx\,ds\Big|.\end{aligned}$$

We write the integrand on the right-hand side of this inequality in the form

$$\begin{aligned}&\Big(a^{ij}_{\rm aux}\Big(\frac{x}{\varepsilon}, \frac{s}{\varepsilon^2}\Big) - p\Big(\frac{x}{\varepsilon}, \frac{s}{\varepsilon^2}\Big)\overline{a}^{ij}\Big)\widetilde{V}^\varepsilon(x,s)\frac{\partial}{\partial x_i}\frac{\partial}{\partial x_j}\widetilde{u}^0_\varepsilon(x,s)\\ &\quad= \Big(\gamma_1^{ij}\Big(\frac{x}{\varepsilon}, \frac{t}{\varepsilon^2}\Big) + \gamma_2^{ij}\Big(\frac{t}{\varepsilon^2}\Big)\Big)\widetilde{V}^\varepsilon(x,s)\frac{\partial}{\partial x_i}\frac{\partial}{\partial x_j}\widetilde{u}^0_\varepsilon(x,t)\\ &\quad= \Big(\varepsilon\operatorname{div}\kappa_1^{ij}\Big(\frac{x}{\varepsilon}, \frac{t}{\varepsilon^2}\Big) + \gamma_2^{ij}\Big(\frac{t}{\varepsilon^2}\Big)\Big)\widetilde{V}^\varepsilon(x,s)\frac{\partial}{\partial x_i}\frac{\partial}{\partial x_j}\widetilde{u}^0_\varepsilon(x,t),\end{aligned}$$

where

$$\begin{aligned}\gamma_2^{ij}(s) &= \int_{\mathbb{T}^d}(a^{ij}_{\rm aux}(y,s) - p(y,s)\overline{a}^{ij})\,ds,\\ \gamma_1^{ij}(y,s) &= (a^{ij}_{\rm aux}(y,s) - p(y,s)\overline{a}^{ij}) - \gamma_2^{ij}(s),\end{aligned}$$

$\kappa^{ij,k}(y,s)$ is periodic with respect to $y$, and $\operatorname{div}_y\kappa(y,s) = \gamma_1(y,s)$. This representation immediately yields the estimate

$$\begin{aligned}&\|\widetilde{V}^\varepsilon(\cdot,t)\|^2_{L_2(\mathbb{R}^d)} + \|\nabla\widetilde{V}^\varepsilon\|^2_{L_2(\mathbb{R}^d\times(0,t))}\\ &\qquad\le C\varepsilon\|\widetilde{V}^\varepsilon\|_{L_2(0,t;H^1(\mathbb{R}^d))}\\ &\qquad\quad + C\Big|\int_0^t\int_{\mathbb{R}^d}\widetilde{V}^\varepsilon(x,s)\gamma_2^{ij}\Big(\frac{s}{\varepsilon^2}\Big)\frac{\partial}{\partial x_i}\frac{\partial}{\partial x_j}\widetilde{u}^0_\varepsilon(x,s)\,dx\,ds\Big|.\end{aligned} \tag{18.105}$$

By construction, the random function $\gamma_2(s)$ is stationary and has zero mean. By the Birkhoff theorem, the functions $\gamma_2(s/\varepsilon^2)$ weakly converge a.s. to zero in $L_2(0,T)$. Changing the variables, we find that

$$\begin{aligned}&\int_0^t\int_{\mathbb{R}^d}\widetilde{V}^\varepsilon(x,s)\gamma_2^{ij}\Big(\frac{s}{\varepsilon^2}\Big)\frac{\partial}{\partial x_i}\frac{\partial}{\partial x_j}\widetilde{u}^0_\varepsilon(x,s)\,dx\,ds\\ &\quad= \int_0^t\int_{\mathbb{R}^d}\widetilde{V}^\varepsilon\Big(x - \frac{s}{\varepsilon}\overline{b} - \frac{1}{\varepsilon}\int_0^s\eta\Big(\frac{\tau}{\varepsilon^2}\Big)\,d\tau, s\Big)\gamma_2^{ij}\Big(\frac{s}{\varepsilon^2}\Big)\frac{\partial}{\partial x_i}\frac{\partial}{\partial x_j}u^0(x,s)\,dx\,ds.\end{aligned}$$

This integral converges to zero in view of the following assertion.

PROPOSITION 18.6. *The family of functions*

$$\check{V}^\varepsilon(x,t) = \widetilde{V}^\varepsilon\Big(x - \frac{s}{\varepsilon}\overline{b} - \frac{1}{\varepsilon}\int_0^s \eta\Big(\frac{\tau}{\varepsilon^2}\Big)\,d\tau, s\Big),$$

$\varepsilon > 0$, $\omega \in \Omega$, *is compact in the space* $L_2(\mathbb{R}^d \times (0,T))$.

The proof will be given later, but now we complete the analysis of the convergence of $\widetilde{V}^\varepsilon$. By (18.105) and Proposition 18.6, the sequence $\widetilde{V}^\varepsilon$ converges a.s. to zero in the $L_2(0,T;H^1(\mathbb{R}^d)) \cap L_\infty(0,T;L_2(\mathbb{R}^d))$-norm. This yields the required assertion of the theorem for the initial data $u_0 \in C_0^\infty(\mathbb{R}^d)$, since, in this case, the $L_\infty(0,T;L_2(\mathbb{R}^d))$-norms of the functions

$$\varepsilon\nabla_x u^0\Big(x + \frac{t}{\varepsilon}\overline{b} + \frac{1}{\varepsilon}\int_0^t \eta\Big(\frac{s}{\varepsilon^2}\Big)\,ds, t\Big)\chi\Big(\frac{x}{\varepsilon}, \frac{t}{\varepsilon^2}\Big)$$

converge to zero.

In the case of general initial data $u_0 \in L_2(\mathbb{R}^d)$, the required assertion can be obtained by approximating $u_0$ by smooth compactly supported functions. Here, the estimate (18.100) is used. The interested reader can carry out the proof. $\square$

We formulate the main results of this section.

THEOREM 18.11. *Let $u^\varepsilon$ be the solution to the problem* (18.73), (18.74), *and let* $\varepsilon \to 0$. *Then the sequence* $u^\varepsilon\big(x - \frac{t}{\varepsilon}\overline{b}, t\big)$ *converges in law in the space* $L_2(\mathbb{R}^d \times (0,T))$ *to a random function* $u^0(x + \sigma W_t, t)$, *where* $W_t$ *is a standard* $n$*-dimensional Wiener process and the matrix* $\sigma$ *is defined by formula* (18.10). *If* $u_0 \in H^1(\mathbb{R}^d)$, *then this function is a solution to the following problem for a stochastic partial differential equation:*

$$\begin{aligned} du &= (\overline{a}^{ij} + (1/2)(\sigma^2)^{ij})\frac{\partial^2}{\partial x_i \partial x_j}u\,dt + \sigma\nabla u\,dW_t, \\ u|_{t=0} &= u_0. \end{aligned} \tag{18.106}$$

*Moreover, the solution is unique and uniquely determines the limit measure.*

PROOF. By Theorem 18.10,

$$\lim_{\varepsilon\to 0}\Big\| u^\varepsilon\Big(x - \frac{t}{\varepsilon}\overline{b}, t\Big) - u^0\Big(x + \frac{1}{\varepsilon}\int_0^t \eta\Big(\frac{s}{\varepsilon^2}\Big)\,ds, t\Big)\Big\|_{L_2(\mathbb{R}^d\times(0,T))} = 0 \quad \text{a.s.}$$

Therefore, to obtain the first assertion, it suffices to prove that the functions

$$u^0\Big(x + \frac{1}{\varepsilon}\int_0^t \eta\Big(\frac{s}{\varepsilon^2}\Big)\,ds, t\Big)$$

converge in law to $u^0(x + \sigma W_t, t)$. This convergence holds in view of Lemma 18.7 and the continuity of the mapping $\Theta : (C(0,T))^d \to L_2(\mathbb{R}^d \times (0,T))$, $\Theta[\psi(\cdot)](x,t) = u^0(x + \psi(t), t)$ (check it!).

To justify the second assertion of the theorem, we apply the Itô–Krylov formula to $u^0(x + \sigma W_t, t)$:

$$\begin{aligned} du^0(x + \sigma W_t, t) = \frac{\partial}{\partial t}u^0(z,t)\Big|_{z = x+\sigma W_t} dt + \sigma\nabla u^0(x + \sigma W_t, t)\,dW_t \\ + (1/2)\sigma\sigma^*\frac{\partial^2}{\partial x_i \partial x_j}u^0(x + \sigma W_t, t). \end{aligned}$$

By (18.102), this equation is equivalent to (18.106).

Finally, the well-posedness of the problem (18.106) is proved in [**48**]. □

It remains to prove Proposition 18.6.

PROOF OF PROPOSITION 18.6. Denote

$$\widetilde{u}^\varepsilon(x,t) = u^\varepsilon\Big(x - \frac{s}{\varepsilon}\bar{b} - \frac{1}{\varepsilon}\int_0^s \eta\Big(\frac{\tau}{\varepsilon^2}\Big)\,d\tau, s\Big).$$

By the definition of $\widetilde{V}^\varepsilon(x,t)$, we have $\widetilde{V}^\varepsilon(x,t) = \widetilde{u}^\varepsilon(x,t) - u^0(x,t) + R_3(x,t)$, where the norms $\|R_3\|_{L_2((0,T)\times\mathbb{R}^d)}$ converge a.s. to zero as $\varepsilon \to 0$. Therefore, it suffices to establish the compactness of the family $\{\widetilde{u}^\varepsilon\}$.

We check that $\widetilde{u}^\varepsilon$ are localized, i.e., for every $\delta > 0$ there is $N = N(\delta) > 0$ such that $\|\widetilde{u}^\varepsilon\|_{L_2(\{|x|>N\}\times(0,T))} \le \delta$. Suppose that $\varphi(s)$ is smooth, $0 \le \varphi(s) \le 1$, $\varphi(s) = 0$ for $s \le 0$, and $\varphi(s) = 1$ for $s \ge 1$. We set

$$\varphi_N(x) = \varphi((|x| - N)/N), \quad \varphi_N^\varepsilon(x,t) = \varphi_N\Big(x - \frac{t}{\varepsilon}\bar{b} - \frac{1}{\varepsilon}\int_0^t \eta\Big(\frac{s}{\varepsilon^2}\Big)\,ds\Big)$$

and multiply equation (18.73) by $u^\varepsilon(x,t)p^\varepsilon(x,t)\varphi_N^\varepsilon(x,t)$, where, as above, $p^\varepsilon(x,t) = p(\frac{x}{\varepsilon}, \frac{t}{\varepsilon^2})$. Integrating the equality obtained and making transformations as above, we find that

$$\begin{aligned}
\int_{\mathbb{R}^d} (u^\varepsilon)^2(x,t)p^\varepsilon(x,t)\varphi_N^\varepsilon(x,t)\,dx &+ \int_0^t\int_{\mathbb{R}^d} |\nabla u^\varepsilon(x,s)|^2 p^\varepsilon(x,s)\varphi_N^\varepsilon(x,s)\,dx\,ds \\
&\le C\int_{\mathbb{R}^d} (u_0)^2(x)\varphi_N(x)\,dx + \frac{C}{N}\|u^\varepsilon\|^2_{L_2(0,t;H^1(\mathbb{R}^d))} \\
&\quad + C\int_0^t |u^\varepsilon(x,s)|(|u^\varepsilon(x,s)| + |\nabla u^\varepsilon(x,s)|)\varphi_N^\varepsilon(x,s)\,dx\,ds.
\end{aligned}$$

Under a suitable choice of $N$, the first two terms on the right-hand side can be made as small as desired. Then the required assertion follows from the Gronwall lemma.

For an arbitrary $\delta > 0$, we fix $N(\delta)$ and choose an orthonormal basis $\{e_j(x)\}$ for the space $L_2(\{|x| < N\})$ consisting of smooth functions vanishing for $|x| = N$. We represent the solution $\widetilde{u}^\varepsilon$ in the form

$$\widetilde{u}^\varepsilon = \sum_{j=1}^\infty \alpha_j^\varepsilon(t)e_j(x) + r^\varepsilon(x,t),$$

where $\alpha_j^\varepsilon$ are the corresponding Fourier coefficients. By the above arguments,

$$\|r^\varepsilon\|_{L_2(\mathbb{R}^d\times(0,T))} \le \delta.$$

Further, by the estimate (18.101), we can choose $M = M(\delta)$ such that

$$\sum_{j=M+1}^\infty \|\alpha_j^\varepsilon e_j\|_{L_2(\mathbb{R}^d\times(0,T))} \le \delta.$$

For every $j$ we have

$$|\alpha_j^\varepsilon(t) - (\widetilde{u}^\varepsilon(t)\widetilde{p}^\varepsilon(t), e_j)| = |(\widetilde{u}^\varepsilon(t)(\widetilde{p}^\varepsilon(t) - 1), e_j)| \le C\varepsilon\|\widetilde{u}^\varepsilon(t)\|_{H^1(\mathbb{R}^d)}\|e_j\|_{H^1(\mathbb{R}^d)},$$

where

$$\widetilde{p}^\varepsilon(x,t) = p^\varepsilon\Big(x - \frac{t}{\varepsilon}\bar{b} - \frac{1}{\varepsilon}\int_0^t \eta\Big(\frac{s}{\varepsilon^2}\Big)\,ds, t\Big).$$

Denote $\widetilde{\alpha}_j^\varepsilon(t) = (\widetilde{u}^\varepsilon(t)\widetilde{p}^\varepsilon(t), e_j)$. The last estimate implies the inequality

$$\sum_{j=1}^{M} \|(\alpha_j^\varepsilon - \widetilde{\alpha}_j^\varepsilon)e_j\|_{L_2(\mathbb{R}^d\times(0,T))} \leq C\varepsilon.$$

Thus, it suffices to verify the compactness of the family of functions $\sum_{j=1}^{M} \widetilde{\alpha}_j^\varepsilon e_j$.

Multiplying (18.73) by $p^\varepsilon(x,t)e_j(x)$ and integrating over $\mathbb{R}^d \times (t_1, t_2)$, we obtain

$$|\widetilde{\alpha}_j^\varepsilon(t_2) - \widetilde{\alpha}_j^\varepsilon(t_1)| \leq C_j \int_{t_1}^{t_2} \|u^\varepsilon(s)\|_{H^1(\mathbb{R}^d)}\, ds \leq C_j\sqrt{|t_2 - t_1|}.$$

Here, we used an *a priori* estimate for $u^\varepsilon$. By the Arzelà theorem, the family of functions $\widetilde{\alpha}_j^\varepsilon$, $j = 1, 2, \ldots, M$, is compact in $(C(0,T))^M$. From this follows the compactness of the family $\sum_{j=1}^{M} \widetilde{\alpha}_j^\varepsilon e_j$ in $L_2(\mathbb{R}^d \times (0,T))$. □

# Bibliography


1. E. Acerbi, V. Chiadò Piat, G. Dal Maso, and D. Percivale, *An extension theorem from connected sets, and homogenization in general periodic domains*, Nonlinear Anal. **18** (1992), no. 5, 481–496.
2. R. Adams, *Sobolev Spaces*, Academic Press, New York, 1975.
3. G. Allaire, *Homogenization and two-scale convergence*, SIAM J. Math. Anal. **23** (1992), no. 6, 1482–1518.
4. ______, *Shape Optimization by the Homogenization Method*, Springer-Verlag, Berlin etc., 2002.
5. T. Arbogast, J. Douglas, and U. Hornung, *Derivation of the double porosity model of single phase flow via homogenization theory*, SIAM J. Math. Anal. **21** (1990), no. 4, 823–836.
6. H. Attouch, *Homogénéisation*, Séminaire Bourbaki, 1987/88, Astérisque (1988), no. 161–162, 7–30.
7. M. Avellaneda and A. Majda, *Superdiffusion in nearly stratified flow*, J. Statist. Phys. **69** (1992), no. 3–4, 689–729.
8. N. S. Bakhvalov, *Averaging characteristics of bodies of periodic structure*, Dokl. Akad. Nauk SSSR **218** (1974), no. 5, 1046–1048. (Russian)
9. ______, *Averaging of partial differential equations with rapidly oscillating coefficients*, Dokl. Akad. Nauk SSSR **221** (1975), no. 3, 516–519; English transl., Sov. Math. Dokl. **16** (1975), 351–355.
10. N. S. Bakhvalov and G. P. Panasenko, *Homogenization: Averaging Processes in Periodic Media*, Nauka, Moscow, 1984; English transl., Kluwer, Dordrecht, 1989.
11. J. M. Ball and F. Murat, *Remarks on rank-one convexity and quasi convexity*, Ordinary and Partial Differential Equations. Vol. III, Proc. 11th Dundee Conf., Dundee/UK 1990, Pitman Res. Notes Math. Ser. 254, 1991, pp. 25–37.
12. A. G. Belyaev, *On singular perturbations of boundary value problems*, Dissertation, Moscow, 1990. (Russian)
13. A. G. Belyaev, A. L. Pyatnitskij, and G. A. Chechkin, *Asymptotic behavior of a solution to a boundary value problem in a perforated domain with oscillating boundary*, Sib. Mat. Zhurn. **39** (1998), no. 4, 730–754; English transl., Sib. Math. J. **39** (1998), no. 4, 621–644.
14. ______, *Averaging in a perforated domain with an oscillating third boundary condition*, Mat. Sb. **192** (2001), no. 7, 3–20; English transl., Sb. Math. **192** (2001), no. 7, 933–949.
15. A. Yu. Belyaev, *Lecture Notes on Homogenization Theory*. `http://ansatte.hin.no/alp/Beliaev/lecture.ps`
16. A. Yu. Belyaev and G. A. Chechkin, *Averaging operators with boundary conditions of fine-scaled structure*, Mat. Zametki **65** (1999), no. 4, 496–510; English transl., Math. Notes **65** (1999), no. 4, 418–429.
17. A. Bensoussan, J.-L. Lions, and G. Papanicolau, *Asymptotic Analysis for Periodic Structures*, North-Holland, Amsterdam, 1978.
18. P. Billingsley, *Convergence of Probability Measures*, Wiley, Chichester, 1999.
19. M. E. Bogovskij, *Solution of the first boundary value problem for the equation of continuity of an incompressible medium*, Dokl. Akad. Nauk SSSR **248** (1979), 1037–1040; English transl., Sov. Math. Dokl. **20** (1979), 1094–1098.
20. N. N. Bogolyubov, *On some statistical methods in mathematical physics*, Izv. Akad. Nauk Ukr. SSR, 1945. (Russian)
21. N. N. Bogolyubov and Yu. A. Mitropol′skii, *Asymptotic Methods in the Theory of Non-Linear Oscillations*, Nauka, Moscow, 1974; English transl., Gordon and Breach Science Publishers, New York, 1961.

22. D. I. Borisov and P. P. Gadyl'shin, *On the spectrum of the Laplacian with frequently alternating boundary conditions*, Teor. Mat. Fiz. **118** (1999), no. 3, 347–353; English transl., Theor. Math. Phys. **118** (1999), no. 3, 272–277.
23. F. Bornemann, *Homogenization in Time of Singularily Perturbed Mechanical Systems*, Lect. Notes Math. **1687**, Springer-Verlag, Berlin etc., 1998.
24. G. Bouchitté and I. Fragalà, *Homogenization of thin structures by two-scale method with respect to measures*, SIAM J. Math. Anal. **32** (2001), no. 6, 1198–1226.
25. G. Bouchitté, G. Buttazzo, and P. Seppecher, *Energies with respect to a measure and applications to low dimensional structures*, Calc. Var. Partial Differential Equations **5** (1997), no. 1, 37–54.
26. A. Bourgeat, G. Chechkin, and A. Piatnitski, *Singular double porosity model*, Appl. Anal. **82** (2003), no. 2, 103–116.
27. A. Bourgeat, M. Goncharenko, M. Panfilov, and A. Pankratov, *A general double porosity model*, C. R. Acad. Sci. Paris Sér. II **327** (1999), 1245–1250.
28. A. Bourgeat, S. Luckhaus, and A. Mikelić, *Convergence of the homogenization process for a double porosity model of immiscible two phase flow*, SIAM J. Appl. Math. **27** (1996), no. 6, 1520–1543.
29. A. Bourgeat, A. Mikelić, and A. L. Piatnitskij, *Modèle de double porosité aléatoire*, C. R. Acad. Sci. Paris, Sér. I **327** (1998), 99–104.
30. A. Braides and A. Defranceschi, *Homogenization of Multiple Integrals*, Clarendon Press, Oxford, 1998.
31. M. Briane, *Homogénéisation de Matériaux Fibres et Multi-Couches*, PhD Thesis, University Paris 6, Paris, 1990.
32. G. Buttazzo and G. Dal Maso, *Γ-limits of integral functionals*, J. Analyse Math. **37** (1980), 145–185.
33. A. P. Calderòn and A. Zygmund, *On singular integrals*, Amer. J. Math. **78** (1956), 289–309.
34. F. Campillo, M. L. Kleptsyna, and A. L. Piatnitskij, *Homogenization of random parabolic operator with large potential*, Stochastic Process. Appl. **93** (2001), 57–85.
35. G. A. Chechkin, *The boundary value problems for second order elliptic equations with oscillating boundary conditions*, Nonclassical Partial Differential Equations, Novosibirsk, IM SO RAN, 1988, pp. 95–104. (Russian)
36. ———, *Averaging of boundary value problems with a singular perturbation of the boundary conditions*, Mat. Sb. **184** (1993), no. 6, 99–150; English transl., Russ. Acad. Sci., Sb. Math. **79** (1994), no. 1, 191–222.
37. ———, *Asymptotic expansion of eigenvalues and eigenfunctions of an elliptic operator in a domain with many "light" concentrated masses closely situated on the boundary. Two-dimensional case*, Izv. Ross. Akad. Nauk Ser. Mat. **69** (2005), no. 4, 161–204; English transl., Izv. RAS **69** (2005), no. 4, 805–846.
38. ———, *Asymptotic expansions of eigenelements of the Laplace operator in a domain with many "light" concentrated masses closely located on the boundary. Multi-dimensional case*, Probl. Mat. Anal. **30** (2005), 87–119; English transl., J. Math. Sci. (New York) **128** (2005), no. 5, 3263–5025.
39. G. A. Chechkin and T. P. Chechkina, *On homogenization of problems in domains of the "infusorium" type*, Tr. Semin. Petrovskogo **23** (2003), 386–407; English transl., **23** (2003), J. Math. Sci. (New York) **120** (2004), no. 3, 1470–1482.
40. G. A. Chechkin and E. I. Doronina, *On the asymptotics of the spectrum of a boundary value problems with nonperiodic rapidly alternating boundary conditions*, Functional Differential Equations, E. Mitidieri, S. Pokhozaev, and A. Skubachevskii eds., **8** (2001), no. 1–2, 111–122, Marcel Dekker, New York.
41. G. A. Chechkin, A. Friedman, and A. L. Piatnitskij, *The boundary value problem in domains with very rapidly oscillating boundary*, J. Math. Anal. Appl. **231** (1999), no. 1, 213–234.
42. G. A. Chechkin, V. V. Jikov, D. Lukkassen, and A. L. Piatnitskij, *On homogenization of networks and junctions*, Asymptotic Anal. **30** (2002), 61–80.
43. G. A. Chechkin and A. L. Piatnitskij, *Homogenization of boundary value problem in a locally periodic perforated domain*, Appl. Anal. **71** (1999), no. 1–4, 215–235.
44. D. Cioranescu and P. Donato, *An Introduction to Homogenization*, Clarendon Press, Oxford, 1999.

45. D. Cioranescu and F. Murat, *Un terme étrange venu d'ailleurs I & II*, Nonlinear Partial Differential Equations and Their Applications. Collège de France Seminar, Vols. II, III, H. Berzis and J. L. Lions, eds., Research Notes in Mathematics, 60 & 70, Pitman, London, 1982, pp. 98–138 & 154–178.
46. D. Cioranescu and A. Piatnitskij, *Homogenization in perforated domains with rapidly pulsing perforations*, ESAIM: Control, Optimisation and Calculus of Variations **9** (2003), 461–484.
47. D. Cioranescu and J. Saint Jean Paulin, *Homogenization of Reticulated Structures*, Springer-Verlag, Berlin etc., 1998.
48. G. Da Prato and J. Zabczyk, *Stochastic Equations in Infinite Dimensions*, Cambridge Univ. Press, Cambridge, 1992.
49. B. Dacorogna, *Weak Continuity and Weak Lower Semicontinuity of Non–Linear Functionals,* Lect. Notes Math. **922**, Springer-Verlag, Berlin etc., 1982.
50. G. Dal Maso, *An Introduction to* $\Gamma$*-Convergence*, Birkhäuser, 1993.
51. A. Damlamian and Ta-Tsien Li (Li Daqian), *Boundary homogenization for elliptic problems*, J. Math. Pure Appl. **66** (1987), 351–361.
52. E. De Giorgi, *Sulla differenziabilita e l'analiticita delle estremali degli integrali multipli regolari*, Mem. Accad. Sci. Torino. Cl. Sci. Fis. Mat. Nat. (3) **3** (1957), 25–43.
53. ______, *Sulla convergenza di alcune successioni d'integrali del tipo dell'area*, Rend. Mat. (6) **8** (1975), 277–294.
54. ______, *Convergence problems for functionals and operators*, Proceedings of the International Meeting on Recent Methods in Nonlinear Analysis (Rome, 1978), Pitagora, Bologna, 1979, pp. 131–188.
55. E. De Giorgi and S. Spagnolo, *Sulla convergenza delli integrali dell'energia per operatori ellitici del secondo ordine*, Boll. Un. Mat. Ital. (4) **8** (1973), 391–411.
56. G. de Rham, *Differentiable Manifolds. Forms, Currents, Harmonic Forms*, Springer-Verlag, Berlin etc., 1984.
57. N. Dunford and J. T. Schwartz, *Linear Operators. Part I: General Theory*, Interscience Publishers, New York–London, 1958.
58. I. Ekeland and R. Temam, *Convex Analysis and Variational Problems*, North-Holland, Amsterdam, 1976.
59. H. Ene and G. Paşa, *Metoda omogenizării, Aplicaţii la teoria materialelor compozite*, Ed. Acad. Rep. Soc. România, Bucharest, 1987.
60. R. R. Gadyl′shin, *Asymptotics of the minimum eigenvalue for a circle with fast oscillating boundary conditions*, C. R. Acad. Sci. Paris, Sér. I **323** (1996), no. 3, 319–323.
61. ______, *On the perturbation of the Laplacian spectrum when the boundary condition type changes on a small part of the boundary*, Zhurn. Vychisl. Mat. Mat. Fiz. **36** (1996), no. 7, 77–88; English transl., Comput. Math. Math. Phys. **36** (1996), no. 7, 889–898.
62. ______, *On the eigenvalue asymptotics for periodically clamped membranes*, Algebra i Anal. **10** (1998), no. 1, 3–19; English transl., St. Petersburg Math. J. **10** (1999), no. 1, 1–14.
63. ______, *Boundary value problem for the Laplacian with rapidly oscillating boundary conditions*, Dokl. Akad. Nauk, Ross. Akad. Nauk **362** (1998), no. 4, 456–459; English transl., Dokl. Math. **58** (1998), no. 2, 293–296.
64. ______, *Asymptotics of the eigenvalues of a boundary value problem with rapidly oscillating boundary conditions*, Differ. Uravn. **35** (1999), no. 4, 540–550; English transl., Differ. Equations **35** (1999), no. 4, 540–551.
65. D. Gilbarg and N. S. Trudinger, *Elliptic Partial Differential Equations of Second Order*, Springer-Verlag, Berlin etc., 1983.
66. *Homogenization. In Memory of Serguei Kozlov.* V. Berdichevsky, V. Jikov, and G. Papanicolau, eds., World Scientific, 1999.
67. *Homogenization and Porous Media*, U. Hornung (ed.), Springer-Verlag, New York, 1997.
68. U. Hornung and W. Jäger, *Diffusion, convection, adsorption, and reaction of chemicals in porous media*, J. Differ. Equations **92** (1991), no. 2, 199–225.
69. A. M. Il′in, *A boundary value problem for the second order elliptic equation in a domain with a narrow slit. I. Two-dimensional case*, Mat. Sb. **99** (1976), 514–537. (Russian)
70. ______, *A boundary value problem for the second order elliptic equation in a domain with a narrow slit. II. Domain with a small cavity*, Mat. Sb. **103** (1977), 265–284; English transl., *Math. USSR-Sb.* **32** (1977), 227–244.

71. ———, *Matching of Asymptotic Expansions of Solutions of Boundary Value Problems*, Nauka, Moscow, 1989; English transl., Amer. Math. Soc., Providence, RI, 1992.
72. M. L. Kleptsyna and A. L. Piatnitskij, *Homogenization of random parabolic operators*, Homogenization and Applications to Material Sciences (D. Cioranescu et al,. eds.), Gakkotosho. GAKUTO Int. Ser. Math. Sci. Appl. **9** (1997), 241–255.
73. ———, *Homogenization of a random non-stationary convection-diffusion problem*, Usp. Mat. Nauk **57** (2002), no. 4, 95–118; English transl., Russ. Math. Surv. **57** (2002), no. 4, 729–751.
74. M. A. Krasnosel′skii, E. A. Lifshits, and A. V. Sobolev, *Positive Linear Systems. The Method of Positive Operators*, Nauka, Moscow, 1985; English transl., Heldermann-Verlag, Berlin, 1989.
75. O. A. Ladyzhenskaya, *Boundary-Value Problems of Mathematical Physics*, Nauka, Moscow, 1973.
76. O. A. Ladyzhenskaya and V. V. Solonnikov, *Some problems of vector analysis and generalized formulations of boundary-value problems for the Navier-Stokes equations*, Zap. Nauch. Sem. LOMI **59** (1976), 81–116; English transl., J. Sov. Math. **8** (1978), 257–286.
77. O. A. Ladyzhenskaya, V. V. Solonnikov, and N. N. Uraltseva, *Linear and Quasilinear Equations of Parabolic Type*, Nauka, Moscow, 1968; English transl. of 1st ed., Am. Math. Soc., Providence, RI, 1967.
78. J.-L. Lions, *Quelqes méthodes de résolution des problèmes aux limites non linéaires*, Dunod, Paris, 1969.
79. ———, *Asymptotic expansions in perforated media with a periodic structure*, Rocky Mountain J. Math. **10** (1980), no. 1, 125–140.
80. J.- L. Lions and E. Magenes, *Non-Homogeneous Boundary-Value Problems and Applications I–II*, Springer-Verlag, Berlin etc., 1972–73.
81. R. Sh. Liptser and A. N. Shiryaev, *Theory of Martingales*, Kluwer, Dordrecht, 1989.
82. M. Lobo and M. E. Pérez, *Asymptotic behavior of an elastic body with a surface having small stack regions*, RAIRO Modél. Math. Anal. Numér. **22** (1988), no. 4, 609–624.
83. D. Lukkassen and P. Wall, *Two-scale convergence with respect to measures and homogenization of monotone operators*, J. Funct. Spaces Appl. **3** (2005), no. 2, 125–161.
84. P. Marcellini, *Convergence of second order linear elliptic operators*, Boll. Un. Mat. Ital. (5) **16** (1979), no. 1, 278–290.
85. V. A. Marchenko and E. Ya. Khruslov, *Boundary Value Problems in Domains with Fine-Scaled Boundary*, Naukova Dumka, Kiev, 1974. (Russian)
86. V. G. Maz′ya, *Sobolev spaces*, Izd. Leningrad. Univ., Leningrad, 1985; English transl., Springer-Verlag, Berlin, 1985.
87. V. G. Maz′ya, S. A. Nazarov, and B. A. Plamenevskii, *Asymptotic expansions of the eigenvalues of boundary value problems for the Laplace operator in domains with small holes*, Izv. Akad. Nauk SSSR Ser. Mat. **48** (1984), no. 2, 347–371; English transl., Math. USSR-Izv. **24** (1985), 321–345.
88. V. P. Mikhailov, *Partial Differential Equations*, Nauka, Moscow, 1983. (Russian)
89. G. Milton, *The Theory of Composites*, Cambridge Univ. Press, Cambridge, 2002.
90. U. Mosco, *Convergence of convex sets and of solutions of variational inequalities*, Adv. Math. **3** (1969), 510–585.
91. F. Murat, *Compacité par compensation*, Ann. Scuola Norm. Sup. Pisa, Cl. Sci. Fis. Mat. **5** (1978), 481–507.
92. ———, *Compacité par compensation II*, Proceedings of the International Meeting on Recent Methods in Nonlinear Analysis (Rome, 1978), Pitagora, Bologna, 1979, pp. 245–256.
93. ———, *Compacité par compensation: condition nécéssaire et suffisante de continuité faible sous une hypothèse de rang constant*, Ann. Scuola Norm. Sup. Pisa, Cl. Sci. Fis. Mat. **8** (1981), 69–102.
94. F. Murat and L. Tartar, *Calcul des variations et homogénéisation.* Preprint, R 84012, Paris, 1984.
95. S. A. Nazarov, *Introduction to Asymptotic Methods of Elasticity Theory*, Izd. Leningrad. Univ., Leningrad, 1983. (Russian)
96. ———, *Asymptotic Expansions of Eigenvalues*, Izd. Leningrad. Univ., Leningrad, 1987. (Russian)

97. ______, *Asymptotic Theory of Thin Plates and Rods*, Nauchnaya Kniga (IDMI), Novosibirsk, 2002. (Russian)
98. S. A. Nazarov and M. V. Paukshto, *Discrete Models and Homogenization of Problems of Elasticity Theory*, Izd. Leningrad. Univ., Leningrad, 1984.
99. V. V. Nemytskii and V. V. Stepanov, *Qualitative Theory of Differential Equations*, Princeton Univ. Press, New Jersey, 1960.
100. G. Nguetseng, *A general convergence result for a functional related to the theory of homogenization*, SIAM J. Math. Anal. **20** (1989), no. 3, 608–623.
101. B. Øksendal, *Stochastic Differential Equations. An Introduction with Applications*, Springer-Verlag, Berlin, 2000.
102. O. A. Oleinik, A. S. Shamaev, and G. A. Yosifian, *Mathematical Problems of the Theory of Strongly Nonhomogeneous Elastic Media*, Izd. Moskov. Univ., Moscow, 1990. (Russian)
103. ______, *Mathematical Problems in Elasticity and Homogenization*, North-Holland, Amsterdam, 1992.
104. O. A. Oleinik and G. A. Chechkin, *On asymptotics of solutions and eigenvalues of the boundary value problems with rapidly alternating boundary conditions for the system of elasticity*, Atti Accad. Naz. Lincei, Cl. Sci. Fis. Mat. Natur., Rend. Lincei (9) Mat. Appl. **7** (1996), no. 1, 5–15.
105. O. A. Oleinik and T. A. Shaposhnikova, *On the homogenization of the Poisson equation in partially perforated domain with the arbitrary density of cavities and mixed conditions on their boundary*, Atti Accad. Naz. Lincei, Cl. Sci. Fis. Mat. Natur., Rend. Lincei (9) Mat. Appl. **7** (1996), no. 3, 129–146.
106. G. P. Panasenko, *Asymptotic Analysis of Rod Structures*, Kluwer, Dordrecht–Boston–London, 2001.
107. A. Pankov, *G-Convergence and Homogenization of Nonlinear Partial Differential Operators*, Kluwer, Dordrecht–Boston–London, 1997.
108. E. Pardoux and A. L. Piatnitskij, *Homogenization of a nonlinear random parabolic partial differential equation*, Stochastic Process. Appl. **104** (2003), no. 1, 1–27.
109. S. E. Pastukhova, *Averaging in a mixed directional derivative problem for an elliptic operator in a perforated domain*, Differ. Uravn. **30** (1994), no. 8, 1445–1456; English transl., Differ. Equations **30** (1994), no. 8, 1341–1351.
110. ______, *On the error of averaging for the Steklov problem in a punctured domain*, Differ. Uravn. **31** (1995), no. 6, 1042–1054; English transl., Differ. Equations **31** (1995), no. 6, 975–986.
111. ______, *Tartar's method of compensated compactness in averaging the spectrum of a mixed problem for an elliptic equation in a perforated domain with third boundary condition*, Mat. Sb. **186** (1995), no. 5, 127–144; English transl., Sb. Math. **186** (1995), no. 5, 753–770.
112. ______, *Homogenization of elasticity problems on periodic composite structures*, Mat. Sb. **196** (2005), no. 7, 101–142. (Russian)
113. L. E. Persson, L. Persson, N. Svanstedt, and J. Wyller, *The Homogenization Method. An Introduction*, Studentlitteratur, Chartwell Bratt, Lund, 1993.
114. M. Reed and B. Simon, *Methods of Modern Mathematical Physics, II. Fourier Analysis. Self-Adjointness*, Academic Press, New York–San Francisco–London, 1975.
115. E. Sanchez-Palencia, *Nonhomogeneous Media and Vibration Theory*, Lect. Notes Phys. **127**, Springer-Verlag, Berlin etc., 1980.
116. ______, *Homogenization Techniques for Composite Media*, Springer-Verlag, Berlin etc., 1987.
117. G. V. Sandrakov, *Averaging of the nonstationary Stokes system with viscosity in a punctured domain*, Izv. Ross. Akad. Nauk Ser. Mat. **61** (1997), no. 1, 113–140; English transl., Izv. Math. **61** (1997), no. 1, 113–141.
118. ______, *Averaging of nonstationary problems in the theory of strongly nonhomogeneous elastic media*, Dokl. Akad. Nauk, Ross. Akad. Nauk **358** (1998), no. 3, 308–311. (Russian)
119. Ya. G. Sinai, *The limit behavior of a one-dimensional random walk in a random environment*, Teor. Veroyatnost. Primenen. **27** (1982), no. 2, 247–258. (Russian)
120. S. L. Sobolev, *Some Applications of Functional Analysis in Mathematical Physics*, 3rd ed., Nauka, Moscow, 1988; English transl., Amer. Math. Soc., Providence, RI, 1991.
121. ______, *Selected Questions of the Theory of Function Spaces and Generalized Functions*, Nauka, Moscow, 1989. (Russian)

122. S. Spagnolo, *Sul limite delle soluzioni di problemi di Cauchy relativi all'equazione del calore*, Ann. Scuola Norm. Sup. Pisa, Cl. Sci. Fis. Mat. **21** (1967), no. 3, 657–699.
123. ———, *Sulla convergenza di soluzioni di equazioni parabolichi ed ellitiche*, Ann. Scuola Norm. Sup. Pisa, Cl. Sci. Fis. Mat. **22** (1968), no. 3, 571–597; errata, ibid. **22** (1968), no. 3, 673.
124. L. Tartar, *Problèmes d'homogénéisation dans les équations aux dérivées partielles, Cours Peccot Collège de France. H-Convergence*, F. Murat, ed., Séminaire d'Analyse Fonctionnelle et Numérique, 1977/78, Université d'Alger, 1978.
125. ———, *Compensated compactness and applications to partial differential equations. Nonlinear analysis and mechanics*, Heriot–Watt Symposium, Vol. IV. Research Notes in Mathematics 39, Pitman, 1979, pp. 136–212.
126. R. Temam, *Navier-Stokes Equations, Theory and Numerical Analysis*, Elsevier, North-Holland, Amsterdam–New York–Oxford, 1984.
127. N. S. Trudinger, *Pointwise estimates and quasilinear parabolic equations, Comm. Pure Appl. Math.* **21** (1968), 205-226.
128. M. I. Vishik and L. A. Lyusternik, *Regular degeneration and boundary layer*, Usp. Mat. Nauk **12** (1957), no. 5, 3–122. (Russian)
129. D. B. Volkov, *Averaging of some boundary value problems in domains with a periodic structure*, Zhurn. Vychisl. Mat. Mat. Fiz. **22** (1982), no. 1, 112–122; English transl., U.S.S.R. Comput. Math. Math. Phys. **22** (1982), no. 1, 117–128.
130. K. Yosida, *Functional Analysis*, Springer-Verlag, Berlin, 1994.
131. V. V. Zhikov, *On the homogenization of the system of Stokes equations in a porous medium*, Dokl. Akad. Nauk, Ross. Akad. Nauk **334** (1994), no. 2, 144–147; English transl., Russ. Acad. Sci. Dokl. Math. **49** (1994), no. 1, 52–57.
132. ———, *Connectedness and homogenization. Examples of fractal conductivity*, Mat. Sb. **187** (1996), no. 8, 3–40; English transl., Sb. Math. **187** (1996), no. 8, 1109–1147.
133. ———, *On a homogenization technique for variational problems*, Funkts. Anal. Prilozh. **33** (1999), no. 1, 14–29; English transl., Funct. Anal. Appl. **33** (1999), no. 1, 11–24.
134. ———, *On an extension of the method of two-scale convergence and its applications*, Mat. Sb. **191** (2000), no. 7, 31–72; English transl., Sb. Math. **191** (2000), no. 7, 973–1014.
135. ———, *Homogenization of elasticity problems on singular structures*, Izv. Ross. Akad. Nauk Ser. Mat. **66** (2002), no. 2, 81–148. (Russian)
136. ——— *On two-scale convergence*, J. Math. Sci. **120** (2004), no. 3, 1328–1352.
137. V. V. Zhikov, S. M. Kozlov, and O. A. Oleinik, *Homogenization of Differential Operators*, Fizmatlit, Moscow, 1993. (Russian)
138. ———, *Homogenization of Differential Operators and Integral Functionals,* Springer-Verlag, Berlin etc., 1994.
139. V. V. Zhikov and S. E. Pastukhova, *Homogenization of elasticity problems on periodic lattices of critical thickness*, Mat. Sb. **194** (2003), no. 5, 61–96. (Russian)

# Index

# Titles in This Series

234 **G. A. Chechkin, A. L. Piatnitski, and A. S. Shamaev,** Homogenization: Methods and applications, 2007

233 **A. Ya. Helemskii,** Lectures and exercises on functional analysis, 2006

232 **O. N. Vasilenko,** Number-theoretic algorithms in cryptography, 2007

231 **Kiyosi Itô,** Essentials of stochastic processes, 2006

230 **Akira Kono and Dai Tamaki,** Generalized cohomology, 2006

229 **Yu. N. Lin′kov,** Lectures in mathematical statistics, 2005

228 **D. Zhelobenko,** Principal structures and methods of representation theory, 2006

227 **Takahiro Kawai and Yoshitsugu Takei,** Algebraic analysis of singular perturbation theory, 2005

226 **V. M. Manuilov and E. V. Troitsky,** Hilbert $C^*$-modules, 2005

225 **S. M. Natanzon,** Moduli of Riemann surfaces, real algebraic curves, and their superanaloges, 2004

224 **Ichiro Shigekawa,** Stochastic analysis, 2004

223 **Masatoshi Noumi,** Painlevé equations through symmetry, 2004

222 **G. G. Magaril-Il'yaev and V. M. Tikhomirov,** Convex analysis: Theory and applications, 2003

221 **Katsuei Kenmotsu,** Surfaces with constant mean curvature, 2003

220 **I. M. Gelfand, S. G. Gindikin, and M. I. Graev,** Selected topics in integral geometry, 2003

219 **S. V. Kerov,** Asymptotic representation theory of the symmetric group and its applications to analysis, 2003

218 **Kenji Ueno,** Algebraic geometry 3: Further study of schemes, 2003

217 **Masaki Kashiwara,** $D$-modules and microlocal calculus, 2003

216 **G. V. Badalyan,** Quasipower series and quasianalytic classes of functions, 2002

215 **Tatsuo Kimura,** Introduction to prehomogeneous vector spaces, 2003

214 **L. Š. Grinblat,** Algebras of sets and combinatorics, 2002

213 **V. N. Sachkov and V. E. Tarakanov,** Combinatorics of nonnegative matrices, 2002

212 **A. V. Mel′nikov, S. N. Volkov, and M. L. Nechaev,** Mathematics of financial obligations, 2002

211 **Takeo Ohsawa,** Analysis of several complex variables, 2002

210 **Toshitake Kohno,** Conformal field theory and topology, 2002

209 **Yasumasa Nishiura,** Far-from-equilibrium dynamics, 2002

208 **Yukio Matsumoto,** An introduction to Morse theory, 2002

207 **Ken'ichi Ohshika,** Discrete groups, 2002

206 **Yuji Shimizu and Kenji Ueno,** Advances in moduli theory, 2002

205 **Seiki Nishikawa,** Variational problems in geometry, 2001

204 **A. M. Vinogradov,** Cohomological analysis of partial differential equations and Secondary Calculus, 2001

203 **Te Sun Han and Kingo Kobayashi,** Mathematics of information and coding, 2002

202 **V. P. Maslov and G. A. Omel′yanov,** Geometric asymptotics for nonlinear PDE. I, 2001

201 **Shigeyuki Morita,** Geometry of differential forms, 2001

200 **V. V. Prasolov and V. M. Tikhomirov,** Geometry, 2001

199 **Shigeyuki Morita,** Geometry of characteristic classes, 2001

198 **V. A. Smirnov,** Simplicial and operad methods in algebraic topology, 2001

197 **Kenji Ueno,** Algebraic geometry 2: Sheaves and cohomology, 2001

196 **Yu. N. Lin′kov,** Asymptotic statistical methods for stochastic processes, 2001
195 **Minoru Wakimoto,** Infinite-dimensional Lie algebras, 2001
194 **Valery B. Nevzorov,** Records: Mathematical theory, 2001
193 **Toshio Nishino,** Function theory in several complex variables, 2001
192 **Yu. P. Solovyov and E. V. Troitsky,** $C^*$-algebras and elliptic operators in differential topology, 2001
191 **Shun-ichi Amari and Hiroshi Nagaoka,** Methods of information geometry, 2000
190 **Alexander N. Starkov,** Dynamical systems on homogeneous spaces, 2000
189 **Mitsuru Ikawa,** Hyperbolic partial differential equations and wave phenomena, 2000
188 **V. V. Buldygin and Yu. V. Kozachenko,** Metric characterization of random variables and random processes, 2000
187 **A. V. Fursikov,** Optimal control of distributed systems. Theory and applications, 2000
186 **Kazuya Kato, Nobushige Kurokawa, and Takeshi Saito,** Number theory 1: Fermat's dream, 2000
185 **Kenji Ueno,** Algebraic Geometry 1: From algebraic varieties to schemes, 1999
184 **A. V. Mel′nikov,** Financial markets, 1999
183 **Hajime Sato,** Algebraic topology: an intuitive approach, 1999
182 **I. S. Krasil′shchik and A. M. Vinogradov, Editors,** Symmetries and conservation laws for differential equations of mathematical physics, 1999
181 **Ya. G. Berkovich and E. M. Zhmud′,** Characters of finite groups. Part 2, 1999
180 **A. A. Milyutin and N. P. Osmolovskii,** Calculus of variations and optimal control, 1998
179 **V. E. Voskresenskiĭ,** Algebraic groups and their birational invariants, 1998
178 **Mitsuo Morimoto,** Analytic functionals on the sphere, 1998
177 **Satoru Igari,** Real analysis—with an introduction to wavelet theory, 1998
176 **L. M. Lerman and Ya. L. Umanskiy,** Four-dimensional integrable Hamiltonian systems with simple singular points (topological aspects), 1998
175 **S. K. Godunov,** Modern aspects of linear algebra, 1998
174 **Ya-Zhe Chen and Lan-Cheng Wu,** Second order elliptic equations and elliptic systems, 1998
173 **Yu. A. Davydov, M. A. Lifshits, and N. V. Smorodina,** Local properties of distributions of stochastic functionals, 1998
172 **Ya. G. Berkovich and E. M. Zhmud′,** Characters of finite groups. Part 1, 1998
171 **E. M. Landis,** Second order equations of elliptic and parabolic type, 1998
170 **Viktor Prasolov and Yuri Solovyev,** Elliptic functions and elliptic integrals, 1997
169 **S. K. Godunov,** Ordinary differential equations with constant coefficient, 1997
168 **Junjiro Noguchi,** Introduction to complex analysis, 1998
167 **Masaya Yamaguti, Masayoshi Hata, and Jun Kigami,** Mathematics of fractals, 1997
166 **Kenji Ueno,** An introduction to algebraic geometry, 1997
165 **V. V. Ishkhanov, B. B. Lur′e, and D. K. Faddeev,** The embedding problem in Galois theory, 1997
164 **E. I. Gordon,** Nonstandard methods in commutative harmonic analysis, 1997
163 **A. Ya. Dorogovtsev, D. S. Silvestrov, A. V. Skorokhod, and M. I. Yadrenko,** Probability theory: Collection of problems, 1997

For a complete list of titles in this series, visit the
AMS Bookstore at **www.ams.org/bookstore/**.